Fuel Cells in Environmental Sustainability

This book provides a comprehensive technical analysis of fuel cell technologies, focusing on chemical reactions, material specifications, and performance metrics. It examines practical adaptations through real-world case studies, showcasing successful implementations across various industries, highlighting efficiency gains and emission reductions. It also looks at advances in fuel cell materials and configurations, spotlighting catalysts, electrolytes, and electrodes developed.

Features:

- Describes different types and sources of fuel cells from a sustainability perspective.
- Includes case studies in real-world applications.
- Provides comprehensive applications of fuel cells in energy production for a sustainable environment.
- Discusses environmental and life cycle assessment.
- Presents emerging innovations in fuel cells production and assessment.

This book is aimed at graduate students and researchers in chemical engineering, renewable energy, and clean technologies.

Fuel Cells in Environmental Sustainability

Edited by
Babafemi Raphael Babaniyi
Sesan Abiodun Aransiola
Naga Raju Maddela

CRC Press
Taylor & Francis Group
Boca Raton London New York

CRC Press is an imprint of the
Taylor & Francis Group, an **informa** business

Designed cover image: © Babafemi Raphael Babaniyi

First edition published 2026
by CRC Press
2385 NW Executive Center Drive, Suite 320, Boca Raton FL 33431

and by CRC Press
4 Park Square, Milton Park, Abingdon, Oxon, OX14 4RN

CRC Press is an imprint of Taylor & Francis Group, LLC

ISBN: 978-1-041-08274-3 (hbk)
ISBN: 978-1-041-08439-6 (pbk)
ISBN: 978-1-003-64540-5 (ebk)

DOI: 10.1201/9781003645405

Typeset in Times
by SPi Technologies India Pvt Ltd (Straive)

Contents

Foreword

In the face of escalating environmental challenges and a rapidly evolving energy landscape, the urgency for clean, efficient, and sustainable energy technologies has never been greater. Fuel cells, with their capacity to convert chemical energy directly into electricity with minimal emissions, stand at the forefront of this transformation. This book, *Fuel Cells in Environmental Sustainability*, offers a timely and comprehensive exploration of the pivotal role fuel cell technologies can play in advancing a cleaner and more resilient global energy system.

The chapters within this volume provide an in-depth analysis of the environmental impact of traditional energy sources, effectively contrasting them with the sustainable promise of fuel cells. It explores material and catalyst innovations essential for scaling up fuel cell deployment, alongside cutting-edge strategies for hydrogen production and the critical environmental considerations that accompany them. Of particular note is the integration of fuel cells with renewable energy systems, showcasing their potential as key enablers of decentralized and low-carbon energy infrastructures. Life cycle analysis further emphasizes their long-term environmental benefits, while emerging technologies—such as microbial fuel cells (MFCs)—demonstrate novel approaches to waste-to-energy conversion, bioremediation, and wastewater treatment. This volume also examines specialized fuel cell types, including Proton Exchange Membrane Fuel Cells (PEMFCs), and their application in remote sensing, agriculture, and environmental monitoring. Equally vital are the discussions on health, safety, policy, and economic considerations, all of which are essential for real-world deployment and acceptance. The concluding chapters address the ongoing challenges and future directions for fuel cell technology, including thermal management and surface heat enhancement for improved environmental sustainability.

This book is an essential resource for researchers, engineers, policymakers, and students seeking to understand both the science and the sustainability promise of fuel cells. It provides the foundation and vision needed to navigate the complex intersection of clean energy, environmental stewardship, and technological innovation. As the world collectively shifts toward greener solutions, the insights provided here will prove invaluable in shaping the role of fuel cells in achieving global sustainability goals.

Dr. O. A. Adelaja
Associate Director, TECHPIC, FUTA
Associate Professor of Environmental
Chemistry and Biotechnology

Preface

The global pursuit of sustainable energy solutions has brought fuel cell technology into sharper focus, not merely as an alternative energy source but as a cornerstone for addressing some of the most pressing environmental and energy challenges of our time. This book, *Fuel Cells in Environmental Sustainability*, is born out of the need to provide a holistic, interdisciplinary, and practical insight into the evolving role of fuel cells in supporting a cleaner, greener, and more sustainable future.

This volume covers a broad spectrum of fuel cell research and applications—from fundamental principles to advanced industrial and environmental use cases. The first section introduces the environmental impact of conventional energy systems and lays the groundwork for understanding how fuel cells can significantly reduce greenhouse gas emissions and environmental degradation. Subsequent chapters evaluate the development of advanced materials and catalysts tailored for industrial-scale applications, hydrogen production pathways, and the integration of fuel cells with renewable energy systems. The book also emphasizes real-world applications, including microbial fuel cells (MFCs) for bioremediation, waste-to-energy solutions, and wastewater treatment. These chapters highlight the dual role of fuel cells in both energy generation and environmental remediation—an area gaining increasing attention in the scientific and industrial communities.

Moreover, emerging areas such as Proton Exchange Membrane Fuel Cells (PEMFCs), remote sensing, climate change mitigation, and sustainable agriculture are explored in depth. Considerations related to policy, economics, safety, and lifecycle assessments are integrated to offer a comprehensive view of fuel cell technology from development to deployment. A dedicated section discusses the surface heat enhancement characteristics and their implications on fuel cell performance and environmental impact—an often-overlooked but critical aspect of improving system efficiency. This book is intended for a wide readership, including researchers, engineers, environmental scientists, graduate students, and policy advocates. Whether you are new to the field or an experienced professional, this book aims to provide a valuable reference that bridges scientific innovation with environmental responsibility. We hope this compilation stimulates further research, collaboration, and innovation toward harnessing the full potential of fuel cells for environmental sustainability.

Babafemi Raphael Babaniyi Ph.D
Bioresources Development Centre,
National Biotechnology Research and Development Agency (NBRDA),
Ogbomoso, Nigeria

Sesan Abiodun Aransiola, Ph.D.
Department of Microbiology, Faculty of Science,
University of Abuja,
Abuja, Nigeria

Naga Raju Maddela, Ph. D.
Departamento de Ciencias Biológicas,
Facultad de Ciencias de la Salud,
Universidad Técnica de Manabí,
Portoviejo 130105,
Ecuador

Acknowledgments

We sincerely thank the contributors of this book for their invaluable insights and timely responses throughout the development of this project. Their enthusiasm, dedication, and collaborative spirit were key to completing this volume within the planned schedule. We are truly grateful for their unwavering support. Our heartfelt appreciation also goes to the anonymous reviewers, whose constructive feedback greatly enhanced the quality of this book. A special note of thanks is extended to CRC Press and the production team for their outstanding guidance and steadfast support. Their professionalism and commitment were instrumental in bringing this project to completion in such a short timeframe. It has been both an honor and a privilege to work with them.

Babafemi Raphael Babaniyi Ph.D
Bioresources Development Centre,
National Biotechnology Research and Development Agency (NBRDA),
Ogbomoso, Nigeria

Sesan Abiodun Aransiola, Ph.D.
Department of Microbiology, Faculty of Science,
University of Abuja,
Abuja, Nigeria

Naga Raju Maddela, Ph. D.
Departamento de Ciencias Biológicas,
Facultad de Ciencias de la Salud,
Universidad Técnica de Manabí,
Portoviejo 130105,
Ecuador

About the Editors

Babafemi Raphael Babaniyi is a distinguished researcher and academic in Environmental Chemistry. He earned a B.Sc. in Industrial Chemistry from Caritas University, Enugu (2008), an M.Tech. in Environmental Chemistry from the Federal University of Technology, Akure (2021), and a Ph.D. in Environmental Chemistry from the same institution in 2025. His work spans Environmental and Green Chemistry, Phytoremediation, Chemical Biology, Bioremediation, Environmental Impact Assessment, and Biodegradable Plastics. With over 40 publications, including international journal articles and book chapters, Babafemi has made significant contributions to global scientific discourse. He serves as an Assistant Chief Research Officer at the National Biotechnology Development Agency, where he leads initiatives in biotechnology research and innovation. His dedication to solving environmental challenges through science marks him as a leading voice in sustainability and environmental health.

Sesan Abiodun Aransiola holds a Ph.D. in Environmental Microbiology from the Federal University of Technology, Minna, Nigeria, where he also earned his B.Tech. and M.Tech. degrees in 2009 and 2014, respectively. He is currently a Lecturer in the Department of Microbiology at the University of Abuja, Nigeria. Dr. Aransiola's research focuses on innovative approaches to soil and water remediation, particularly through vermicomposting (vermitechnology), phytoremediation, and microbial biotechnology. His work has explored the use of vermicasts and biofertilizers such as vinasse to enhance soil recovery from heavy metal and plastic pollution. His expertise extends to environmental health, marine resources, and sustainable waste management. An award-winning scientist, Dr. Aransiola has authored over 90 publications, including high-impact journal articles, reviews, and book chapters. He has co-edited several globally recognized scientific books with CRC Press and other international publishers, covering topics such as microbial biotechnology, soil microbiomes, marine bioprospecting, phytoremediation, and food safety.

Previously, he served as Assistant Chief Scientific Officer at the National Biotechnology Development Agency. In addition to his academic and research achievements, he actively engages in community service and sustainability outreach. Dr. Aransiola is a member of the Nigerian Society for Microbiology and the American Society for Microbiology, among other professional bodies.

Naga Raju Maddela is a Full Professor of Microbiology at the Facultad de Ciencias de la Salud, Universidad Técnica de Manabí, Portoviejo, Ecuador. He earned his M.Sc. (1996–1998) and Ph.D. (2012) in Microbiology from Sri Krishnadevaraya University, Anantapuramu, India, with doctoral research focused on the impact of industrial effluents and insecticides on soil microorganisms. With over two decades of teaching experience since 1998, Dr. Maddela has mentored both undergraduate and postgraduate students in microbiology. He served as a Prometeo Investigator at Universidad Estatal Amazónica, Ecuador (2013–2015), received a Postdoctoral Fellowship from Sun Yat-sen

University, China (2016–2018), and secured funding from the China Postdoctoral Science Foundation (2017) and Universidad Técnica de Manabí (2020). He has actively participated in international conferences across China, Cuba, Ecuador, India, and Singapore. A prolific researcher, Dr. Maddela has authored over 120 scientific articles, 25 books, and 50 book chapters, making significant contributions to environmental microbiology and related fields.

Contributors

S.A. Aasa
Department of Mechanical and Mechatronic Engineering
Tshwane University of Technology
Pretoria, South Africa

Falese Babatunde Abimbola
Biomolecular Structure and Dynamics Unit
Department of Biochemistry
Federal University of Technology
Akure, Nigeria

Aisha Eniola Adegbite
Department of Microbiology
Faculty of Science
University of Abuja
Abuja, Nigeria

Sharafadeen Adetunji Adeniji
Department of Physics
Nile University of Nigeria
Abuja, Nigeria.

Olarewaju John Adunbi
Bioresources Development Center
National Biotechnology Research and Development Agency (NBRDA)
Ogbomoso, Nigeria

Bolaji Oluwatosin Akanbi
Department of Microbiology
Faculty of Science
University of Abuja
Abuja, Nigeria

Chima Bartholomew Akpa
GT, Nigerian Content Development and Management Board
Yenegoa, Nigeria

Joshua Ogboghena Akhigbe
Department of Microbiology
Faculty of Science
University of Abuja
Abuja, Nigeria

Ulelu Jessica Akor
Department of Animal Production
University of Ilorin
Ilorin, Nigeria

Sesan Abiodun Aransiola
Department of Microbiology
Faculty of Science
University of Abuja
Abuja, Nigeria

Babafemi Raphael Babaniyi
Bioresources Development Center
National Biotechnology Research and Development Agency (NBRDA)
Ogbomoso, Nigeria

Gabriel Gbenga Babaniyi
Department of Agricultural Development and Management, Agricultural and Rural Management Training Institute (ARMTI)
Ilorin, Nigeria

Mukul Machhindra Barwant
Department of Botany
Sanjivani Arts Commerce and Science College
Kopargoan, India

Olafare Biola Blessing
Enzyme Biotechnology and Environmental Health Unit
Department of Biochemistry
Federal University of Technology
Akure, Nigeria

Mauricio Colpari-Pozzo
Pontificia Universidad Católica del Ecuador
Ecuador, Portoviejo

D.A. Desai
Department of Mechanical and Mechatronic Engineering
Tshwane University of Technology
Pretoria, South Africa

Namla Djadjitti
Department of Biological Sciences
Faculty of Science
King Abdulaziz University
Jeddah, Saudi Arabia

Rodrigo Duarte-Casar
Pontificia Universidad Católica del Ecuador
Ecuador, Portoviejo

R.K. Dunne
Department of Mechanical and Mechatronic Engineering
Tshwane University of Technology
Pretoria, South Africa

Martina Onyinye Eze
Department of Geology
Michael Okpara University of Agriculture
Umudike, Nigeria

Gloria Ginikanwa Ezeagu
Department of Biology
Microbiology & Science Laboratory Technology
Faculty of Science
Nile University of Nigeria
Abuja, Nigeria

Abisola Abiodun Ibikunle
Department Of Chemistry
School of Physical Sciences
Federal University of Technology
Akure, Nigeria

Aisha Umar Ibrahim
Department of Biology
Microbiology & Science Laboratory Technology
Faculty of Science
Nile University of Nigeria
Abuja, Nigeria

Oluwafemi Christopher Igbasan
Department Of Chemistry
School of Physical Sciences
Federal University of Technology
Akure

Emmanuel Ojo Ige
Department of Chemistry
School of Physical Sciences
Federal University of Technology
Akure, Nigeria

Odangowei Inetiminebi Ogidi
Department of Biochemistry
Bayelsa Medical University
Yenagoa, Nigeria

Olumide Ayobami Iyiola
Department of Rural Development and Gender Issues
Agricultural & Rural Management Training Institute (ARMTI)
Ilorin, Nigeria

Johnson Jayeola
Fring Associate Inc.
Toronto, Canada

Agabi Enaji Kanu
Department of Biology, Microbiology & Science Laboratory Technology
Faculty of Science
Nile University of Nigeria
Abuja, Nigeria.

Anand Kumar
College of Food Science and Technology
Guangdong Provincial Key Laboratory of Aquatic Product Processing and Safety
Guangdong Ocean University
Zhanjiang, China

John A. Lawal
Department of Chemical Sciences
Achievers University
Owo, Nigeria.

Olusola T. Lawal
Medical Biochemistry,
School of Basic Medical Sciences Department
Federal University of Technology
Akure, Nigeria

Debasree Lodh
Department of Botany
Holy Cross College
Agartala, India

Naga Raju Maddela
Departamento de Ciencias Biológicas
Facultad la Ciencias de la Salud
Universidad Técnica de Manabí
Portoviejo, Manabí, Ecuador

Josephine Madukaihe Madu
Department of Food and Industrial Biotechnology
National Biotechnology Research and Development
Abuja, Nigeria.

Collins Madojemu
Department of Biology
Microbiology & Science Laboratory Technology
Faculty of Science
Nile University of Nigeria
Abuja, Nigeria.

Sani Sambo Datsugwai Mohammed
Department of Biology
Microbiology & Science Laboratory Technology
Faculty of Science
Nile University of Nigeria
Abuja, Nigeria

Darío Mondavi-Sobby
Pontificia Universidad Católica del Ecuador
Manta

Marell Navarro-Rojas
Universidad Politécnica Salesiana
Cuenca, Ecuador

Monday Peter Obagunwa
Bioresources Development Center, National Biotechnology Research and Development Agency (NBRDA)
Ogbomoso, Nigeria

Olusola D. Ogundele
Department of Chemical Sciences
Achievers University
Owo, Nigeria

Emmanuel Adesuyi Omole
Department of Marine Biology
University of Padova
Padova, Italy

Olaewe Tunde Olaniyi
Enzyme Biotechnology and Environmental Health Unit
Department of Biochemistry
Federal University of Technology
Akure, Nigeria

Constance Omono Omokhodion
Department of Microbiology
Faculty of Science
University of Abuja
Abuja, Nigeria

Beatrice Ovute Onodagu
Department of Microbiology
University of Nigeria Teaching Hospital
Ituku/Ozalla, Enugu

Tatiana Ordóñez-Zambrano
Pontificia Universidad Católica del Ecuador
Portoviejo

Oluwatosin D. Owolala
Department of Pharmaceutical Science
Robert Gordon University
Aberdeen

Marlene Rojas-Le-Fort
Pontificia Universidad Católica del Ecuador
Ecuador, Portoviejo

Juan Carlos Romero-Benavides
Universidad Técnica Particular de Loja
Loja, Ecuador

Dipanwita Chaudhuri Sil
Department of Botany
Holy Cross College
Agartala, India

Balwant Singh
Department of Botany
K. S. Saket PG College
Dr. Ram Manohar Lohia Avadh University
Ayodhya, India

Kafilat Olayemi Suleiman
Department of Biology
Microbiology & Science Laboratory Technology
Faculty of Science
Nile University of Nigeria
Abuja, Nigeria

Amir Vazirizadeh
Persian Gulf Research and Studies Institute
Fishery and Marine Biology Department
Persian Gulf University
Bushehr, Iran

Amina Yahaya
Department of Microbiology
Faculty of Science
University of Abuja
Abuja, Nigeria

Mehtab Yasmeen
Department of Botany
Veeranari Chakali Ilamma Women's University
Koti, Hyderabad, India

1 Overview of Fuel Cell Sustainability

Abisola Abiodun Ibikunle, Babafemi Raphael Babaniyi, and Monday Peter Obagunwa

1.1 INTRODUCTION

Fuel sustainability refers to the long-term, responsible management and use of energy resources, especially fuels, such that they meet present energy demands without compromising the ability of future generations to meet theirs (Dincer & Acar 2017). It encompasses environmental, economic, and social dimensions, including minimizing carbon emissions, reducing dependence on non-renewable resources, ensuring equitable access to energy, and preserving ecological systems (Kabeyi, & Olanrewaju 2022). Fuel sustainability involves shifting from fossil fuels (coal, oil, and gas) to cleaner, renewable energy sources such as biofuels, hydrogen, solar, wind, and hydroelectric power, as well as improving energy efficiency (Abbasi et al., 2022). Fuel sustainability is critical to navigating the global energy transition. It offers pathways to mitigate climate change, protect finite resources, and ensure a just and equitable energy future (Saleh & Hassan 2024; Sreekanth 2024). Advancing research, investing in renewable fuels, and formulating integrated policies are essential to accelerate this transition (Onabowale 2024). It plays a vital role in addressing two of the most pressing global challenges: climate change and long-term energy security (Elkhatat & Al-Muhtaseb 2024). As fossil fuel dependence continues to strain environmental systems and global energy markets, a shift toward sustainable fuel alternatives has become essential (Londoño-Pulgarin et al., 2021). Many nations, especially Nigeria, rely heavily on imported fossil fuels, which makes them vulnerable to disruptions and price fluctuations (olujobi et al., 2023).

1.2 CATEGORIES OF FUEL SUSTAINABILITY

1.2.1 Non-renewable Energy

1.2.1.1 Fossil Fuel

Fossil fuels are non-renewable because they take millions of years to form and are being consumed far faster than they are replenished (Freedman et al., 2024). Rising global energy demand, driven by population growth and developing economies, compounds this issue. Their use also contributes significantly to environmental problems, particularly greenhouse gas emissions (Omer 2009). Despite efforts to expand low-carbon energy sources, fossil fuels will remain essential in the near term, necessitating emission reduction strategies (Prasad et al., 2020). As reserves decline, supporting the Hubbert Peak Theory alternative technologies are needed to extend fossil fuel use and mitigate future shortages (Deffeyes 2008). Fossil fuels are finite resources. Fossil fuels have long been central to industrial and consumer energy needs, and despite increasing modifications to meet evolving demands, they remain the dominant energy source (Zuo et al., 2016). Refining processes have become more advanced to produce fuels that meet modern standards, but rising operational costs and difficulty in fully utilizing refinery outputs persist (Gudde 2018). Meanwhile, fossil fuel reserves, once thought abundant, are depleting rapidly, prompting interest in alternative technologies to extend the viability of traditional fuels and address anticipated shortages, as predicted by the Hubbert

DOI: 10.1201/9781003645405-1

Peak Theory (Olev, 2012). This theory, proposed by M. King Hubbert, suggests fossil fuel production follows a bell-shaped curve, peaking before entering irreversible decline, underscoring the urgency for renewable energy development, though widespread adoption remains limited (Roos, 2021).

1.2.1.2 Coal

Coal, formed from preserved plant matter in swamp environments, is the most abundant and globally distributed fossil fuel (Arnold 2023). It exists in various forms: lignite, subbituminous, bituminous, and anthracite, each with increasing carbon content and energy potential. Coal is primarily used for electricity generation and industrial heating, especially in developing countries (Kerimray et al., 2017). Technological improvements have kept coal competitive, and its liquefaction into synthetic fuels offers compatibility with existing fuel infrastructure (Vasireddy et al., 2011). However, coal-to-liquid processes emit significant CO_2, raising environmental concerns. Despite this, coal gasification and liquefaction remain practical alternatives amid fluctuating oil prices and vast coal reserves (Al-Kuwari, 2023). Coal bed methane, generated during coal formation, is mainly extracted in the U.S., accounting for about 9% of its gas output. Globally, its share remains minor (Kholod et al., 2020). Tight gas, found in low-permeability sandstones, has low natural flow rates but is increasingly accessed using hydraulic fracturing (Boak & Kleinberg 2020).

Coal, a sedimentary rock used widely as fuel, is mined in over 50 countries. Of 3,600 global coal fields, 75 countries hold coal resources, totaling 14.81 billion tons, with recoverable reserves at 1.25 billion tons (Afonja 2022). Most high-quality coal is extracted via underground mining, and leading producers also dominate reserve rankings (Cameron and Stanley 2017).

1.2.1.3 Petroleum

Petroleum has powered industries and transportation for over a century. Non-OPEC producers are facing production declines, and forecasts suggest global oil production may have already peaked or will do so soon (Al-Fattah 2020). Rising crude oil prices have made alternatives like coal liquefaction economically appealing. Refining technology has advanced to meet consumer demands, though cost and complexity challenges remain (Asghar et al., 2022). Refineries vary in configuration, producing a wide range of products: gases, fuels, lubricants, and petrochemicals. Petroleum also yields coke, used in refinery operations. Industry pressures product demand, feedstock changes, environmental regulations, and technological innovation have driven refining evolution (Griffiths et al., 2022). Petroleum continues to dominate global energy supply and serves as a raw material for countless industrial products. Petroleum, a naturally occurring flammable liquid composed mainly of hydrocarbons, is found in underground geological formations (Pang et al., 2015). Although over 50,000 oil fields exist, 94% of known reserves are concentrated in fewer than 1,500 major fields, mainly in the Middle East, the CIS, and Venezuela. Other regions like Europe and Oceania have minimal reserves (Tiratsoo 2022). Between 2005 and 2009, large deepwater oil discoveries were made in several countries, including Brazil and the U.S., with deepwater fields yielding much larger reserves than onshore ones (Anjos et al., 2024). About 70% of oil reserves lie on continental shelves. Offshore oil production has risen steadily, from 28% in 2000 to nearly 35% in 2008, with major contributions from Venezuela, the U.S., and Gulf Arab states (liu et al., 2019).

1.2.1.4 Other Oil Sources

Unconventional oil sources, such as bituminous sand and shales, are also significant. Bitumen is mined in 23 countries, with some deposits being extremely large. Composed mainly of methane, natural gas is found in gas fields, oil fields, and coal beds (Tolmachev et al., 2020). It must be refined to remove impurities and can yield useful by-products like ethane, propane, and sulfur. Though extraction technologies lag oil, natural gas is recognized as a clean, abundant alternative. Challenges include costly conversion to liquids and long-distance transport, but strategic advantages exist, especially near production sites (Safari et al., 2019). Its use spans residential heating, power generation, and as a feedstock for various products. Burning natural gas emits less CO_2 and pollutants than coal or oil, but methane leaks and environmental impacts from drilling remain concerns (Alagoz et al., 2023). Advances in drilling and sensing technology have reduced its environmental footprint. Today, natural gas plays a crucial role in the global energy mix and is expected to grow as cleaner technologies emerge (liang et al., 2012). Natural gas reserves are heavily concentrated in Russia, Iran, and Qatar. Offshore gas discoveries have grown in recent years, especially in Australia, Israel, and Venezuela. By 2000, offshore gas accounted for 31% of global production (Bernardino et al., 2020) Unconventional gas sources include gas hydrates, coal bed methane, and tight gas. Gas hydrates, mainly methane trapped in ice-like structures under high pressure and low temperature, occur in oceanic sediments but are not yet commercially exploited (Xiao et al., 2023).

1.2.2 Environmental Impacts of Fossil Fuel Use

1.2.2.1 Environmental Effect of Coal

Coal use involves mining, transport, and processing, with combustion releasing numerous pollutants including sulfur dioxide (SO_2), particulates, radium, and carbon dioxide (CO_2), a key greenhouse gas (Pandey et al., 2018). Coal power plants emit more pollutants than oil or gas plants, except for NO_2, aldehydes, and hydrocarbons. Surface mining disturbs more land and water, while deep mining poses higher safety risks (Hannun, & Razzaq 2022). Acid mine drainage, mainly from underground mines, pollutes water sources. Surface mining leads to erosion, sediment pollution, and flooding unless reclaimed (Adeniyi et al., 2022). Rail transport and electricity transmission require extensive land. Coal plants also discharge large amounts of waste heat.

Technological improvements aim to reduce air pollution. Baseline systems use flue gas cleanup to lower SO_2 and particulates but increase costs and reduce efficiency (Cui et al., 2018). Advanced combustion and fuel conversion technologies (e.g., fluidized bed boilers, gasifiers, and liquefaction) can meet U.S. air quality standards for SO_2 and particulates but struggle with NO_x control (Hou 2024). However, all coal technologies emit toxic compounds such as polycyclic aromatic hydrocarbons and heavy metals, with little progress on CO_2 reduction (Yi et al., 2015).

1.2.2.2 Environmental Effect of Petroleum

Petroleum extraction causes land disturbance, brine disposal, and oil spills—especially offshore. Most marine oil pollution comes from tankers, bilges, and runoffs (Bhattacharjee & Dutta 2022). Oil pipelines can rupture, contaminate soil and groundwater. Refineries emit pollutants and transporting refined products risks spills. Combustion of oil releases fewer pollutants than coal, with less environmental impact overall (Hannun & Razzaq 2022). However, unconventional sources like oil shale disturb large areas and produce toxic by-products and waste, requiring heavy water use and raising further pollution concerns (Pal & Sen 2024).

1.2.2.3 Environmental Effect of Natural Gas

Natural gas extraction resembles oil production, often from shared wells. It affects land through brine disposal and pipeline construction (Cordes et al. 2016). Processing and combustion contribute to air, water, and heat pollution, especially through high NO_x emissions. Liquefied natural gas poses fire and explosion risks (Ozgen et al., 2021). Human activities influence climate mainly through heat release and emissions of particles and trace gases. Fossil fuel combustion is the primary source of both (Nica et al., 2019). Waste heat from fossil fuel use alters local climates, especially in urbanized and industrialized areas where energy sources are concentrated (Firth et al., 2019). Though global climate models show that large-scale heat emissions may cause only minor atmospheric changes within model noise, regional models suggest localized warming—up to 1.2°C in winter and 0.3°C in summer—within shallow atmospheric layers (Zhao et al., 2021). Large energy clusters could significantly affect temperature, soil moisture, cloud formation, and precipitation, possibly increasing the frequency of severe weather events. However, global-scale impacts remain uncertain due to model limitations. Particles from fossil fuel combustion, agriculture, and natural sources influence climate by interacting with solar radiation and cloud formation (Zhang 2020). While total global particle emissions are around 3000 million tons per year, about 100 million tons come from fossil fuels (Vohra et al., 2021). Due to short atmospheric residence times, these particles concentrate near cities and industrial zones (Zhang et al., 2015). Air quality controls have reduced emissions in developed nations, and further increases in fossil fuel use may not significantly raise global particle levels (Shindell & Smith 2019). Initially thought to cause cooling by reflecting sunlight, particles are now recognized for their complex effects, including both warming and cooling depending on absorption properties and atmospheric conditions (Li et al., 2020). Overall, particles likely contribute to a net warming effect, but their full climatic impact remains under investigation. Human activities such as fossil fuel combustion, deforestation, cement production, and gas flaring have steadily increased atmospheric CO_2 since the Industrial Revolution. Oceans, the primary CO_2 sink, cannot absorb the excess quickly enough, leading to a global distribution of CO_2 due to its long atmospheric residence time (~28 years). CO_2 traps outgoing infrared radiation, creating a greenhouse effect that warms the lower atmosphere and cools the stratosphere (Irfan et al., 2024). Climate models consistently project global warming, especially amplified in polar regions. However, uncertainties in feedback mechanisms, biosphere interactions, and future fossil fuel use complicate projections, and long transition times to alternative energy limit rapid policy action (Moriarty & Honnery 2019).

1.2.2.4 Effects from Carbon Emission

Carbon emissions from fossil fuels rose from 0.1 billion tons in 1860 to over 5 billion tons by 1980, with atmospheric CO_2 increasing from ~290 ppm to ~335 ppm (Kojima 2019). Annual increases have accelerated, exceeding 1 ppm/year, and are globally uniform. If current energy trends persist, CO_2 could reach ~380 ppm by 2000 and double pre-industrial levels by 2050, with potential peaks at 4–8 times those levels over the next few centuries (Adamo et al., 2021). Once elevated, concentrations will remain high for over a millennium. Doubling atmospheric CO_2 is expected to cause a global temperature rise of about 3°C (±1.5°C), with 3–4 times greater warming in the Arctic (Huang et al., 2017). Although regional climate change details remain uncertain, all major models agree on overall warming (James et al., 2017). A hypothetical scenario comparing fossil fuel versus alternative energy use confirms that fossil-based strategies significantly elevate CO_2 and global temperatures (Johnsson et al., 2019).

1.2.2.5 Global Warming

Possible consequences of global warming include disrupted weather patterns, altered precipitation, shifted agricultural zones, freshwater shortages, and cryosphere changes, such as irreversible conditions (Jha & Dev 2024). Arctic ice loss, sea level rise, disturbed ocean circulation and fisheries, and wide-ranging ecological, societal, and ethical impacts (Mondal & Lee 2025). While precise regional effects are unclear, the risks of large-scale disruption are substantial (Chang & Tanner, 2022).

1.2.3 Renewable Energy

Renewable and alternative energy sources present the most viable path to addressing the dual challenges of energy security and climate change, despite requiring significant upfront investment (Elkhatat & Al-Muhtaseb 2024). Shifting sustainable resources capable of providing 150 petawatt hours (PWh) of energy will demand not only large-scale investment but also breakthrough technologies (lodhi et al., 2024). Future solutions may involve fuel systems based on CO_2 and water (H_2O), supporting both climate mitigation and the broader energy transition (Schiemann et al., 2022). A balanced energy mix that includes fossil fuels, hydrogen, biofuels, and renewables could offer a practical strategy moving forward (Chen et al., 2023). Transitioning to sustainable fuels, particularly biofuels derived from renewable biomass, can substantially reduce these emissions, especially in high-impact sectors like transportation (Bonenkamp et al., 2020). Renewable fuels like second-generation biofuels have the potential to cut lifecycle CO_2 emissions by as much as 60–80% compared to fossil fuels (Kannan et al., 2022). Their adoption is therefore crucial for achieving global net-zero targets by 2050, alongside innovations in energy storage and clean transport.

1.2.3.1 Solar Energy

Engineers have worked for decades to harness solar thermal energy, and the technology is now approaching large-scale viability (Hasan et al., 2023; Nwokediegwu et al., 2024). The Earth receives over 10,000 times more solar energy each year than the world consumes, and each day's solar input exceeds annual global energy use, making the sun a highly promising energy source (Hanna et al., 2019). The challenge remains in reducing the cost of converting sunlight into electricity. Solar energy, though only a fraction of what reaches Earth, is sufficient to meet global energy needs. It's been used historically for heating and, more recently, for electricity through photovoltaic cells (Hasan et al., 2023). In 2017, solar made up 6% of U.S. renewable energy use (<1% of total energy) (Herche 2017). However, solar power is intermittent, available only when the sun shines and is unevenly distributed geographically, limiting its reliability (Ukoba et al., 2024). Despite this, solar is a popular supplemental energy source, especially with growing rooftop installations.

1.2.3.2 Biomass

Biomass, made up of organic matter from plants, animals, and their by-products, releases heat when burned or metabolized (Tursi 2019). When refined, it becomes biofuel—usable forms like ethanol, methane, biodiesel, or charcoal. Presently, commercial biofuels are mostly derived from food crops such as corn, soy, and palm, sparking concerns over food versus fuel competition (Naylor & Higgins 2018). To ensure sustainability, future biofuel production must rely on non-food sources. Starch is also being tested as a fuel due to its low cost and ease of processing (Morais et al., 2019).

1.2.3.3 Wind Energy

Wind energy, created by solar-driven atmospheric temperature differences, has gained momentum since the 1970s energy crises (Yıldız & MacEachern 2018). Modern wind turbines, now exceeding 4 MW, are highly efficient and expected to become a key contributor to future electricity supply and emissions reduction (Bošnjaković et al., 2022). Though they alter natural landscapes, their role in mitigating climate risks outweighs aesthetic drawbacks. Wind energy harnesses the movement of air caused by uneven solar heating to generate electricity (Abdali et al., 2019). Wind turbines, modern versions of traditional windmills, convert wind's kinetic energy into electricity (Gupta 2015). Offshore and high-altitude sites offer better potential. Wind power is non-polluting and water-free. In 2017, U.S. capacity rose from 40,181 MW in 2010 to 82,183 MW. Turbines, which take up little space, are often located on farmland, supporting rural economies. Like solar, wind energy is intermittent and best used as a supplemental source (Poggi et al., 2018).

1.2.3.4 Ocean Energy

Ocean energy sourced from wind-generated waves holds great potential but is unevenly distributed (Neill, & Hashemi). Technologies like tidal barrages, tidal and wave turbines are advancing, but further innovation is required. Additional opportunities exist in salinity gradient power, also known as blue energy, which uses osmosis or electrodialysis to generate electricity from salt concentration differences (Jia et al., 2014).

1.2.3.5 Hydropower

Hydropower, also known as hydroelectric power, is one of the oldest and most widely used forms of renewable energy (Bagher et al., 2015). It harnesses the kinetic and potential energy of moving or falling water to turn turbines, which in turn drive generators to produce electricity. Globally, hydropower is the second-largest source of renewable electricity after biomass, and it accounts for a significant portion of the energy mix in many countries due to its reliability and maturity (Yang, 2024). Hydropower systems can be classified into three main types: storage (reservoir-based), pumped-storage, and run-of-river systems, each with distinct operational characteristics and environmental impacts.

Storage hydropower involves the construction of dams to create large reservoirs. These reservoirs store water that can be released as needed to generate electricity, offering flexibility in responding to peak demand (Luo et al., 2025). This form is highly effective for grid stability and energy security but often comes with significant ecological and social costs, such as habitat disruption, land submersion, and displacement of local communities (Kabeyi, & Olanrewaju 2022).

Pumped-storage hydropower functions like a large battery. It pumps water from a lower elevation to a higher reservoir during periods of low electricity demand (or excess generation from sources like wind and solar) (Blakers et al., 2021). When electricity demand rises, the stored water is released back down to generate power. This system is essential for energy storage and balancing variable renewables, making it a key component in integrating solar and wind into modern power grids (Worku 2022).

Run-of-river systems divert a portion of a river's flow through a channel or pipeline to spin turbines without significantly altering the river's natural path or creating large reservoirs (Suder 2024). These systems are generally smaller in scale and less disruptive to aquatic ecosystems, but they depend heavily on consistent water flow, making them less reliable during dry seasons or droughts.

1.2.3.6 Geothermal Energy

Geothermal energy harnesses heat stored beneath the Earth's surface, a renewable resource generated by the natural decay of radioactive elements in the Earth's core (Naveen Kumar & Vigneshwaran, 2023). This geothermal heat manifests at the surface through features such as geysers, hot springs, and fumaroles, indicating regions with high geothermal potential (Rop et al., 2024). To access this energy, deep wells are drilled into geologically active areas—often near tectonic plate boundaries or volcanic zones—where temperatures are sufficiently high (Jolie et al., 2021).

Once tapped, geothermal heat can be used in several ways. High-temperature sources are primarily utilized for generating electricity through steam-driven turbines in geothermal power plants (Zhang, & Taleghani 2024). Medium- and low-temperature geothermal resources are ideal for direct-use applications, such as district heating, greenhouse agriculture, and industrial processes. Additionally, geothermal heat pumps use the relatively constant underground temperatures just a few meters below the surface for energy-efficient space heating and cooling in residential and commercial buildings (Gharbia et al. 2024).

As of 2015, the U.S. had approximately 21,000 megawatts (MW) of installed geothermal capacity, making it one of the global leaders in this technology. This capacity includes both electricity generation and direct-use systems, primarily concentrated in the western states like California and Nevada (Dadashi et al., 2022).

Geothermal energy is considered highly sustainable and environmentally friendly. Its lifecycle greenhouse gas emissions are significantly lower than those of fossil fuel-based power generation. In most systems, emissions such as carbon dioxide or hydrogen sulfide are minimal and can be effectively removed using gas scrubbers. Moreover, geothermal power plants have a small land footprint and require less surface disturbance compared to wind or solar farms (Rahman et al., 2022).

One of the most compelling advantages of geothermal systems is their stability and reliability. Unlike wind and solar, geothermal energy provides continuous baseload power that is not subject to weather fluctuations. This makes it a crucial component of a diversified and resilient renewable energy portfolio (xu et al., 2022).

1.2.4 Environmental Impact of Renewable Energy

1.2.4.1 Wind Energy

Offshore wind turbines generally have fewer environmental effects than land-based ones, though all wind turbines face some challenges. They may cause visual impacts, rare fires, or fluid leaks. Noise pollution can affect humans and animals, but offshore locations reduce this (Grimsrud et al., 2024). Wind turbines can kill birds and bats, especially those on migratory routes, which is a greater concern for threatened species (Guest et al., 2022). Research is ongoing to reduce this impact. Construction involves limited disruption from roads, turbine production, and foundation concrete (Mathern et al., 2021). But life cycle analysis shows turbines generate far more energy than they consume during production and installation.

1.2.4.2 Solar Energy

Solar energy's environmental impact depends on location. Manufacturing photovoltaic cells produces hazardous waste and emits greenhouse gases due to fossil fuel use (Tawalbeh et al., 2021). While solar panels are often installed on rooftops or built into structures, large solar farms can harm fragile desert ecosystems. Some solar thermal systems use hazardous heat-transfer fluids and require water for cleaning and cooling, which can strain ecosystems, especially in arid regions (Coimbra, 2025).

1.2.4.3 Hydropower

Hydropower, though renewable and emission-free during operation, can cause major ecological damage. Dams and reservoirs obstruct fish migration, harm aquatic life, and require fish ladders to mitigate some effects (Girgiri et al., 2023). Storage hydropower destroys land ecosystems, disrupts plant and animal life, floods natural and cultural sites, and may displace communities, as seen with the Three Gorges Dam (Kuban 2021). Reservoirs alter water flow, chemistry, and temperature, changing river ecosystems. Stagnant water can emit methane and carbon dioxide, sometimes equaling or exceeding emissions from fossil fuels. Despite its many advantages including low operational emissions, high efficiency, and long lifespans, hydropower is limited by geographic constraints (Gemechu & Kumar, 2022). Suitable locations require specific topographic and hydrological conditions, such as elevation changes and consistent water flow. Moreover, seasonal variations in rainfall and snowmelt, as well as long-term impacts from climate change, can affect water availability and reduce generation capacity over time (Wasti et al., 2022). Hydropower plays a crucial role in decarbonizing electricity systems, but sustainable development practices are necessary to mitigate its environmental and social impacts. Modern approaches increasingly emphasize small-scale, modular hydropower, improved fish passage systems, and multi-purpose water management strategies that balance energy generation with ecosystem preservation and community needs (Paraschiv & Paraschiv, (2023).

1.2.4.4 Geothermal Energy

Geothermal energy is a renewable resource that harnesses heat stored beneath the Earth's surface. This subterranean heat originates from the planet's formation and the ongoing radioactive decay of minerals in the Earth's crust and mantle (Yadav, et al., 2022). In geothermal power systems, wells are drilled into hot rock formations to access steam or hot water, which is then used to drive turbines and generate electricity, or to provide direct heating for buildings, greenhouses, and industrial processes.

Despite its sustainability and minimal carbon footprint, geothermal energy presents several challenges that affect its long-term viability and expansion:

1. Sustainable Use of Groundwater: Traditional hydrothermal geothermal systems rely on naturally occurring reservoirs of water and steam (Sharmin et al., 2023). These systems are only sustainable if the rate of water extraction does not exceed the natural recharge rate.

Excessive withdrawal can lead to declining pressure and temperature within the reservoir, reducing energy output and, in severe cases, forcing plants to shut down. To combat this, many modern geothermal facilities re-inject used water back into the ground to maintain reservoir pressure and extend system longevity (Davoodi et al., 2024).

2. Geochemical and Mechanical Challenges: Geothermal fluids are often rich in dissolved minerals, including silica, sulfur, and heavy metals. These minerals can precipitate out at surface temperatures, leading to scaling, corrosion, and clogging of pipes and turbines (Duku et al., 2022). This increases maintenance costs and shortens the equipment lifespan. Advanced materials and chemical treatments are required to mitigate these effects, but they add complexity and expense to operations.
3. Geographic Limitations: Conventional geothermal power generation is highly site-specific. Productive geothermal resources are typically found in tectonically active regions—such as along the Pacific Ring of Fire, where magma lies relatively close to the Earth's surface (Livescu et al. 2023). This restricts the widespread deployment of geothermal energy to a limited number of countries like the U.S., Iceland, Indonesia, the Philippines, and Kenya. In areas without natural hydrothermal activity, geothermal energy is currently not economically viable without significant technological enhancement (Sharmin et al., 2023).
4. Enhanced Geothermal Systems (EGS): To overcome geographic limitations, researchers and developers are advancing enhanced geothermal systems, which involve drilling deep into hot dry rock and injecting high-pressure water to create artificial fractures that facilitate heat exchange (Akhigbe 2025). While EGS could dramatically expand the availability of geothermal energy, it carries geophysical risks, most notably induced seismicity (Zhou et al., 2024). The process of injecting water into subsurface rock formations can cause small earthquakes, especially if faults are reactivated. Though these quakes are typically minor, they raise public safety and regulatory concerns, particularly in populated areas. Geothermal energy offers stable, low-emission power generation with a small land footprint and high reliability (Im et al., 2021).

Most renewable energy sources solar, wind, hydropower, and biomass are derived directly or indirectly from solar energy. Since the sun will shine for billions of years, it represents an inexhaustible energy source. Renewable technologies enable decentralized power generation, allowing homes and communities to produce energy independently of the grid. This contrasts with fossil fuels like coal, which require large, centralized facilities. Small-scale solar panels and wind turbines are feasible for remote or underdeveloped areas lacking grid infrastructure. Renewables also avoid environmental hazards like oil spills and reduce emissions that contribute to climate change and air pollution, improving public health.

1.3 BOOSTING ENERGY EFFICIENCY FOR CARBON EMISSION REDUCTION

Enhancing energy efficiency also plays a key role in emission reduction. Studies show that a 10% improvement in efficiency can lead to an 8.8% decline in CO_2 emissions over time, making it a powerful complement to renewable energy adoption (Min et al., 2022). Fuel sustainability is not just about switching energy sources; it also involves managing land, water, and ecosystems responsibly (Guddaraddi et al., 2023). The production of biofuels, for example, must be balanced with the preservation of biodiversity, protection of food systems, and the conservation of water resources (Barwant et al., 2025). Waste-to-energy approaches further support sustainability goals by using agricultural, municipal, and industrial waste as fuel feedstock, reinforcing circular economy practices. Effective implementation of fuel sustainability requires integrated policies that align energy, agriculture, environment, and industry sectors. Coordinated planning ensures that sustainability efforts are systemic rather than fragmented (Raman et al., 2024). By embracing sustainable fuels, nations can also mitigate the risks associated with fossil fuel market volatility and geopolitical tensions. Investment in clean technologies such as solar, wind, and green hydrogen not only addresses environmental concerns but also enhances national resilience and long-term energy security (Ndigwe et al., 2025). Investing in sustainable fuels such as biofuels and hydrogen especially those produced locally can strengthen energy independence and stability. Sustainable energy also offers equitable solutions. Cost-effective and decentralized renewable systems can expand access to energy in rural or underserved areas, fostering social inclusion and economic development (Rumbayan et al., 2025).

1.4 ALIGNING ENERGY POLICY AND SUSTAINABLE FINANCE FOR A JUST ENERGY TRANSITION

The global shift to renewable energy, driven by climate change mitigation, demands coordinated action from policymakers and financial institutions (Adewumi et al., 2024). To meet Paris Agreement goals, renewables must exceed 60% of global power generation by 2030, requiring over $4 trillion in annual investments (Amir Fikry 2025). This transition involves more than technological change; it necessitates deep economic restructuring. Energy policies and financial mechanisms are interdependent policies that create stable markets, while sustainable finance provides essential capital. Instruments like feed-in tariffs, carbon pricing, and renewable portfolio standards reduce investment risk and signal long-term support (Onabowale 2024). In turn, sustainable finance tools green bonds, climate funds, and ESG-integrated investments channel capital toward clean energy. The European Green Deal exemplifies this synergy, combining regulatory frameworks with financial taxonomies to accelerate decarbonization and manage high upfront costs. Sustainable finance plays a central role in meeting climate targets, not only funding infrastructure but also backing innovation in technologies like hydrogen and energy storage. It also requires mainstream financial institutions to embed ESG criteria to ensure alignment with climate goals (Mahmood et al., 2024). Strong policies reinforce this by ensuring market predictability, boosting investment, and promoting inclusive energy access. Key policy instruments such as FITs, net metering, and renewable energy auctions have proven effective globally. Germany and Japan used FITs to drive wind and solar growth, while net metering in the U.S. and India encouraged rooftop solar adoption (Kishore et al., 2025). Competitive auctions in Brazil and South Africa lowered costs and enhanced transparency. Successful national strategies, such as Denmark's wind policy or China's renewable mandates highlight the importance of stability and clear targets. Public-private partnerships (PPPs) further bridge funding gaps by combining public support with private expertise. Morocco's Noor Solar Complex and Kenya's Lake Turkana Wind Power Project demonstrate how PPPs can deliver large-scale renewable energy in developing markets (Ogunsola et al., 2024). These collaborative models are vital for scaling clean energy deployment, especially where capital is scarce, and contribute meaningfully to national development and global sustainability goals.

1.5 BENEFITS OF ENERGY POLICY AND SUSTAINABLE FINANCE INTEGRATION

Aligning energy policies with financial incentives accelerates renewable energy adoption by reducing investment barriers (Obuseh et al., 2025). Tools like feed-in tariffs, tax credits, and grants such as the U.S. Investment Tax Credit lower upfront costs, making projects more viable. When paired with stable policy frameworks, as seen in Germany's Renewable Energy Sources Act, these incentives drive large-scale deployment. China and India exemplify this approach, achieving rapid growth in wind and solar through subsidies, auctions, and concessional financing. Integrated energy policy and finance also mitigate climate change by promoting low-carbon energy (Baziar & Parsa, 2024. Carbon pricing and emission targets disincentivize fossil fuels, while sustainable finance, including green bonds, channels capital into clean energy. The European Investment Bank's green bond program, for example, has funded extensive renewable development. Furthermore, sustainable finance builds climate resilience by supporting adaptive infrastructure like microgrids. Blended finance initiatives in countries such as Bangladesh have brought solar power to millions of off-grid households (Hassan et al., 2022). Economically, renewable energy drives job creation and GDP growth. The sector employed over 12 million people globally in 2020, with countries like Brazil and Germany showing strong employment through supportive policies. Renewable infrastructure also stimulates investment and economic diversification. Projects like Morocco's Noor Solar Complex illustrate how renewables can boost GDP and foster industrial development (Benbba et al., 2024). In developed economies, renewable investment—such as the $50 billion in U.S. private funding in 2021—has proven a key growth engine. This underscores the importance of aligning finance with policy for sustainable development.

1.6 CHALLENGES IN IMPLEMENTING ENERGY POLICY AND SUSTAINABLE FINANCE IN DEVELOPING ECONOMIES

The lack of unified global energy policies hinders renewable energy expansion by causing inefficiencies and inconsistent investment flows (Dai et al., 2025). Disparities in incentives, regulatory timelines, and carbon pricing systems such as the EU's Emissions Trading System versus weaker or absent frameworks elsewhere create uncertainty and discourage cross-border cooperation. Regulatory misalignment also delays international projects due to incompatible standards and permitting processes. Collaborative platforms like IRENA and UNFCCC are essential for promoting policy harmonization and supporting global market development (Latilo et al., 2024).

High upfront costs and limited access to affordable financing remain major obstacles, particularly in emerging markets where capital is scarce, and risk perceptions are high. Local developers often lack financial capacity, and many financial institutions are hesitant to fund renewable projects due to regulatory uncertainty or lack of expertise. Innovative tools like blended finance, public-private partnerships, and credit enhancements (e.g., guarantees, insurance) help de-risk investments. Programs like the Green Climate Fund and MIGA have enabled project financing in high-risk countries such as Kenya, Nigeria, and Indonesia (Brown, 2022).

Market volatility, especially fluctuations in fossil fuel and electricity prices, threatens the competitiveness and revenue stability of renewable projects. Without long-term contracts like power purchase agreements (PPAs), developers face financial risk. Policy measures such as carbon pricing and renewable portfolio standards help stabilize market conditions by internalizing environmental costs (Yadav 2025). Uncertainties around new technologies also deter investment. Limited performance data for innovations like energy storage or floating wind turbines raises concerns about long-term reliability. Government-backed guarantees and demonstration funding, such as through the EU's Horizon Europe program, are key to supporting the deployment of emerging technologies and mitigating associated financial risks.

1.7 CONCLUSION

Achieving sustainable energy systems demands a strategic transition from fossil fuels to low-carbon and renewable alternatives. This includes scaling up technologies like solar, wind, hydropower, and biofuels, while enhancing energy efficiency and embedding circular economic practices. To overcome structural and financial barriers, especially in developing economies, coordinated global policies and sustainable finance mechanisms—such as feed-in tariffs, green bonds, and public-private partnerships—are essential. By aligning innovation, regulation, and investment, the world can reduce environmental harm, strengthen energy security, and ensure an inclusive, climate-resilient future.

REFERENCES

Abbasi, K. R., Shahbaz, M., Zhang, J., Irfan, M., & Alvarado, R. (2022). Analyze the environmental sustainability factors of China: The role of fossil fuel energy and renewable energy. *Renewable Energy*, *187*, 390–402. https://doi.org/10.1016/j.renene.2022.01.066

Abdali, L. M. A., Yakimovich, B. A., & Kuvshinov, V. V. (2019). Hybrid power generation by using solar and wind energy. In Экологическая, промышленная и энергетическая безопасность-*2019* (pp. 26–31).

Adamo, N., Al-Ansari, N., & Sissakian, V. (2021). Review of climate change impacts on human environment: Past, present and future projections. *Engineering*, *13*(11), 605–630. https://doi.org/10.4236/eng.2021.1311044

Adeniyi, A. G., Emenike, E. C., Iwuozor, K. O., Okoro, H. K., & Ige, O. O. (2022). Acid mine drainage: The footprint of the Nigeria mining industry. *Chemistry Africa*, *5*(6), 1907–1920. https://doi.org/10.1007/s42250-022-00493-3

Adewumi, A., Olu-Lawal, K. A., Okoli, C. E., Usman, F. O., & Usiagu, G. S. (2024). Sustainable energy solutions and climate change: A policy review of emerging trends and global responses. *World Journal of Advanced Research and Reviews*, *21*(2), 408–420. https://doi.org/10.30574/wjarr.2024.21.2.0474

Afonja, A. A. (2022). *Basic coal science and technology*. Chudace Publishing.

Akhigbe, E. E. (2025). Advancing geothermal energy: A review of technological developments and environmental impacts. *Gulf Journal of Advance Business Research*. https://doi.org/10.51594/gjabr.v3i2.104

Alagoz, E., Alghawi, Y., & Ergul, M. S. (2023). Innovation in exploration and production: How technology is changing the oil and gas landscape. *Journal of Energy and Natural Resources*, *12*(3), 25–29.

Al-Fattah, S. M. (2020). Non-OPEC conventional oil: Production decline, supply outlook and key implications. *Journal of Petroleum Science and Engineering*, *189*, 107049. https://doi.org/10.1016/j.petrol.2020.107049

Al-Kuwari, O. (2023). *The future of liquified natural gas (LNG) in the energy transition: Options and implications for the LNG industry in a decarbonising world* (Doctoral dissertation, UCL (University College London)).

Amir Fikry, A. A. (2025). Investing in renewable energy: Challenges and proposed solutions to achieve the sustainable development goals of vision 2030. Российские регионы в фокусе перемен: сборник докладов–*2024*.–Екатеринбург, *2025*, 953–957. http://elar.urfu.ru/handle/10995/142836

Anjos, S. M., Sombra, C. L., & Spadini, A. R. (2024). Petroleum exploration and production in Brazil: From onshore to ultra-deepwaters. *Petroleum Exploration and Development*, *51*(4), 912–924. https://doi.org/10.1016/S1876-3804(24)60515-X

Arnold, B. J. (2023). Coal formation. In *The coal handbook* (pp. 3–21). Woodhead Publishing. https://doi.org/10.1016/B978-0-12-824328-2.00014-5

Asghar, A., Sairash, S., Hussain, N., Baqar, Z., Sumrin, A., & Bilal, M. (2022). Current challenges of biomass refinery and prospects of emerging technologies for sustainable bioproducts and bioeconomy. *Biofuels, Bioproducts and Biorefining*, *16*(6), 1478–1494. https://doi.org/10.1002/bbb.2403

Bagher, A. M., Vahid, M., Mohsen, M., & Parvin, D. (2015). Hydroelectric energy advantages and disadvantages. *American Journal of Energy Science*, *2*(2), 17–20. https://www.researchgate.net/publication/275094706

Barwant, M. M., Sharma, S. A., Banerjee, S., Lodh, D., & Fulari, P. (2025). Impact of Biofuels on the Environment, Biodiversity, Agriculture, and Water Resources: Potential Solutions. In *Recent trends in lignocellulosic biofuels and bioenergy* (pp. 65–92). Singapore: Springer. https://doi.org/10.1007/978-981-96-4636-4_3

Baziar, A., & Parsa, N. (2024). Legal challenges in renewable energy development: A comparative study of China and selected countries. *arXiv preprint arXiv:2412.10203*. https://doi.org/10.48550/arXiv.2412.10203

Benbba, R., Barhdadi, M., Ficarella, A., Manente, G., Romano, M. P., El Hachemi, N., … & Outzourhit, A. (2024). Solar energy resource and power generation in morocco: Current situation, potential, and future perspective. *Resources*, *13*(10), 140. https://doi.org/10.3390/resources13100140

Bernardino, A. F., Cordes, E. E., & Schlacher, T. A. (2020). The natural capital of offshore oil, gas, and methane hydrates in the World. *Natural capital and exploitation of the deep ocean*, *64*(683), 111.

Bhattacharjee, A., & Dutta, S. (2022). A Unique Collaborative Perspective on the Utilisation of Biochar in Accelerated Biodegradation of Discharge from Factories. In *Biochar and its Application in Bioremediation* (pp. 345–361). Singapore: Springer Nature Singapore. https://doi.org/10.1007/978-981-16-4059-9_16

Blakers, A., Stocks, M., Lu, B., & Cheng, C. (2021). A review of pumped hydro energy storage. *Progress in Energy*, *3*(2), 022003. https://doi.org/10.1088/2516-1083/abeb5b

Boak, J., & Kleinberg, R. (2020). Shale gas, tight oil, shale oil and hydraulic fracturing. In *Future energy*, Trevor M. Letch (ed.), Third Edition (pp. 67–95). Elsevier.

Bonenkamp, T. B., Middelburg, L. M., Hosli, M. O., & Wolffenbuttel, R. F. (2020). From bioethanol containing fuels towards a fuel economy that includes methanol derived from renewable sources and the impact on European Union decision-making on transition pathways. *Renewable and Sustainable Energy Reviews, 120*, 109667. https://doi.org/10.1016/j.rser.2019.109667

Bošnjaković, M., Katinić, M., Santa, R., & Marić, D. (2022). Wind turbine technology trends. *Applied Sciences, 12*(17), 8653. https://doi.org/10.3390/app12178653

Brown, L. H. (2022). Addressing the adaptation finance gap: Scaling up investment in climate change adaptation using blended finance solutions. http://hdl.handle.net/11427/39047

Cameron, P. D., & Stanley, M. C. (2017). *Oil, gas, and mining: A sourcebook for understanding the extractive industries*. World Bank Publications.

Chang, S. E., & Tanner, A. (2022). A community impact scale for regional disaster planning with transportation disruption. *Natural Hazards Review, 23*(3), 04022019. https://doi.org/10.1061/(ASCE)NH.1527-6996.0000563

Chen, S., Zhang, C., & Lu, X. (2023). Energy conversion from Fossil fuel to renewable energy. In *Handbook of air quality and climate change* (pp. 1–44). Singapore: Springer Nature Singapore. https://doi.org/10.1007/978-981-15-2527-8_42-1

Coimbra, C. F. (2025). Energy meteorology for the evaluation of solar farm thermal impacts on desert habitats. *Advances in Atmospheric Sciences, 42*(2), 313–326. https://doi.org/10.1007/s00376-024-4242-3

Cordes, E. E., Jones, D. O., Schlacher, T. A., Amon, D. J., Bernardino, A. F., Brooke, S., … & Witte, U. (2016). Environmental impacts of the deep-water oil and gas industry: A review to guide management strategies. *Frontiers in Environmental Science, 4*, 58. https://doi.org/10.3389/fenvs.2016.00058

Cui, L., Li, Y., Tang, Y., Shi, Y., Wang, Q., Yuan, X., & Kellett, J. (2018). Integrated assessment of the environmental and economic effects of an ultra-clean flue gas treatment process in coal-fired power plant. *Journal of Cleaner Production, 199*, 359–368. https://doi.org/10.1016/j.jclepro.2018.07.174

Dadashi, Z., Mahmoudi, A., & Rashidi, S. (2022). Capacity and strategies of energy production from renewable sources in Arab countries until 2030: A review from renewable energy potentials to environmental issues. *Environmental Science and Pollution Research, 29*(32), 47837–47866. https://doi.org/10.1007/s11356-022-20544-z

Dai, J., Farooq, U., & Alam, M. M. (2025). Navigating energy policy uncertainty: Effects on fossil fuel and renewable energy consumption in G7 economies. *International Journal of Green Energy, 22*(2), 239–252. https://doi.org/10.1080/15435075.2024.2413676

Davoodi, S., Al-Shargabi, M., Wood, D. A., Slivkin, S., Shishaev, G., & Rukavishnikov, V. S. (2024). Geothermal energy recovery from abandoned petroleum wells: A review of the challenges and opportunities. *Sustainable Energy Technologies and Assessments, 68*, 103870. https://doi.org/10.1016/j.seta.2024.103870

Deffeyes, K. S. (2008). *The future of fossil fuels: From Hubbert's peak* (Vol. 4). Princeton University Press.

Dincer, I., & Acar, C. (2017). Smart energy systems for a sustainable future. *Applied energy, 194*, 225–235. https://doi.org/10.1016/j.apenergy.2016.12.058

Duku, E. O., Ongarora, B. G., & Tanui, P. (2022). Effect of boiling and cooling of geothermal fluids on precipitation of secondary minerals: A case study of Olkaria Fields, Kenya. *Journal of Geoscience and Environment Protection, 10*(9), 251–270. https://doi.org/10.4236/gep.2022.109015

Elkhatat, A., & Al-Muhtaseb, S. (2024). Climate change and energy security: A comparative analysis of the role of energy policies in advancing environmental sustainability. *Energies, 17*(13), 3179. https://doi.org/10.3390/en17133179

Firth, A., Zhang, B., & Yang, A. (2019). Quantification of global waste heat and its environmental effects. *Applied Energy, 235*, 1314–1334. https://doi.org/10.1016/j.apenergy.2018.10.102

Freedman, B., Dorsey, W., Frazier, A., Kambhampati, M., Galiotos, J., & Mukherjee, S. (2024). Renewable and non-renewable energy sources. *Environmental Science*.

Gemechu, E., & Kumar, A. (2022). A review of how life cycle assessment has been used to assess the environmental impacts of hydropower energy. *Renewable and Sustainable Energy Reviews, 167*, 112684. https://doi.org/10.1016/j.rser.2022.112684

Gharbia, Y., Derakhshandeh, J. F., Amer, A. M., & Dinc, A. (2024). Geothermal heat pump for space cooling and heating in Kuwaiti climate. *Processes, 12*(5), 910. https://doi.org/10.3390/pr12050910

Girgiri, A., Baba, M. M., Alibe, A. M., & Kashim, B. G. (2023). Elucidation of the impact of renewable energy power plants on the environment. *Fane-Fane International Multi-Disciplinary Journal, 7*(1, June), 17–25.

Griffiths, S., Sovacool, B. K., Kim, J., Bazilian, M., & Uratani, J. M. (2022). Decarbonizing the oil refining industry: A systematic review of sociotechnical systems, technological innovations, and policy options. *Energy Research & Social Science*, *89*, 102542. https://doi.org/10.1016/j.erss.2022.102542

Grimsrud, K., Hagem, C., Haaskjold, K., Lindhjem, H., & Nowell, M. (2024). Spatial trade-offs in national land-based wind power production in times of biodiversity and climate crises. *Environmental and Resource Economics*, *87*(2), 401–436. https://doi.org/10.1007/s10640-023-00764-8

Guddaraddi, A., Singh, A., Amrutha, G., Saikanth, D. R. K., Kurmi, R., Singh, G., … & Singh, B. V. (2023). Sustainable biofuel production from agricultural residues an eco-friendly approach: A review. *Int J Environ Clim Change*, *13*(10), 2905–2914. https://doi.org/10.9734/IJECC/2023/v13i102956

Gudde, N. J. (2018). Adaptation of oil refineries to make modern fuels. *Proceedings of the Institution of Mechanical Engineers, Part D: Journal of Automobile Engineering*, *232*(1), 5–21. https://doi.org/10.1177/0954407016680522

Guest, E. E., Stamps, B. F., Durish, N. D., Hale, A. M., Hein, C. D., Morton, B. P., … & Fritts, S. R. (2022). An updated review of hypotheses regarding bat attraction to wind turbines. *Animals*, *12*(3), 343. https://doi.org/10.3390/ani12030343

Gupta, A. K. (2015). Efficient wind energy conversion: Evolution to modern design. *Journal of Energy Resources Technology*, *137*(5), 051201. https://doi.org/10.1115/1.4030109

Hanna, S. H. S., Obiomon, P., Osborne-Lee, I. W., Cesaretti, G. P., & Misso, R. (2019). Global Solar Energy availability model and use in relationship to Ecological Human imprint: Economic Sustainability Impact and Assessment. *Renewable Energy and Power Quality Journal*, *17*, 522–529.

Hannun, R. M., & Razzaq, A. H. A. (2022, March). Air pollution resulted from coal, oil and gas firing in thermal power plants and treatment: A review. In *IOP conference series: Earth and environmental science* (Vol. 1002, No. 1, p. 012008). IOP Publishing. https://doi.org/10.1088/1755-1315/1002/1/012008

Hasan, M. M., Hossain, S., Mofijur, M., Kabir, Z., Badruddin, I. A., Yunus Khan, T. M., & Jassim, E. (2023). Harnessing solar power: A review of photovoltaic innovations, solar thermal systems, and the dawn of energy storage solutions. *Energies*, *16*(18), 6456. https://doi.org/10.3390/en16186456

Hassan, R., Das, B. K., & Hasan, M. (2022). Integrated off-grid hybrid renewable energy system optimization based on economic, environmental, and social indicators for sustainable development. *Energy*, *250*, 123823. https://doi.org/10.1016/j.energy.2022.123823

Herche, W. (2017). Solar energy strategies in the US utility market. *Renewable and Sustainable Energy Reviews*, *77*, 590–595. https://doi.org/10.1016/j.rser.2017.04.028

Hou, Y. (2024). A comprehensive review of fuels used in combustion processes. https://urn.fi/URN:NBN:fi-fe2024043024187

Huang, J., Zhang, X., Zhang, Q., Lin, Y., Hao, M., Luo, Y., … & Zhang, J. (2017). Recently amplified arctic warming has contributed to a continual global warming trend. *Nature Climate Change*, *7*(12), 875–879. https://doi.org/10.1038/s41558-017-0009-5

Im, D. H., Chung, J. B., Kim, E. S., & Moon, J. W. (2021). Public perception of geothermal power plants in Korea following the Pohang earthquake: A social representation theory study. *Public Understanding of Science*, *30*(6), 724–739. https://doi.org/10.1177/09636625211012551

Irfan, M., Musarat, M. A., Alaloul, W. S., & Ghufran, M. (2024). Radiative forcing on climate change: Assessing the effect of greenhouse gases on energy balance of Earth. In *Advances and technology development in greenhouse gases: Emission, capture and conversion* (pp. 137–167). Elsevier. https://doi.org/10.1016/B978-0-443-19066-7.00012-6\

James, R., Washington, R., Schleussner, C. F., Rogelj, J., & Conway, D. (2017). Characterizing half-a-degree difference: A review of methods for identifying regional climate responses to global warming targets. *Wiley Interdisciplinary Reviews: Climate Change*, *8*(2), e457. https://doi.org/10.1002/wcc.457

Jha, M. K., & Dev, M. (2024). Impacts of climate change. In *Smart internet of things for environment and healthcare* (pp. 139–159). Cham: Springer Nature Switzerland. https://doi.org/10.1007/978-3-031-70102-3_10

Jia, Z., Wang, B., Song, S., & Fan, Y. (2014). Blue energy: Current technologies for sustainable power generation from water salinity gradient. *Renewable and Sustainable Energy Reviews*, *31*, 91–100. https://doi.org/10.1016/j.rser.2013.11.049

Johnsson, F., Kjärstad, J., & Rootzén, J. (2019). The threat to climate change mitigation posed by the abundance of fossil fuels. *Climate Policy*, *19*(2), 258–274. https://doi.org/10.1080/14693062.2018.1483885

Jolie, E., Scott, S., Faulds, J., Chambefort, I., Axelsson, G., Gutiérrez-Negrín, L. C., … & Zemedkun, M. T. (2021). Geological controls on geothermal resources for power generation. *Nature Reviews Earth & Environment*, *2*(5), 324–339. https://doi.org/10.1038/s43017-021-00154-y

Kabeyi, M. J. B., & Olanrewaju, O. A. (2022). Sustainable energy transition for renewable and low carbon grid electricity generation and supply. *Frontiers in Energy research*, *9*, 743114. https://doi.org/10.3389/fenrg.2021.743114

Kannan, R., Panos, E., Hirschberg, S., & Kober, T. (2022). A net-zero Swiss energy system by 2050: Technological and policy options for the transition of the transportation sector. *Futures & Foresight Science*, *4*(3–4). https://doi.org/10.1002/ffo2.126

Kerimray, A., Rojas-Solórzano, L., Torkmahalleh, M. A., Hopke, P. K., & Gallachóir, B. P. Ó. (2017). Coal use for residential heating: Patterns, health implications and lessons learned. *Energy for Sustainable Development*, *40*, 19–30. https://doi.org/10.1016/j.esd.2017.05.005

Kholod, N., Evans, M., Pilcher, R. C., Roshchanka, V., Ruiz, F., Coté, M., & Collings, R. (2020). Global methane emissions from coal mining to continue growing even with declining coal production. *Journal of Cleaner Production*, *256*, 120489.

Kishore, T. S., Kumar, P. U., & Ippili, V. (2025). Review of global sustainable solar energy policies: Significance and impact. *Innovation and Green Development*, *4*(2), 100224. https://doi.org/10.1016/j.igd.2025.100224

Kojima, T. (Ed.). (2019). *Carbon dioxide problem: Integrated Energy and environmental policies for the 21st century*. Routledge.

Kuban, N. (2021). Hydroelectric Plants and Dams as Industrial Heritage in the Context of Nature-Culture Interrelation: An Overview of Examples in Turkey. *Energies*, *14*(5), 1281. https://doi.org/10.3390/en14051281

Latilo, A., Uzougbo, N. S., Ugwu, M. C., Oduro, P., & Aziza, O. R. (2024). Developing legal frameworks for successful engineering, procurement, and construction projects. *International Journal of Applied Reserch in Social Sciences*, *6*(8), 1868–1883. https://doi.org/10.51594/ijarss.v6i8.1430

Li, W., Li, Y., & Shah, K. W. (2020). A materials perspective on radiative cooling structures for buildings. *Solar Energy*, *207*, 247–269. https://doi.org/10.1016/j.solener.2020.06.095

Liang, F. Y., Ryvak, M., Sayeed, S., & Zhao, N. (2012). The role of natural gas as a primary fuel in the near future, including comparisons of acquisition, transmission and waste handling costs of as with competitive alternatives. *Chemistry Central Journal*, *6*, 1–24. https://doi.org/10.1186/1752-153X-6-S1-S4

Liu, Z., Wang, H., Blackbourn, G., Ma, F., He, Z., Wen, Z., ... & Wu, Z. (2019). Heavy oils and oil sands: Global distribution and resource assessment. *Acta Geologica Sinica-English Edition*, *93*(1), 199–212. https://doi.org/10.1111/1755-6724.13778

Livescu, S., Dindoruk, B., Schulz, R., Boul, P., Kim, J., & Wu, K. (2023). Geothermal and electricity production: Scalable geothermal concepts in the future of geothermal in Texas: Contemporary prospects and perspectives

Lodhi, S. K., Gill, A. Y., & Hussain, I. (2024). Renewable energy technologies: Present patterns and upcoming paths in ecological power production. *Global Journal of Universal Studies*, *1*(1), 108–131.

Londoño-Pulgarin, D., Cardona-Montoya, G., Restrepo, J. C., & Muñoz-Leiva, F. (2021). Fossil or bioenergy? Global fuel market trends. *Renewable and Sustainable Energy Reviews*, *143*, 110905. https://doi.org/10.1016/j.rser.2021.110905

Luo, X., Huang, L., & Sui, X. (2025). Capacity optimization of existing reservoir hydropower expansion and its impact on power system flexibility. *Applied Energy*, *377*, 124549. https://doi.org/10.1016/j.apenergy.2024.124549

Mahmood, S., Sun, H., Iqbal, A., Alhussan, A. A., & El-Kenawy, E. S. M. (2024). Green finance, sustainable infrastructure, and green technology innovation: Pathways to achieving sustainable development goals in the belt and road initiative. *Environmental Research Communications*, *6*(10), 105036. https://doi.org/10.1088/2515-7620/ad898f

Mathern, A., von der Haar, C., & Marx, S. (2021). Concrete support structures for offshore wind turbines: Current status, challenges, and future trends. *Energies*, *14*(7), 1995. https://doi.org/10.3390/en14071995

Min, J., Yan, G., Abed, A. M., Elattar, S., Khadimallah, M. A., Jan, A., & Ali, H. E. (2022). The effect of carbon dioxide emissions on the building energy efficiency. *Fuel*, *326*, 124842. https://doi.org/10.1016/j.fuel.2022.124842

Mondal, S., & Lee, M. A. (2025). Impact of global warming on fisheries. In *Food security, Nutrition and sustainability through aquaculture technologies* (pp. 227–253). Cham: Springer Nature Switzerland.

Morais, R. R., Pascoal, A. M., Pereira-Júnior, M. A., Batista, K. A., Rodriguez, A. G., & Fernandes, K. F. (2019). Bioethanol production from Solanum lycocarpum starch: A sustainable non-food energy source for biofuels. *Renewable Energy*, *140*, 361–366. https://doi.org/10.1016/j.renene.2019.02.056

Moriarty, P., & Honnery, D. (2019). Ecosystem maintenance energy and the need for a green EROI. *Energy Policy*, *131*, 229–234. https://doi.org/10.1016/B978-0-12-814104-5.00006-5

Naveen Kumar, K. A., & Vigneshwaran, A. (2023). Renewable energy resources and their types. In *AI Techniques for renewable source integration and battery charging methods in electric vehicle applications* (pp. 116–135). IGI Global.

Naylor, R. L., & Higgins, M. M. (2018). The rise in global biodiesel production: Implications for food security. *Global food security*, *16*, 75–84. https://doi.org/10.1016/j.gfs.2017.10.004

Ndigwe, C. I., Erondu, I. N., Nnamdi, K. C., & Ekpemuaka, E. (2025). Clean energy transition for the green economy: A pathway to a sustainable green economy. In *Securing sustainable futures through blue and green economies* (pp. 427–454). IGI Global Scientific Publishing.

Nica, A., Popescu, A., & Ibanescu, D. C. (2019). Human influence on the climate system. *Current Trends in Natural Sciences*, *8*(15), 209–215.

Nwokediegwu, Z. Q. S., Ibekwe, K. I., Ilojianya, V. I., Etukudoh, E. A., & Ayorinde, O. B. (2024). Renewable energy technologies in engineering: A review of current developments and future prospects. *Engineering Science & Technology Journal*, *5*(2), 367–384.

Obuseh, E., Eyenubo, J., Alele, J., Okpare, A., & Oghogho, I. (2025). A Systematic Review of Barriers to Renewable Energy Integration and Adoption. *Journal of Asian Energy Studies*, *9*, 26–45. https://doi.org/10.24112/jaes.090002

Ogunsola, O. Y., Adebayo, Y. A., Dienagha, I. N., Ninduwezuor-Ehiobu, N., & Nwokediegwu, Z. S. (2024). Public-private partnership models for financing renewable energy and infrastructure development in Sub-Saharan Africa. *Gulf Journal of Advance Business Research*, *2*(6), 483–492. https://doi.org/10.51594/gjabr.v2i6.61

Olev, D. (2012). *Peak Oil and growth: The challenges and opportunities posed by the finite character of the fossil fuels to the world economic order* (Doctoral dissertation).

Olujobi, O. J., Okorie, U. E., Olarinde, E. S., & Aina-Pelemo, A. D. (2023). Legal responses to energy security and sustainability in Nigeria's power sector amidst fossil fuel disruptions and low carbon energy transition. *Heliyon*, *9*(7).

Omer, A. M. (2009). Energy use and environmental impacts: A general review. *Journal of renewable and Sustainable Energy*, *1*(5). https://doi.org/10.1063/1.3220701

Onabowale, O. (2024). Energy policy and sustainable finance: Navigating the future of renewable energy and energy markets. *World Journal of Advanced Research and Reviews*, *25*, 2235–2252. https://doi.org/10.30574/wjarr.2025.25.1.0319

Ozgen, S., Cernuschi, S., & Caserini, S. (2021). An overview of nitrogen oxides emissions from biomass combustion for domestic heat production. *Renewable and Sustainable Energy Reviews*, *135*, 110113. https://doi.org/10.1016/j.rser.2020.110113

Pal, D., & Sen, S. (2024). In-depth coverage of petroleum waste sources, characteristics, environmental impact, and sustainable remediation process. In *Impact of petroleum waste on environmental pollution and its sustainable management through circular economy* (pp. 1–38). Cham: Springer Nature Switzerland. https://doi.org/10.1007/978-3-031-48220-5_1

Pandey, B., Gautam, M., & Agrawal, M. (2018). Greenhouse gas emissions from coal mining activities and their possible mitigation strategies. In *Environmental carbon footprints* (pp. 259–294). Butterworth-Heinemann. https://doi.org/10.1016/B978-0-12-812849-7.00010-6

Pang, X. Q., Jia, C. Z., & Wang, W. Y. (2015). Petroleum geology features and research developments of hydrocarbon accumulation in deep petroliferous basins. *Petroleum Science*, *12*, 1–53. https://doi.org/10.1007/s12182-015-0014-0

Paraschiv, L. S., & Paraschiv, S. (2023). Contribution of renewable energy (hydro, wind, solar and biomass) to decarbonization and transformation of the electricity generation sector for sustainable development. *Energy Reports*, *9*, 535–544. https://doi.org/10.1016/j.egyr.2023.07.024

Poggi, F., Firmino, A., & Amado, M. (2018). Planning renewable energy in rural areas: Impacts on occupation and land use. *Energy*, *155*, 630–640.

Prasad, S., Venkatramanan, V., & Singh, A. (2020). Renewable energy for a low-carbon future: Policy perspectives. In *Sustainable bioeconomy: Pathways to sustainable development goals* (pp. 267–284). Singapore: Springer Singapore. https://doi.org/10.1007/978-981-15-7321-7_12

Rahman, A., Farrok, O., & Haque, M. M. (2022). Environmental impact of renewable energy source based electrical power plants: Solar, wind, hydroelectric, biomass, geothermal, tidal, ocean, and osmotic. *Renewable and sustainable energy reviews*, *161*, 112279. https://doi.org/10.1016/j.rser.2022.112279

Raman, R., Gunasekar, S., Dávid, L. D., Rahmat, A. F., & Nedungadi, P. (2024). Aligning sustainable aviation fuel research with sustainable development goals: Trends and thematic analysis. *Energy Reports*, *12*, 2642–2652. https://doi.org/10.1016/j.egyr.2024.08.076

Roos, A. (2021). *Renewing power: Including global asymmetries within the system boundaries of solar photovoltaic technology*. Lund University.

Rop, B. K., Anyona, S., Kipseba, E. K., Namwiba, W. H., & Kipkiror, L. J. (2024). Review of geothermal resources as manifestations of volcanism in Eastern Africa. http://localhost/xmlui/handle/123456789/6461

Rumbayan, M., Kindangen, J., Sambul, A., Sompie, S., & Cross, J. (2025). Solar energy implementation in rural communities and its contributions to SDGs: A systematic literature review. *Unconventional Resources*, 100180. https://doi.org/10.1016/j.uncres.2025.100180

Safari, A., Das, N., Langhelle, O., Roy, J., & Assadi, M. (2019). Natural gas: A transition fuel for sustainable energy system transformation?. *Energy Science & Engineering*, *7*(4), 1075–1094. https://doi.org/10.1002/ese3.380

Saleh, H. M., & Hassan, A. I. (2024). The challenges of sustainable energy transition: A focus on renewable energy. *Applied Chemical Engineering*, *7*(2), 2084. https://doi.org/10.59429/ace.v7i2.2084

Schiemann, M., Böhm, B., Chirone, R., Senneca, O., Ströhle, J., Umeki, K., & Vujanovic, M. (2022). Technical solutions to foster the global energy transition: Special issue on clean fuel conversion technologies for carbon dioxide and pollutant reduction. *Renewable and Sustainable Energy Reviews*, *154*, 111770. https://doi.org/10.1016/j.rser.2021.111770

Sharmin, T., Khan, N. R., Akram, M. S., & Ehsan, M. M. (2023). A state-of-the-art review on geothermal energy extraction, utilization, and improvement strategies: Conventional, hybridized, and enhanced geothermal systems. *International Journal of Thermofluids*, *18*, 100323. https://doi.org/10.1016/j.ijft.2023.100323

Shindell, D., & Smith, C. J. (2019). Climate and air-quality benefits of a realistic phase-out of fossil fuels. *Nature*, *573*(7774), 408–411. https://doi.org/10.1038/s41586-019-1554-z

Sreekanth, K. J. (2024). Energy policy analyses, energy transition and sustainability. In: K. J. Sreekanth (Eds) *Energy efficiency improvements with emission abatement for energy sustainability*. Springer, Singapore. https://doi.org/10.1007/978-981-97-9480-5_8

Suder, P. (2024). Run-of-River Device for Hydropower from the Crum Creek.

Tawalbeh, M., Al-Othman, A., Kafiah, F., Abdelsalam, E., Almomani, F., & Alkasrawi, M. (2021). Environmental impacts of solar photovoltaic systems: A critical review of recent progress and future outlook. *Science of The Total Environment*, *759*, 143528. https://doi.org/10.1016/j.scitotenv.2020.143528

Tiratsoo, E. N. (2022). World oilfields–Past, present, future. In *Seventy five years of progress in oil field science and technology* (pp. 39–47). Routledge.

Tolmachev, O., Urunov, A., Muminova, S., Dvoichenkova, G., & Davydov, I. (2020). Review of unconventional hydrocarbon resources: Production technologies and opportunities for development. *Mining of Mineral Deposits*. https://doi.org/10.33271/mining14.04.113

Tursi, A. (2019). A review on biomass: Importance, chemistry, classification, and conversion. *Biofuel Research Journal*, *6*(2), 962–979. https://doi.org/10.18331/BRJ2019.6.2.3

Ukoba, K., Yoro, K. O., Eterigho-Ikelegbe, O., Ibegbulam, C., & Jen, T. C. (2024). Adaptation of solar energy in the Global South: Prospects, challenges and opportunities. *Heliyon*, *10*(7).

Vasireddy, S., Morreale, B., Cugini, A., Song, C., & Spivey, J. J. (2011). Clean liquid fuels from direct coal liquefaction: Chemistry, catalysis, technological status and challenges. *Energy & Environmental Science*, *4*(2), 311–345.

Vohra, K., Vodonos, A., Schwartz, J., Marais, E. A., Sulprizio, M. P., & Mickley, L. J. (2021). Global mortality from outdoor fine particle pollution generated by fossil fuel combustion: Results from GEOS-Chem. *Environmental research*, *195*, 110754. https://doi.org/10.1016/j.envres.2021.110754

Wasti, A., Ray, P., Wi, S., Folch, C., Ubierna, M., & Karki, P. (2022). Climate change and the hydropower sector: A global review. *Wiley Interdisciplinary Reviews: Climate Change*, *13*(2), e757. https://doi.org/10.1002/wcc.757

Worku, M. Y. (2022). Recent advances in energy storage systems for renewable source grid integration: A comprehensive review. *Sustainability*, *14*(10), 5985. https://doi.org/10.3390/su14105985

Xiao, P., Li, J. J., Chen, W., Pang, W. X., Peng, X. W., Xie, Y., … & Chen, G. J. (2023). Enhanced formation of methane hydrate from active ice with high gas uptake. *Nature Communications*, *14*(1), 8068. https://doi.org/10.1038/s41467-023-43487-6

Xu, D., Yuan, Z. L., Bai, Z., Wu, Z., Chen, S., & Zhou, M. (2022). Optimal operation of geothermal-solar-wind renewables for community multi-energy supplies. *Energy*, *249*, 123672. https://doi.org/10.1016/j.energy.2022.123672

Yadav, A. (2025). Promoting economic stability: The role of renewable energy transition in mitigating global volatility. *International Journal of Energy Sector Management*. https://doi.org/10.1108/IJESM-06-2024-0032

Yadav, A., Agrahari, G. K., & Agrahari, S. (2022). Geothermal energy: Exploration, exploitation, and production. In *Status and future challenges for non-conventional energy sources volume 1* (pp. 79–109). Singapore: Springer Singapore. https://doi.org/10.1007/978-981-16-4505-1_5

Yang, P. (2024). Hydropower. In *Renewable energy: challenges and solutions* (pp. 109–138). Cham: Springer International Publishing. https://doi.org/10.1007/978-3-031-49125-2_4

Yi, Q., Li, W., Feng, J., & Xie, K. (2015). Carbon cycle in advanced coal chemical engineering. *Chemical Society Reviews*, *44*(15), 5409–5445. https://doi.org/10.1039/C4CS00453A

Yıldız, İ., & MacEachern, C. (2018). 1.2 historical aspects of energy. *Comprehensive Energy Systems*, *1*(2.2), 24.

Zhang, B. (2020). The effect of aerosols to climate change and society. *Journal of Geoscience and Environment Protection*, *8*(08), 55. http://www.scirp.org/journal/Paperabs.aspx?PaperID=102481

Zhang, Q., & Taleghani, A. D. (2024). Intelligent fluid flow management for enhanced geothermal system in fractured horizontal wells. *Applied Thermal Engineering*, *239*, 122191. https://doi.org/10.1016/j.applthermaleng.2023.122191

Zhang, R., Wang, G., Guo, S., Zamora, M. L., Ying, Q., Lin, Y., … & Wang, Y. (2015). Formation of urban fine particulate matter. *Chemical reviews*, *115*(10), 3803–3855. https://doi.org/10.1021/acs.chemrev.5b00067

Zhao, L., Oleson, K., Bou-Zeid, E., Krayenhoff, E. S., Bray, A., Zhu, Q., … & Oppenheimer, M. (2021). Global multi-model projections of local urban climates. *Nature Climate Change*, *11*(2), 152–157. https://doi.org/10.1038/s41558-020-00958-8

Zhou, W., Lanza, F., Grigoratos, I., Schultz, R., Cousse, J., Trutnevyte, E., … & Wiemer, S. (2024). Managing induced seismicity risks from enhanced geothermal systems: A good practice guideline. *Reviews of Geophysics*, *62*(4), e2024RG000849. https://doi.org/10.1029/2024RG000849

Zou, C., Zhao, Q., Zhang, G., & Xiong, B. (2016). Energy revolution: From a fossil energy era to a new energy era. *Natural Gas Industry B*, *3*(1), 1–11. https://doi.org/10.1016/j.ngib.2016.02.001

2 Environmental Impact of Traditional Energy Sources

Marlene Rojas-Le-Fort, Mauricio Colpari-Pozzo, Marell Navarro-Rojas, Juan Carlos Romero-Benavides, Tatiana Ordóñez-Zambrano, Darío Mondavi-Sobby, and Rodrigo Duarte-Casar

2.1 INTRODUCTION

The accelerating impacts of climate change are compelling societies worldwide to reassess fundamental aspects of human activity, from agricultural practices to energy production systems. At the heart of this transition lies the urgent need to address the environmental footprint of traditional energy sources (Rezaei et al., 2023; Tol, 2018). Fossil fuels—coal, oil, and natural gas—have powered industrial development for over a century, but their continued use comes at an increasingly unsustainable ecological cost. The combustion of these energy sources accounts for approximately 75% of global greenhouse gas emissions, driving climate change while simultaneously degrading air quality, water resources, and terrestrial ecosystems (Soeder, 2021).

Recent research shows the multidimensional environmental consequences of fossil fuel dependence. Beyond their well-documented role in climate change, traditional energy systems contribute to widespread ecological damage through extraction processes, transportation, and waste disposal (Shamoon et al., 2022). Surface mining operations permanently alter landscapes and destroy habitats, while oil spills and pipeline leaks have caused catastrophic damage to marine and freshwater ecosystems (Hettithanthri et al., 2024). Air pollution from fossil fuel combustion has been linked to millions of premature deaths annually, creating a global public health crisis that disproportionately affects vulnerable populations in both developed and developing nations (Lelieveld et al., 2019).

The environmental impacts of conventional energy systems extend across spatial and temporal scales. Locally, extraction activities (of fossil fuels but also minerals) contaminate groundwater and disrupt communities (Mukhopadhyay et al., 2022; Shiquan et al., 2022). Regionally, acid rain from power plant emissions damages forests and aquatic systems (Sanders & Barreca, 2022). Globally, the cumulative effects of carbon emissions threaten to destabilize climate systems that have remained relatively stable throughout human civilization (Asif et al., 2024). These interconnected challenges highlight the need for comprehensive energy transitions that address both mitigation and adaptation imperatives.

Legal and policy frameworks have struggled to keep pace with the growing recognition of these environmental costs. While some jurisdictions have implemented carbon pricing mechanisms and pollution regulations, the lack of coordinated global action has allowed many environmental externalities to remain unaccounted for in energy markets. This regulatory gap perpetuates the artificial economic advantage of fossil fuels over cleaner alternatives, delaying the necessary transition to sustainable energy systems (Abbass et al., 2022). Recent studies suggest that fully accounting for the costs of environmental damages would fundamentally reshape energy economics, making renewable sources immediately competitive with conventional options (Oliver Yébenes, 2024).

This chapter examines the complex environmental legacy of traditional energy systems through an interdisciplinary lens. It analyzes the lifecycle impacts of fossil fuel use, from extraction to

 DOI: 10.1201/9781003645405-2

consumption, while considering the social and ecological consequences of delayed action. The discussion emphasizes how these environmental burdens are distributed unevenly across populations and generations, raising important questions about energy justice and intergenerational equity. By synthesizing current scientific understanding with policy analysis, this chapter provides a foundation for rethinking humanity's energy future in an era of ecological constraints. The transition away from fossil fuels represents both a formidable challenge and a historic opportunity. As we stand at the crossroads of energy system transformation, understanding the full environmental costs of conventional energy sources becomes essential for making informed decisions about our collective future. The following analysis aims to provide that understanding while setting the stage for discussions of sustainable alternatives in subsequent chapters.

2.2 TRADITIONAL AND ALTERNATIVE ENERGY SOURCES

Energy production has historically relied on conventional fossil fuel sources, such as coal, oil, and natural gas (Zhironkin & Abu-Abed, 2024). However, the transition toward sustainable energy requires a shift to less polluting, alternative sources that reduce environmental impact and enhance energy security, addressing climate change and long-term reliability (Karduri & Ananth, 2023). This section examines both traditional (and primary) and alternative energy sources, highlighting their characteristics, benefits, and challenges.

2.2.1 Traditional Energy Sources

Traditional—or primary—energy sources primarily consist of fossil fuels and biomass, which have historically supplied the majority of global energy (Mohammad et al., 2021). These energy carriers have powered civilization, industrialization, and modern economies, but pose significant environmental and sustainability concerns.

Coal has been a dominant energy source for centuries due to its abundance and cost-effectiveness, making it a foundational fuel for industrialization and electricity generation (Kosowski, 2024). It is primarily used for electricity generation and industrial processes. However, coal combustion releases high levels of carbon dioxide (CO_2), sulfur dioxide (SO_2), nitrogen oxides (NO_x), and particulate matter, contributing to air pollution, acid rain, and climate change.

Oil is nowadays a crucial global energy resource, widely used in transportation, heating, and industrial applications, due to its high energy density and ease of storage and transportation (Bebbington et al., 2020). Despite its high energy density and versatility, oil extraction and consumption result in greenhouse gas emissions, environmental degradation, and geopolitical tensions due to uneven global distribution.

Natural gas is often considered a cleaner alternative to coal and oil because it emits lower CO_2 and pollutant levels when burned, making it a key transitional fuel in many national energy strategies (Webb, 2015). It is widely used for electricity generation, heating, and as a feedstock in chemical industries. However, concerns about methane leaks during extraction and transportation, as well as reliance on finite reserves, limit its long-term sustainability.

Biomass has been used as a primary source of energy for centuries, particularly in traditional societies where it serves as the main fuel for cooking, heating, and small-scale industrial activities. Derived from organic materials such as wood, charcoal, crop residues, and animal dung, biomass is widely available and accessible, making it a crucial energy source in many developing regions.

Traditional biomass combustion takes place in open fires or simple reactors, which often operate with low efficiency and produce high emissions of CO_2 and other toxic gases. While it offers an affordable and renewable energy source, its widespread use presents significant environmental and health challenges, including deforestation, air pollution, and greenhouse gas emissions, which have long been cause for concern (Crutzen & Andreae, 1990). Despite these drawbacks, biomass remains a vital energy source for millions of people, underscoring the need for improved technologies and sustainable management practices to mitigate its negative impacts.

Biomass, however, is nowadays also used as an alternative energy source in the shape of biofuels (Malik et al., 2024). More about this will be detailed in the next section.

2.3 ALTERNATIVE ENERGY SOURCES

Alternative energy sources encompass any energy source that can replace fossil fuels, including both renewable sources and some non-renewable sources like nuclear energy. The main goal of the renewable energy branch of alternative energies is to offer sustainable solutions to meet global energy demands from sources that are inexhaustible and can regenerate naturally while reducing environmental impacts. These include solar, wind, hydro, geothermal, wave, tidal, and bioenergy.

2.3.1 Solar Energy

Solar energy harnesses the sun's radiation through photovoltaic (PV) panels or solar thermal systems. It is an abundant and renewable resource with minimal emissions. Advancements in solar panel efficiency and energy storage technologies have made it a viable alternative; however, intermittency and land use remain challenges (Ukoba et al., 2024).

2.3.2 Wind Energy

Wind power is generated by converting kinetic energy from the wind into electricity using wind turbines. It is a cost-effective and rapidly expanding renewable energy source. However, variability in wind speeds, potential impacts on wildlife, and land use considerations pose challenges to its widespread adoption (Boadu & Otoo, 2024).

2.3.3 Hydropower

Hydropower utilizes flowing or falling water to generate electricity, making it one of the most established renewable energy sources. Large-scale hydropower projects provide significant electricity generation capacity, but they can have ecological and social impacts, including habitat disruption and displacement of communities. Small-scale and run-of-river systems offer more sustainable alternatives. Droughts brought about by climate change remain a challenge for hydropower, as is the case in Ecuador and Brazil (de Jong et al., 2018; Dupuits et al., 2024).

2.3.4 Geothermal Energy

Geothermal energy utilizes heat from the Earth's interior to generate electricity and supply direct heating. It provides a reliable and continuous energy source with low emissions. However, geothermal resources are dependent on location, and drilling costs can be high (Shah et al., 2024). Geothermal energy can be recovered from abandoned oil wells and can integrate with green hydrogen production, providing a new avenue for sustainable energy transformation (Davoodi et al., 2024; Hamlehdar et al., 2024).

2.3.5 Bioenergy

Bioenergy is derived from organic materials, including biomass, agricultural waste, and biofuels. It can replace fossil fuels in electricity generation, heating, and transportation. While bioenergy is renewable, concerns about land use competition, carbon neutrality, and resource availability require careful management (Raina et al., 2024).

2.3.6 Wave Energy

Wave energy captures the energy transported by surface ocean waves, primarily generated by wind interactions with the sea surface. Some technologies include point absorbers, oscillating water columns, and attenuators, all of which convert the mechanical motion of waves into electricity through hydraulic systems, air turbines, or linear generators. Ongoing research focuses on improving efficiency, survivability, and environmental integration, making wave energy an increasingly viable component of diversified renewable energy portfolios, especially for island and coastal nations (Chen & Wu, 2024).

2.3.7 Tidal Energy

Tidal energy harnesses the predictable movement of ocean tides, primarily driven by the gravitational interactions between the Earth, Moon, and Sun. It is one of the most consistent forms of renewable energy, offering a reliable generation profile with minimal greenhouse gas emissions. Although the environmental footprint of tidal energy is generally low, installations can impact marine habitats, sediment transport, and aquatic life if not carefully planned. However, when integrated with ecological monitoring and adaptive design, tidal energy represents a sustainable and scalable complement to variable sources like solar and wind in coastal energy strategies (Chen & Wu, 2024).

The transition from traditional to alternative energy sources is essential for sustainable energy production. While fossil fuels have historically driven economic growth, their environmental impacts necessitate a shift toward renewable alternatives. Advances in technology, policy support, and infrastructure development will determine the success of integrating alternative energy sources into global energy systems.

2.4 THE IMPORTANCE OF EVALUATING ENVIRONMENTAL IMPACT

The transition from traditional to alternative energy sources must consider their respective environmental impacts to ensure sustainable energy solutions. Even though renewable energy sources generally reduce greenhouse gas emissions and pollution, their deployment and long-term sustainability require careful evaluation of ecological, social, and economic factors. A comprehensive comparison of solar, wind, biomass, hydropower, and pumped storage renewable energy sources revealed that while these alternatives have lower environmental costs, their impacts vary significantly, emphasizing the need for thorough sustainability assessments (ARIDI et al., 2025).

Integrating renewable energy with digital innovations has been shown to drive green economic growth, showing the importance of policy and technological advancements in ensuring a just and efficient energy transition (Hwang, 2023). Mammadov and Akbarova further underscore that energy transition research increasingly focuses on environmental benefits, sustainable development, and technological innovations, reinforcing the necessity for continuous optimization (Mammadov & Akbarova, 2025).

2.5 RELEVANCE TO THE TRANSITION TOWARD FUEL CELLS

Fuel cells represent a promising technology in transitioning toward cleaner and more sustainable energy systems. By converting energy directly into electricity through electrochemical reactions, fuel cells offer a highly efficient and low-emission alternatives to traditional combustion-based power generation (de Sá, 2024). Their relevance in the energy transition lies in their potential to complement renewable energy sources by addressing intermittency issues. Electrolyzers can store excess energy from renewables as green hydrogen, which fuel cells can later convert back into electricity (Bianchi et al., 2023). This synergy improves grid stability (Kosowski, 2024) and reduces dependency on fossil fuels, as mentioned by de Vasconcelos and Lavoie (2019).

2.6 FOSSIL FUELS AS TRADITIONAL ENERGY SOURCES

Fossil fuels have been the cornerstone of global energy production for over a century, powering industrialization, economic growth, and modern society. Their dominance stems from their high energy density, relative abundance, and established infrastructure, making them the most accessible and cost-effective energy sources for meeting the world's growing demands. Fossil fuels have enabled technological advancements, urbanization, and improved living standards, but their widespread use has come at a significant cost to both society and the environment. The combustion of fossil fuels is the largest contributor to greenhouse gas emissions, driving climate change, air pollution, and environmental degradation (Ramanathan & Feng, 2009). Additionally, their extraction and use have led to habitat destruction, water contamination, and social conflicts over resource control. Despite these challenges, fossil fuels remain deeply entrenched in global energy systems due to their historical role, economic inertia, and the lack of widespread alternatives. This reliance sets the stage for a deeper questioning of global dependence and historical context, as well as the urgent need to transition to cleaner, more sustainable energy sources to mitigate their profound impacts on the planet and future generations.

2.7 CLASSIFICATION OF PRIMARY ENERGY SOURCES

Coal, oil, natural gas, and biomass are the primary energy sources that have powered human civilization for centuries. Their origin and uses are summarized in Figure 2.1. Each of these fuels differs in its physical state, formation process, energy content, and applications, shaping the global energy landscape. Coal is a solid fossil fuel from ancient plant material subjected to high pressure and temperature over millions of years. It is classified into different types based on carbon content and energy density, with anthracite being the most energy-rich and lignite the least efficient (Kosowski, 2024). Oil, or petroleum, is a liquid hydrocarbon mixture derived from the remains of microscopic marine organisms buried under sedimentary rock layers and transformed over millions of years due to the severe pressure and temperature conditions. Through refining, crude oil is converted into various products, including gasoline, diesel, and jet fuel, making it a dominant force in transportation and industrial applications (Bebbington et al., 2020). Natural gas, primarily composed of methane, is often associated with oil reservoirs or shale formations. It burns cleaner than coal and oil, making it a preferred choice for electricity generation and residential heating (Webb, 2015).

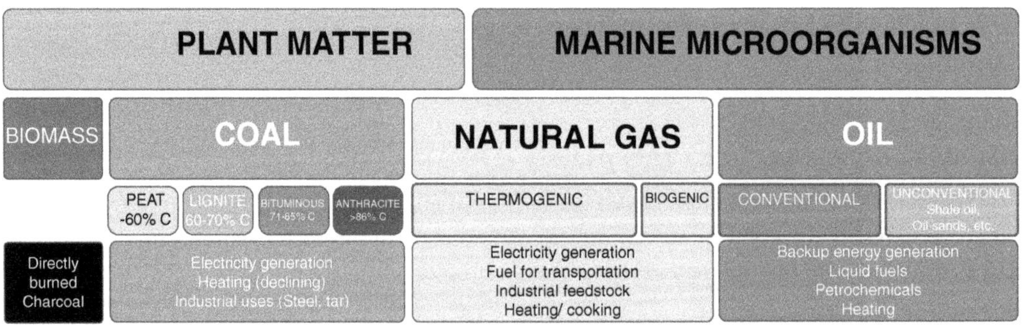

FIGURE 2.1 Scheme of primary energy sources, their origin and uses.

2.8 GLOBAL DEPENDENCE ON FOSSIL FUELS

The global dependence on fossil fuels has been a defining feature of industrialization and economic development since the 19th century. Coal, oil, and natural gas have powered industries, transportation, and households, enabling unprecedented growth in global GDP and living standards. The inertia of existing infrastructure and economic systems has perpetuated this reliance, making the transition to alternative energy sources a complex and challenging endeavor.

The environmental consequences of global fossil fuel dependence are profound and far-reaching. Coal, oil, and gas combustion are the primary sources of greenhouse gas emissions, contributing significantly to global warming and climate change. Rising temperatures, melting ice caps, and extreme weather events are direct outcomes of this reliance, threatening ecosystems, biodiversity, and human livelihoods. These impacts highlight the urgent need to reduce fossil fuel consumption and transition to cleaner energy systems, such as fuel cells, which offer a sustainable alternative with a minimal environmental footprint.

The global dependence on fossil fuels has also shaped geopolitical dynamics and economic structures, often leading to instability and conflict. Countries rich in fossil fuel reserves, such as those in the Middle East, Russia, and Venezuela, have significantly influenced global energy markets, created power imbalances, and fostered dependency among energy-importing nations. This reliance has made economies vulnerable to price volatility and supply disruptions, as seen during the oil crises of the 1970s and more recent fluctuations in oil and gas markets. Furthermore, the fossil fuel industry has historically received substantial subsidies, diverting resources from developing renewable energy technologies. Shifting away from fossil fuels could reduce these geopolitical tensions and create more resilient and diversified energy systems.

2.9 EMERGING ALTERNATIVES

Renewables and fuel cells represent a promising solution to mitigate global dependence on fossil fuels by providing a clean, efficient, and versatile energy source. Unlike traditional combustion-based energy systems, fuel cells generate electricity through electrochemical reactions, producing only water and heat as byproducts. This technology can be applied across various sectors, including transportation, industry, and residential energy systems, offering a pathway to decarbonization. By integrating fuel cells with renewable energy sources such as hydrogen produced from wind or solar power, societies can reduce their reliance on fossil fuels while addressing climate change and energy security concerns. However, the widespread adoption of fuel cells requires significant investment, policy support, and technological advancements to overcome current barriers and accelerate the transition to a sustainable energy future.

2.10 ENVIRONMENTAL CONSEQUENCES OF FOSSIL FUEL USE

It has already been mentioned that indiscriminate fossil fuel use has several environmental consequences and is a primary contributor to the climate crisis (Jiang et al., 2024). This section expands on the consequences and dangers.

2.10.1 Greenhouse Gas Emissions and Climate Change

Using fossil fuels has had profound environmental consequences, with greenhouse gas emissions among the most significant. The combustion of coal, oil, and natural gas releases large quantities of carbon dioxide (CO_2), methane (CH_4), and nitrogen oxides (NO_x) into the atmosphere. These gases trap heat within the lower atmospheric layers, contributing to the greenhouse effect and driving global warming. CO_2 is the primary contributor, accounting for most emissions, while CH_4, though less abundant, is far more potent in its heat-trapping capacity. NO_x emissions not only contribute to global warming but also play a role in forming ground-level ozone and smog, exacerbating air quality issues (Montzka et al., 2011; Ramanathan & Feng, 2009). These emissions have led to a steady rise in global temperatures, resulting in climate variability, more frequent and severe weather events, and long-term disruptions to ecosystems and human societies.

The impact of fossil fuel-derived greenhouse gas emissions on global warming and climate variability is well-documented and alarming. Rising global temperatures have caused polar ice caps and glaciers to melt, leading to rising sea levels that threaten coastal communities and ecosystems. Climate variability has increased the frequency and intensity of extreme weather events, such as hurricanes, droughts, and heat waves, which disrupt agriculture, water supplies, and infrastructure. The warming and acidification of the oceans have led to coral bleaching and the loss of marine biodiversity, further destabilizing ecosystems (Figueroa-Pico et al., 2021). These changes not only pose environmental risks but also have far-reaching socio-economic consequences, including the displacement of populations, food insecurity, and increased health risks. Reducing fossil fuel use and transitioning to cleaner energy sources, such as fuel cells, is essential to mitigate these impacts and safeguard the planet for future generations.

In addition to ecological and human health consequences, climate change threatens the durability and performance of built infrastructure. Concrete structures such as bridges, seawalls, and columns are increasingly vulnerable to accelerated degradation due to climate-induced stressors (Boccacci et al., 2023; Orcesi et al., 2022). Rising temperatures contribute to thermal expansion and cracking, while elevated humidity and atmospheric CO_2 levels promote carbonation, which leads to steel reinforcement corrosion (Bastidas-Arteaga et al., 2013; de Larrard et al., 2014). In coastal regions, salt-laden winds and rising sea levels accelerate chloride ingress, weakening the structural core (Amleh et al., 2024; Sannasiraj et al., 2025).

A local example of this phenomenon is visible in a deteriorated concrete column along the perimeter fence of a nearby university (see Figure 2.2). The image shows advanced spalling and corrosion of steel reinforcement, likely exacerbated by prolonged exposure to moisture and atmospheric pollutants. Studies have projected that under intensified environmental exposure, infrastructure such as bridges and retaining walls may experience a 30–40% reduction in service life due to the combined effects of carbonation, chloride attack, and freeze–thaw cycles (Amleh et al., 2024; Sannasiraj et al., 2025; Stewart et al., 2011; Val et al., 2025). Such examples underscore the urgent need for emissions reduction and adaptive strategies in infrastructure design and maintenance.

2.10.1.1 Air Pollution and Public Health

The combustion of fossil fuels releases, apart from greenhouse gases, harmful pollutants such as particulate matter (PM), sulfur dioxide (SO_2), and nitrogen oxides (NO_x). The health impacts of particulate matter, SO_2, and NO_x emissions are severe and widespread, particularly for vulnerable populations such as children, older people, and those with pre-existing conditions. Particulate matter, especially fine particles ($PM_{2.5}$), can penetrate deep into the lungs and even enter the bloodstream, causing various health problems. SO_2 and NO_x, which in turn are also strongly linked to respiratory diseases, including asthma, chronic obstructive pulmonary disease (COPD), and lung cancer (Amster et al., 2014). Fine particulate matter and NO_x have been associated with cardiovascular diseases, such as heart attacks, strokes, and hypertension. SO_2, when inhaled, can irritate the respiratory system and exacerbate conditions like bronchitis. On the other hand, they are key contributors to the formation of secondary pollutants like acid rain and ground-level ozone, which harm ecosystems, degrade soil quality, and damage vegetation. These pollutants are particularly prevalent in urban areas with high traffic and industrial activity, where fossil fuel consumption is concentrated. The World Health Organization (WHO) estimates that air pollution, primarily driven by fossil fuel combustion, is responsible for millions of premature deaths annually (Irfan, 2024; Khaltaev & Axelrod, 2024). These health burdens also translate into significant economic costs, including healthcare expenses and lost productivity. By replacing fossil fuels with cleaner energy technologies

FIGURE 2.2 Exposed steel and spalling in a concrete fence column, showing visible signs of climate-related deterioration such as corrosion and loss of concrete cover.

like fuel cells, societies can reduce air pollution, alleviate the burden of respiratory and cardiovascular diseases, and improve overall quality of life.

2.10.1.2 Water Pollution and Resource Depletion

The extraction and use of fossil fuels, like any human-made industry, may cause disasters if proper measures and maintenance are not regularly made; the lack of these procedures has led to numerous environmental disasters, with oil spills, fracking, and thermal pollution being among the most damaging (Singh, 2024). Oil spills, whether from offshore drilling accidents or pipeline leaks, have devastating effects on marine and coastal ecosystems, killing wildlife, destroying habitats, and contaminating water supplies. Fracking, or hydraulic fracturing, used to extract natural gas and oil from shale formations, poses significant risks to freshwater ecosystems due to the release of toxic chemicals and the potential for groundwater contamination (Ramanathan & Feng, 2009). Additionally, thermal pollution from power plants, which discharge heated water into rivers and lakes, disrupts aquatic ecosystems by reducing oxygen levels and altering the habitats of fish and other organisms (Edo et al., 2024). These activities highlight the extensive environmental costs of fossil fuel dependence, underscoring the urgent need for cleaner energy alternatives.

Freshwater ecosystems are especially vulnerable to fossil fuel extraction and use impacts, compounded by pesticide residues (Duarte-Casar et al., 2025; Mishra, 2023). Oil spills can contaminate rivers and lakes, harming aquatic life and making water unsafe for human consumption. Fracking operations often require large amounts of water, depleting local water resources and introducing pollutants that disrupt the delicate balance of freshwater ecosystems (Vieira da Silva Filho et al., 2022). Thermal pollution from power plants further exacerbates these issues by raising water temperatures, which can lead to the death of temperature-sensitive species and promote the growth of harmful algal blooms. These combined effects threaten biodiversity, reduce the availability of clean water, and compromise the ecological services that freshwater systems provide, such as water filtration and habitat support. Transitioning to sustainable energy sources like fuel cells, which have minimal water requirements and produce no thermal pollution, is essential to protect and preserve these vital ecosystems for future generations.

2.10.1.3 Land Degradation and Biodiversity Loss

The extraction and use of fossil fuels often come with hidden environmental costs, including deforestation, mining, and habitat destruction, which have far-reaching consequences for ecosystems and biodiversity. Large-scale deforestation frequently occurs to make way for oil and gas exploration, coal mining, and the construction of infrastructure such as pipelines and access roads, reducing carbon sequestration capacity and displacing countless species, leading to habitat fragmentation and loss of biodiversity. Mining activities, particularly for coal and tar sands, degrade landscapes, pollute waterways, and disrupt ecosystems, causing long-term environmental damage. The infrastructure associated with fossil fuel extraction, such as drilling sites and refineries, further encroaches on natural habitats, threatening wildlife and disrupting ecological balance. These activities exemplify the extensive environmental footprint of fossil fuels, emphasizing the need for a transition to cleaner energy sources like fuel cells, which have a significantly lower impact on land use and ecosystems, helping preserve the planet's natural heritage. Examples of the above are: deforestation, faunal and societal damage in the Amazon due to oil exploration, particularly in the Yasuní area of Ecuador (Center for Economic and Social Rights, 1994; Finer et al., 2008); and the habitat destruction brought about by Tar sands mining in Canada (Kelly et al., 2010).

2.11 COMPARATIVE ENVIRONMENTAL IMPACT: FOSSIL FUELS VERSUS ALTERNATIVE ENERGY

The environmental impact of energy sources is a critical factor in evaluating their sustainability and long-term viability. However, each alternative energy source comes with its own set of environmental trade-offs, making it essential to carefully assess their overall impact.

When comparing energy efficiency and carbon footprints, alternative energy sources consistently outperform fossil fuels. Solar and wind energy, for example, generate electricity with minimal greenhouse gas emissions once the infrastructure is in place (Wevers et al., 2020). Nuclear energy, despite concerns about waste and safety, produces virtually no CO_2 during operation and has a high energy output relative to its input (Burchart & Przytuła, 2025). In contrast, fossil fuels are inherently inefficient, as a significant portion of their energy content is lost as waste heat during combustion (Sarkar et al., 2023). Fuel cells, particularly hydrogen fuel cells, stand out for their high efficiency and low emissions. They convert chemical energy directly into electricity with efficiencies often exceeding 50%, compared to the 30–40% efficiency of traditional fossil fuel power plants (Medrano-García et al., 2022). Moreover, when hydrogen is produced using renewable energy, fuel cells can operate with near-zero carbon emissions, making them a promising solution for reducing the carbon footprint of energy systems.

While renewable energy sources and nuclear power offer significant environmental benefits, they are not without trade-offs. Solar and wind energy require large land areas and can impact local ecosystems, particularly if not sited responsibly (van Zalk & Behrens, 2018). The production and disposal of solar panels and wind turbines also involve resource extraction and waste management challenges (Gibon et al., 2017). Hydropower, while renewable, can disrupt aquatic ecosystems and displace communities (Gerbens-Leenes et al., 2020). Nuclear energy, despite its low emissions, raises concerns about radioactive waste storage, potential for catastrophic accidents, and high capital costs (Sasse & Trutnevyte, 2020). These trade-offs highlight the need for careful planning and innovation to minimize negative impacts while maximizing the benefits of alternative energy sources.

Fuel cells offer a unique set of advantages in reducing emissions and addressing the limitations of other alternative energy sources. Unlike combustion-based systems, fuel cells produce electricity through electrochemical reactions, emitting only water and heat as byproducts (Imdahl et al., 2021). This makes them an exceptionally clean technology, particularly when paired with green hydrogen produced from renewable energy (Naseri et al., 2024). Fuel cells are also highly versatile, capable of powering everything from vehicles to industrial processes to residential energy systems (Shari et al., 2024). Their ability to provide reliable, decentralized energy makes them an ideal complement to intermittent renewables like solar and wind, enhancing grid stability and energy security (Arshad et al., 2025). Integrating fuel cells into energy systems can substantially reduce dependence on fossil fuels and help meet global climate targets.

A comparative analysis of fossil fuels, renewable energy, and fuel cells regarding their environmental impact is presented in Table 2.1.

2.12 POLICY, REGULATIONS, AND GLOBAL INITIATIVES

The transition to sustainable energy systems is heavily influenced by policy frameworks, regulations, and global initiatives that aim to reduce greenhouse gas emissions and promote cleaner energy technologies. At the forefront of these efforts is the Paris Agreement, a landmark international accord adopted in 2015, which seeks to limit global warming to well below 2°C above pre-industrial levels, with efforts to cap the increase at 1.5°C (He et al., 2022; Larch & Wanner, 2024). The agreement has galvanized countries to set ambitious emissions reduction targets and implement national strategies to achieve them. These strategies often include incentives for renewable energy adoption, carbon pricing mechanisms, and stricter emissions standards for industries and vehicles. Additionally, global initiatives such as the United Nations Sustainable Development Goals (SDGs), particularly Goal 7 (Affordable and Clean Energy), provide a broader framework for addressing energy access, efficiency, and sustainability. These international efforts are complemented by regional and local policies that tailor solutions to specific contexts, ensuring that the transition to cleaner energy is inclusive and equitable. However, the effectiveness of these policies depends on robust enforcement, transparent monitoring, and the willingness of nations to collaborate and share best practices.

While policy and regulation are critical in driving the energy transition, technological advancements and market dynamics are equally important in accelerating the shift toward cleaner energy (Hwang, 2023). Innovations in renewable energy technologies, energy storage, and fuel cells have significantly reduced costs and improved efficiency, making them increasingly competitive with

TABLE 2.1
Comparison of Environmental Impacts: Fossil Fuels vs. Alternative Energy Sources

Aspect	Fossil Fuels	Renewable Energy	Fuel Cells
Greenhouse Gas Emissions	Very high (CO_2, CH_4, NO_x)	Very low (near-zero in operation)	Near-zero if using green hydrogen
Air Pollution	High (SO_2, particulate matter, smog)	Minimal (dust during construction)	Very low
Water Usage	High (cooling, fracking, spills)	Moderate (mainly hydropower impact)	Low (only water output)
Land Impact	Mining, drilling, and habitat destruction	Solar and wind need land, but are less invasive if planned carefully	Minimal (compact systems)
Resource Depletion	Non-renewable, finite resources	Renewable, replenishable	Renewable if hydrogen is green
Waste Production	Toxic waste (e.g., ash, sludge)	Equipment disposal (solar panels, turbines)	Water vapor only
Reliability	High but environmentally costly	Depends on location and weather	High when coupled with storage

fossil fuels. For instance, the declining costs of solar panels and wind turbines, coupled with advancements in battery storage, have enabled the rapid deployment of renewable energy systems worldwide. Similarly, fuel cell technology has seen significant progress, particularly in the transportation and industrial sectors, where it offers a viable alternative to fossil fuels. Market-driven shifts, such as the growing demand for electric vehicles (EVs) and corporate commitments to achieving net-zero emissions, further propel the adoption of clean energy solutions. Companies are investing heavily in sustainable energy projects, recognizing both the environmental imperative and the economic opportunities they present. These technological and market-driven changes create a positive feedback loop, where increased adoption leads to further innovation and cost reductions, making clean energy more accessible and attractive. Policy frameworks and market forces shape a future where sustainable energy systems replace fossil fuels, paving the way for a cleaner, more resilient, and equitable energy landscape.

2.13 THE NEED FOR SUSTAINABLE ENERGY TRANSITIONS

The world is at a critical juncture where the need for sustainable energy transitions has never been more urgent. Fossil fuels, which have powered global development for over a century, are now recognized as a major driver of climate change, environmental degradation, and public health crises. The increasing frequency of extreme weather events, rising global temperatures, and the depletion of natural resources underscore the necessity of shifting to cleaner, renewable energy systems. Sustainable energy transitions are not just about reducing carbon emissions; they are about creating resilient, equitable, and future-proof energy systems supporting economic growth while preserving the planet for future generations. This transition requires a collective effort from governments, industries, and individuals to embrace innovative technologies and policies prioritizing sustainability over short-term gains (Sachs et al., 2019).

Sustainable energy solutions can potentially drive local development by creating jobs, improving energy access, and fostering economic resilience. In many parts of the world, especially in developing regions, communities still lack reliable access to electricity and rely on inefficient and polluting energy sources, such as kerosene or diesel generators. Transitioning to sustainable energy systems—such as solar, wind, and fuel cells—can provide clean and affordable power to these communities, enhancing quality of life and enabling economic opportunities. Localized energy systems, such as microgrids powered by fuel cells, can reduce dependence on centralized fossil fuel-based infrastructure, enhancing energy security and empowering communities to take control of their energy needs (Akinyele et al., 2020). This decentralized approach supports environmental goals and promotes social and economic equity.

Fuel cells are emerging as a key technology in the global effort to decarbonize energy systems. As mentioned, fuel cells produce electricity through electrochemical reactions, emitting only water and heat as byproducts. This makes them an environmentally friendly alternative to fossil fuels, particularly in sectors that are difficult to electrify, such as heavy industry and long-haul transportation. Hydrogen fuel cells, in particular, have gained attention for their potential to store and deliver clean energy at scale (Kanani & Zahedi, 2024). When hydrogen is produced using renewable energy sources, fuel cells can play a pivotal role in achieving net-zero emissions. By integrating fuel cells into energy grids, transportation networks, and industrial processes, societies can significantly reduce their carbon footprint and move closer to a sustainable energy future.

Modern biomass is a versatile and sustainable energy source used to generate electricity, heat, and biofuels while reducing reliance on fossil fuels. Through technologies like anaerobic digestion (biogas), gasification, and direct combustion, organic materials—such as agricultural residues, wood pellets, and energy crops—are converted into clean energy. Advanced biofuels (e.g., cellulosic ethanol) and biomass co-firing in power plants further reduce carbon emissions. As a renewable resource, biomass supports circular economies by repurposing waste and promoting carbon neutrality, though sustainability depends on responsible sourcing to avoid deforestation and land-use conflicts (Mignogna et al., 2024). Its integration with carbon capture (BECCS) could even facilitate negative emissions, aiding climate efforts goals.

Despite the clear benefits of sustainable energy technologies like fuel cells, replacing fossil fuels is not without challenges. The existing global infrastructure heavily relies on fossil fuels, and transitioning to new energy systems requires massive investments in technology, infrastructure, and workforce training. Additionally, the intermittent nature of renewable energy sources like solar and wind poses challenges for energy storage and grid stability, which fuel cells can help address but not entirely solve. Economic and political barriers, such as fossil fuel subsidies and resistance from entrenched industries, further complicate the transition. Overcoming these challenges requires coordinated efforts, including policy support, public-private partnerships, and international cooperation to create an enabling environment for sustainable energy adoption.

While the challenges are significant, the opportunities presented by replacing fossil fuels are immense. The shift to sustainable energy systems can drive innovation, create new industries, and generate millions of jobs worldwide. Fuel cells, for example, are at the forefront of this transformation, offering solutions for clean transportation, efficient power generation, and energy storage. The growing demand for renewable energy technologies also opens opportunities for research and development, fostering advancements in materials science, engineering, and energy management (Shamoon et al., 2022). Moreover, reducing dependence on fossil fuels can enhance energy security by diversifying energy sources and reducing vulnerability to price volatility and supply disruptions (Khan et al., 2024). These opportunities highlight the potential for a just and inclusive energy transition that benefits both people and the planet.

Looking ahead, the future of energy sustainability will depend on the widespread adoption of innovative technologies, supportive policies, and a shared commitment to combating climate change. Fuel cells are expected to play a central role in this transition, particularly as advancements in hydrogen production, storage, and distribution make them more accessible and cost-effective (Kanani & Zahedi, 2024). The integration of fuel cells with other renewable energy systems, such as solar and wind, will be critical for creating resilient and flexible energy grids (Kanani & Zahedi, 2024). Additionally, fostering international collaboration and knowledge-sharing will be essential to accelerate the global energy transition. By prioritizing sustainability, investing in clean technologies, and addressing the social and economic dimensions of energy systems, the world can move toward a future where energy is not only clean and abundant but also equitable and accessible to all.

2.14 CONCLUSIONS

The mounting evidence of climate change and the well-documented environmental and public health consequences of fossil fuel reliance underscore the urgent need for a systemic transition to sustainable energy systems. This analysis has clarified not only the ecological and societal burdens imposed by conventional energy sources but also the transformative potential of renewable energy technologies in mitigating these impacts.

The shift toward clean energy extends beyond technological innovation; it represents a moral obligation to promote intergenerational equity and planetary stewardship. Effective policy frameworks, bolstered by targeted investments in infrastructure and research, are crucial for accelerating this transition. Furthermore, integrating principles of social equity into energy planning is essential to ensure that sustainability benefits, including energy access, security, and resilience, are distributed fairly across communities. At this critical juncture, the imperative for action is clear. The adoption of low-carbon technologies, such as hydrogen fuel cells powered by renewable hydrogen, must be prioritized to reduce greenhouse gas emissions while ensuring economic stability. Achieving this goal will require unprecedented collaboration among governments, industries, and civil society to tackle the technical, economic, and institutional barriers to change.

Reconfiguring global energy systems is essential to preserve ecological integrity, protect public health, and promote a just and sustainable future. The decisions made today will shape the trajectory of environmental, social, and economic outcomes for decades to come. It is crucial for all stakeholders to act with determination, ensuring that sustainability serves as the foundation of energy policy and practice moving forward.

REFERENCES

Abbass, K., Qasim, M. Z., Song, H., Murshed, M., Mahmood, H., & Younis, I. (2022). A review of the global climate change impacts, adaptation, and sustainable mitigation measures. *Environmental Science and Pollution Research, 29*(28), 42539–42559. https://doi.org/10.1007/s11356-022-19718-6

Akinyele, D., Olabode, E., & Amole, A. (2020). Review of fuel cell technologies and applications for sustainable microgrid systems. *Inventions, 5*(3), Article 3. https://doi.org/10.3390/inventions5030042

Amleh, L., Hassan, M., & Hussein, L. (2024). Influence of climate change on the probability of chloride-induced corrosion initiation for RC bridge decks made of geopolymer concrete. *Sustainability, 16*(18), 8200. https://doi.org/10.3390/su16188200

Amster, E. D., Haim, M., Dubnov, J., & Broday, D. M. (2014). Contribution of nitrogen oxide and sulfur dioxide exposure from power plant emissions on respiratory symptom and disease prevalence. *Environmental Pollution, 186*, 20–28. https://doi.org/10.1016/j.envpol.2013.10.032

Aridi, R., Aridi, M., Marie-Lise, P., & Lemenand, Thierry. (2025). Eco-environmental, and social impacts of producing electricity with various renewable energy sources. *Energy*, 135139. https://doi.org/10.1016/j.energy.2025.135139

Arshad, M. Y., Hessel, V., Halog, A., Lewis, D., & Tran, N. N. (2025). Decentralisation transition in the chemical, energy, and waste management sectors: Innovations, opportunities, and sustainable pathways – A review. *Sustainable Energy Technologies and Assessments, 76*, 104307. https://doi.org/10.1016/j.seta.2025.104307

Asif, M., Sahin, G., & Khalid, M. (2024). *Handbook of energy and environment in the 21st century: Technology and policy dynamics*. CRC Press.

Bastidas-Arteaga, E., Schoefs, F., Stewart, M. G., & Wang, X. (2013). Influence of global warming on durability of corroding RC structures: A probabilistic approach. *Engineering Structures, 51*, 259–266. https://doi.org/10.1016/j.engstruct.2013.01.006

Bebbington, J., Schneider, T., Stevenson, L., & Fox, A. (2020). Fossil fuel reserves and resources reporting and unburnable carbon: Investigating conflicting accounts. *Critical Perspectives on Accounting, 66*, 102083. https://doi.org/10.1016/j.cpa.2019.04.004

Bianchi, F. R., Bosio, B., Conte, F., Massucco, S., Mosaico, G., Natrella, G., & Saviozzi, M. (2023). Modelling and optimal management of renewable energy communities using reversible solid oxide cells. *Applied Energy, 334*, 120657. https://doi.org/10.1016/j.apenergy.2023.120657

Boadu, S., & Otoo, E. (2024). A comprehensive review on wind energy in Africa: Challenges, benefits and recommendations. *Renewable and Sustainable Energy Reviews, 191*, 114035. https://doi.org/10.1016/j.rser.2023.114035

Boccacci, G., Frasca, F., Bartolucci, B., Vergelli, L., Bertolin, C., & Siani, A. M. (2023). Climate-induced conservation risks of historic reinforced concrete buildings: Preliminary results from literature review. *The International Archives of the Photogrammetry Remote Sensing and Spatial Information Sciences, XLVIII-M-2–2023*, 227–234. https://doi.org/10.5194/isprs-archives-xlviii-m-2-2023-227-2023

Burchart, D., & Przytuła, I. (2025). Review of environmental life cycle assessment for fuel cell electric vehicles in road transport. *Energies, 18*(5), 1229. https://doi.org/10.3390/en18051229

Center for Economic and Social Rights. (1994). Rights violations in the ecuadorian Amazon: The human consequences of oil development. *Health and Human Rights, 1*(1), 82–100. https://doi.org/10.2307/4065263

Chen, P., & Wu, D. (2024). A review of hybrid wave-tidal energy conversion technology. *Ocean Engineering, 303*, 117684. https://doi.org/10.1016/j.oceaneng.2024.117684

Crutzen, P. J., & Andreae, M. O. (1990). Biomass burning in the tropics: Impact on atmospheric chemistry and biogeochemical cycles. *Science, 250*(4988), 1669–1678. https://doi.org/10.1126/science.250.4988.1669

Vieira da Silva Filho, S., Peyerl, D., & Moutinho dos Santos, E. (2022). Early insights on the fracking impacts to the water-energy nexus in Brazil: Is there a risk of water scarcity in the shale gas prospective areas? *Journal of Cleaner Production, 336*, 130390. https://doi.org/10.1016/j.jclepro.2022.130390

Davoodi, S., Al-Shargabi, M., Wood, D. A., Slivkin, S., Shishaev, G., & Rukavishnikov, V. S. (2024). Geothermal energy recovery from abandoned petroleum wells: A review of the challenges and opportunities. *Sustainable Energy Technologies and Assessments, 68*, 103870. https://doi.org/10.1016/j.seta.2024.103870

de Jong, P., Tanajura, C. A. S., Sánchez, A. S., Dargaville, R., Kiperstok, A., & Torres, E. A. (2018). Hydroelectric production from Brazil's São Francisco River could cease due to climate change and inter-annual variability. *Science of The Total Environment*, *634*, 1540–1553. https://doi.org/10.1016/j.scitotenv.2018.03.256

de Larrard, T., Bastidas-Arteaga, E., Duprat, F., & Schoefs, F. (2014). Effects of climate variations and global warming on the durability of RC structures subjected to carbonation. *Civil Engineering and Environmental Systems*, *31*(2), 153–164. https://doi.org/10.1080/10286608.2014.913033

de Sá, M. H. (2024). Electrochemical devices to power a sustainable energy transition—an overview of green hydrogen contribution. *Applied Sciences*, *14*(5), 2168. https://doi.org/10.3390/app14052168

de Vasconcelos, B. R., & Lavoie, J.-M. (2019). Recent advances in power-to-X technology for the production of fuels and chemicals. *Frontiers in Chemistry*, *7*, 392. https://doi.org/10.3389/fchem.2019.00392

Duarte-Casar, R., Romero-Benavides, J. C., Bailon-Moscoso, N., Navarro-Rojas, M., & Rojas-Le-Fort, M. (2025). Synergies between green chemistry principles and the sustainable development goals (SDGs) in developing eco-friendly pesticides. In B. R. Babaniyi & E. E. Babaniyi (Eds.), *The interplay of pesticides and climate change: environmental dynamics and challenges* (pp. 291–312). Springer Nature Switzerland. https://doi.org/10.1007/978-3-031-81669-7_11

Dupuits, E., Garcés, A., Llambí, L. D., & Bustamante, M. (2024). Strategies for monitoring and evaluation of climate change adaptation: Localizing global approaches into Andean realities. *Npj Climate Action*, *3*(1), 1–11. https://doi.org/10.1038/s44168-023-00097-8

Edo, G. I., Itoje-Akpokiniovo, L. O., Obasohan, P., Ikpekoro, V. O., Samuel, P. O., Jikah, A. N., Nosu, L. C., Ekokotu, H. A., Ugbune, U., Oghroro, E. E. A., Emakpor, O. L., Ainyanbhor, I. E., Mohammed, W. A.-S., Akpoghelie, P. O., Owheruo, J. O., & Agbo, J. J. (2024). Impact of environmental pollution from human activities on water, air quality and climate change. *Ecological Frontiers*, *44*(5), 874–889. https://doi.org/10.1016/j.ecofro.2024.02.014

Figueroa-Pico, J., Tortosa, F. S., & Carpio, A. J. (2021). Natural and anthropogenic-induced stressors affecting the composition of fish communities on the rocky reefs of Ecuador. *Marine Pollution Bulletin*, *164*, 112018. https://doi.org/10.1016/j.marpolbul.2021.112018

Finer, M., Jenkins, C. N., Pimm, S. L., Keane, B., & Ross, C. (2008). Oil and gas projects in the western amazon: Threats to wilderness, biodiversity, and indigenous peoples. *PLoS ONE*, *3*(8), e2932. https://doi.org/10.1371/journal.pone.0002932

Gerbens-Leenes, W., Vaca-Jiménez, S., & Mekonnen, M. (2020). Burning water, overview of the contribution of arjen hoekstra to the water energy nexus. *Water*, *12*(10), 2844. https://doi.org/10.3390/w12102844

Gibon, T., Arvesen, A., & Hertwich, E. G. (2017). Life cycle assessment demonstrates environmental co-benefits and trade-offs of low-carbon electricity supply options. *Renewable and Sustainable Energy Reviews*, *76*, 1283–1290. https://doi.org/10.1016/j.rser.2017.03.078

Hamlehdar, M., Beardsmore, G., & Narsilio, G. A. (2024). Hydrogen production from low-temperature geothermal energy – A review of opportunities, challenges, and mitigating solutions. *International Journal of Hydrogen Energy*, *77*, 742–768. https://doi.org/10.1016/j.ijhydene.2024.06.104

He, R., Luo, L., Shamsuddin, A., & Tang, Q. (2022). Corporate carbon accounting: A literature review of carbon accounting research from the Kyoto Protocol to the Paris Agreement. *Accounting & Finance*, *62*(1), 261–298. https://doi.org/10.1111/acfi.12789

Hettithanthri, O., Nguyen, T. B. T., Fiedler, T., Phan, C., Vithanage, M., Pallewatta, S., Nguyen, T. M. L., Nguyen, P. Q. A., & Bolan, N. (2024). A review of oil spill dynamics: Statistics, impacts, countermeasures, and weathering behaviors. *Asia-Pacific Journal of Chemical Engineering*, *19*(6), e3128. https://doi.org/10.1002/apj.3128

Hwang, Y. K. (2023). The synergy effect through combination of the digital economy and transition to renewable energy on green economic growth: Empirical study of 18 Latin American and caribbean countries. *Journal of Cleaner Production*, *418*, 138146. https://doi.org/10.1016/j.jclepro.2023.138146

Imdahl, C., Blume, C., Blume, S., Zellmer, S., Gensicke, M., & Herrmann, C. (2021). Potentials of hydrogen technologies for sustainable factory systems. *Procedia CIRP*, *98*, 583–588. https://doi.org/10.1016/j.procir.2021.01.158

Irfan, H. (2024). Air pollution and cardiovascular health in South Asia: A comprehensive review. *Current Problems in Cardiology*, *49*(2), 102199. https://doi.org/10.1016/j.cpcardiol.2023.102199

Jiang, T., He, X., Su, B., Havea, P. H., Wei, K., Kundzewicz, Z. W., & Liu, D. (2024). COP 28: Challenge of coping with climate crisis. *The Innovation*, *5*(1). https://doi.org/10.1016/j.xinn.2023.100559

Kanani, B., & Zahedi, A. (2024). Step toward sustainable development through the integration of renewable energy systems with fuel cells: A review. *Sustainable Energy Technologies and Assessments*, *70*, 103935. https://doi.org/10.1016/j.seta.2024.103935

Karduri, R. K. R., & Ananth, C. (2023). The Role of Policy in Accelerating the Energy Transition. *SSRN Electronic Journal*. https://doi.org/10.2139/ssrn.4637804

Kelly, E. N., Schindler, D. W., Hodson, P. V., Short, J. W., Radmanovich, R., & Nielsen, C. C. (2010). Oil sands development contributes elements toxic at low concentrations to the Athabasca River and its tributaries. *Proceedings of the National Academy of Sciences*, *107*(37), 16178–16183. https://doi.org/10.1073/pnas.1008754107

Khaltaev, N., & Axelrod, S. (2024). Cardiovascular disease mortality and air pollution in countries with different socioeconomic status. *Chronic Diseases and Translational Medicine*, *10*(03), 247–255. https://doi.org/10.1002/cdt3.116

Khan, K., Khurshid, A., Cifuentes-Faura, J., & Xianjun, D. (2024). Does renewable energy development enhance energy security? *Utilities Policy*, *87*, 101725. https://doi.org/10.1016/j.jup.2024.101725

Kosowski, P. (2024). From fossil fuels to renewables: Clustering european primary energy production from 1990 to 2022. *Energies*, *17*(22), 5596. https://doi.org/10.3390/en17225596

Larch, M., & Wanner, J. (2024). The consequences of non-participation in the Paris Agreement. *European Economic Review*, *163*, 104699. https://doi.org/10.1016/j.euroecorev.2024.104699

Lelieveld, J., Klingmüller, K., Pozzer, A., Burnett, R. T., Haines, A., & Ramanathan, V. (2019). Effects of fossil fuel and total anthropogenic emission removal on public health and climate. *Proceedings of the National Academy of Sciences*, *116*(15), 7192–7197. https://doi.org/10.1073/pnas.1819989116

Malik, K., Capareda, S. C., Kamboj, B. R., Malik, S., Singh, K., Arya, S., & Bishnoi, D. K. (2024). Biofuels Production: A Review on Sustainable Alternatives to Traditional Fuels and Energy Sources. *Fuels*, *5*(2), Article 2. https://doi.org/10.3390/fuels5020010

Mammadov, N., & Akbarova, S. (2025). Research Areas on Transition to Renewable Energy Sources in Bibliographic Publications. *BIO Web of Conferences*, *151*, 02007. https://doi.org/10.1051/bioconf/202515102007

Medrano-García, J. D., Charalambous, M. A., & Guillén-Gosálbez, G. (2022). Economic and environmental barriers of CO2-based fischer–tropsch electro-diesel. *ACS Sustainable Chemistry & Engineering*, *10*(36), 11751–11759. https://doi.org/10.1021/acssuschemeng.2c01983

Mignogna, D., Szabó, M., Ceci, P., & Avino, P. (2024). Biomass energy and biofuels: Perspective, potentials, and challenges in the energy transition. *Sustainability*, *16*(16), Article 16. https://doi.org/10.3390/su16167036

Mishra, R. K. (2023). Fresh Water availability and Its Global challenge. *British Journal of Multidisciplinary and Advanced Studies*, *4*(3), Article 3. https://doi.org/10.37745/bjmas.2022.0208

Mohammad, N., Ishak, W. W. M., Mustapa, S. I., & Ayodele, B. V. (2021). Natural Gas as a Key Alternative Energy Source in Sustainable Renewable Energy Transition: A Mini Review. *Frontiers in Energy Research*, *9*, 625023. https://doi.org/10.3389/fenrg.2021.625023

Montzka, S. A., Dlugokencky, E. J., & Butler, J. H. (2011). Non-CO2 greenhouse gases and climate change. *Nature*, *476*(7358), 43–50. https://doi.org/10.1038/nature10322

Mukhopadhyay, A., Duttagupta, S., & Mukherjee, A. (2022). Emerging organic contaminants in global community drinking water sources and supply: A review of occurrence, processes and remediation. *Journal of Environmental Chemical Engineering*, *10*(3), 107560. https://doi.org/10.1016/j.jece.2022.107560

Naseri, N., Hani, S. E., Machmoum, M., Elbouchikhi, E., & Daghouri, A. (2024). Energy management strategy for a net zero emission islanded photovoltaic microgrid-based green hydrogen system. *Energies*, *17*(9), 2111. https://doi.org/10.3390/en17092111

Oliver Yébenes, M. (2024). Climate change, ESG criteria and recent regulation: Challenges and opportunities. *Eurasian Economic Review*, *14*(1), 87–120. https://doi.org/10.1007/s40822-023-00251-x

Orcesi, A., O'Connor, A., Bastidas-Arteaga, E., Stewart, M. G., Imam, B., Kreislova, K., Schoefs, F., Markogiannaki, O., Wu, T., Li, Y., Salman, A., Hawchar, L., & Ryan, P. C. (2022). Investigating the effects of climate change on material properties and structural performance. *Structural Engineering International*, *32*(4), 577–588. https://doi.org/10.1080/10168664.2022.2107468

Raina, N., Chuetor, S., Elalami, D., Tayibi, S., & Barakat, A. (2024). Biomass valorization for bioenergy production: Current techniques, challenges, and pathways to solutions for sustainable bioeconomy. *BioEnergy Research*, *17*(4), 1999–2028. https://doi.org/10.1007/s12155-024-10792-x

Ramanathan, V., & Feng, Y. (2009). Air pollution, greenhouse gases and climate change: Global and regional perspectives. *Atmospheric Environment*, *43*(1), 37–50. https://doi.org/10.1016/j.atmosenv.2008.09.063

Rezaei, E. E., Webber, H., Asseng, S., Boote, K., Durand, J. L., Ewert, F., Martre, P., & MacCarthy, D. S. (2023). Climate change impacts on crop yields. *Nature Reviews Earth & Environment*, *4*(12), 831–846. https://doi.org/10.1038/s43017-023-00491-0

Sachs, J. D., Schmidt-Traub, G., Mazzucato, M., Messner, D., Nakicenovic, N., & Rockström, J. (2019). Six Transformations to achieve the sustainable development goals. *Nature Sustainability*, *2*(9), 805–814. https://doi.org/10.1038/s41893-019-0352-9

Sanders, N. J., & Barreca, A. I. (2022). Adaptation to environmental change: Agriculture and the unexpected incidence of the acid rain program. *American Economic Journal: Economic Policy*, *14*(1), 373–401. https://doi.org/10.1257/pol.20190060

Sannasiraj, R. D. A., Shi, S., Liu, X., Gryllias, K., Vandepitte, D., Chronopoulos, D., & Zhang, L. (2025). A fully coupled depth-dependent corrosion model for reinforced concrete piles under marine environmental conditions. *Construction and Building Materials*, *472*, 140795. https://doi.org/10.1016/j.conbuildmat.2025.140795

Sarkar, O., Modestra, J. A., Rova, U., Christakopoulos, P., & Matsakas, L. (2023). Waste-derived renewable hydrogen and methane: Towards a potential energy transition solution. *Fermentation*, *9*(4), 368. https://doi.org/10.3390/fermentation9040368

Sasse, J.-P., & Trutnevyte, E. (2020). Regional impacts of electricity system transition in Central Europe until 2035. *Nature Communications*, *11*(1), 4972. https://doi.org/10.1038/s41467-020-18812-y

Shah, M., Prajapati, M., Yadav, K., & Sircar, A. (2024). A comprehensive review of geothermal energy storage: Methods and applications. *Journal of Energy Storage*, *98*, 113019. https://doi.org/10.1016/j.est.2024.113019

Shamoon, A., Haleem, A., Bahl, S., Javaid, M., Bala Garg, S., Chandmal Sharma, R., & Garg, J. (2022). Environmental impact of energy production and extraction of materials—A review. *Materials Today: Proceedings*, *57*, 936–941. https://doi.org/10.1016/j.matpr.2022.03.159

Shari, B. E., Moumouni, Y., Ohunakin, O. S., Blechinger, P., Madougou, S., & Rabani, A. (2024). Exploring the role of green hydrogen for distributed energy access planning towards net-zero emissions in Nigeria. *Sustainable Energy Research*, *11*(1), 16. https://doi.org/10.1186/s40807-024-00107-1

Shiquan, D., Amuakwa-Mensah, F., Deyi, X., Yue, C., & Yue, C. (2022). The impact of mineral resource extraction on communities: How the vulnerable are harmed. *The Extractive Industries and Society*, *10*, 101090. https://doi.org/10.1016/j.exis.2022.101090

Singh, V. (2024). Water pollution. In V. Singh (Ed.), *Textbook of environment and ecology* (pp. 253–266). Springer Nature. https://doi.org/10.1007/978-981-99-8846-4_17

Soeder, D. J. (2021). Fossil fuels and climate change. In D. J. Soeder (Ed.), *Fracking and the environment: A scientific assessment of the environmental risks from hydraulic fracturing and fossil fuels* (pp. 155–185). Springer International Publishing. https://doi.org/10.1007/978-3-030-59121-2_9

Stewart, M. G., Wang, X., & Nguyen, M. N. (2011). Climate change impact and risks of concrete infrastructure deterioration. *Engineering Structures*, *33*(4), 1326–1337. https://doi.org/10.1016/j.engstruct.2011.01.010

Tol, R. S. J. (2018). The economic impacts of climate change. *Review of Environmental Economics and Policy*, *12*(1), 4–25. https://doi.org/10.1093/reep/rex027

Ukoba, K., Yoro, K. O., Eterigho-Ikelegbe, O., Ibegbulam, C., & Jen, T.-C. (2024). Adaptation of solar energy in the Global South: Prospects, challenges and opportunities. *Heliyon*, *10*(7). https://doi.org/10.1016/j.heliyon.2024.e28009

Val, D. V., Andrade, C., Sykora, M., Stewart, M. G., Bastidas-Arteaga, E., Mlcoch, J., Truong, Q. C., & Soueidy, C.-P. E. (2025). Probabilistic modelling of deterioration of reinforced concrete structures. *Structural Safety*, *113*, 102454. https://doi.org/10.1016/j.strusafe.2024.102454

van Zalk, J., & Behrens, P. (2018). The spatial extent of renewable and non-renewable power generation: A review and meta-analysis of power densities and their application in the U.S. *Energy Policy*, *123*, 83–91. https://doi.org/10.1016/j.enpol.2018.08.023

Webb, R. (2015). Safety first, environment last: Improving regulation of gas pipeline leaks. *SSRN Electronic Journal*. https://doi.org/10.2139/ssrn.2671847

Wevers, J. B., Shen, L., & van der Spek, M. (2020). What does it take to go net-zero-CO2? a life cycle assessment on long-term storage of intermittent renewables with chemical energy carriers. *Frontiers in Energy Research*, *8*, 104. https://doi.org/10.3389/fenrg.2020.00104

Zhironkin, S., & Abu-Abed, F. (2024). Fossil fuel prospects in the energy of the future (energy 5.0): A review. *Energies*, *17*(22), 5606. https://doi.org/10.3390/en17225606

3 Materials and Catalysts Development for Industrial-Scale Fuel Cells

Olusola D. Ogundele, John A. Lawal, Oluwatosin D. Owolala, and Johnson Jayeola

3.1 INTRODUCTION

The escalating concerns over climate change, environmental degradation, and the finite nature of fossil fuels have intensified the global pursuit of sustainable and clean energy solutions (Singh, 2021). Among the myriad alternatives, fuel cells have emerged as a promising technology due to their high efficiency, scalability, and minimal environmental impact. Unlike traditional combustion-based power generation, fuel cells convert chemical energy directly into electrical energy through electrochemical reactions, resulting in higher efficiencies and lower emissions (Kartha & Grimes, 1994). Fuel cells are versatile in their applications, ranging from small-scale portable devices to large-scale power generation systems (Wang *et al.*, 2023). Their ability to operate on various fuels, including hydrogen, natural gas, and methanol, further enhances their adaptability across different sectors. Notably, when hydrogen is used as the primary fuel, the only by-products are water and heat, making it an environmentally benign energy source (Younas *et al.*, 2022).

The significance of fuel cells extends beyond environmental benefits, Table 3.1 shows the key applications of fuel cells. They offer energy security by reducing dependence on imported fuels and contribute to economic growth through the creation of new industries and job opportunities (Stambouli, 2011). Moreover, their quiet operation and modular design make them suitable for integration into urban environments and remote locations alike. Fuel cells have found applications across various industries, demonstrating their versatility and potential for large-scale deployment (Qasem & Abdulrahman, 2024). In the transportation sector, fuel cell electric vehicles (FCEVs) offer a zero-emission alternative to internal combustion engine vehicles. Major automotive manufacturers have developed FCEVs that provide comparable performance and range to conventional vehicles, with the added benefit of rapid refueling times (Singh *et al.*, 2023).

In the realm of stationary power, fuel cells are employed for both primary and backup power generation (Abdin, 2024). They are particularly valuable in critical infrastructure, such as hospitals, data centers, and telecommunication facilities, where an uninterrupted power supply is essential. Their high efficiency and reliability make them an attractive option for distributed generation and grid stabilization (Kubert, 2010). Portable fuel cells cater to the growing demand for reliable and long-lasting power sources in consumer electronics and military applications (Nguyen, 2006). They offer extended operation times compared to traditional batteries and can be rapidly recharged by refueling, thus enhancing the usability of portable devices in remote or off-grid locations (Gupta *et al.*, 2024).

The performance, durability, and cost-effectiveness of fuel cells are intrinsically linked to the materials and catalysts employed in their construction. The development of advanced materials is crucial for enhancing the efficiency and longevity of fuel cells, thereby facilitating their widespread

DOI: 10.1201/9781003645405-3

TABLE 3.1
Key Applications of Fuel Cells (Pardhi *et al.*, 2022; Fakhreddine *et al.*, 2023; McNicol *et al.*, 2001)

Application Sector	Examples	Benefits
Transportation	Fuel cell cars, buses, trucks, trains	Zero emissions, quick refueling, long driving range
Stationary Power	Backup power for hospitals, data centers, residential use	High efficiency, reliability, grid independence
Portable Devices	Laptops, military equipment, remote sensors	Extended operation, rapid refueling, portability

adoption (He *et al.*, 2020). Catalysts play a pivotal role in the electrochemical reactions within fuel cells. Traditionally, platinum-based catalysts have been the standard due to their excellent activity. However, the high cost and limited availability of platinum present significant barriers to large-scale commercialization. Consequently, research has intensified into alternative catalysts, including non-precious metal catalysts and nanostructured materials, which offer comparable performance at a reduced cost (Huang *et al.*, 2021).

Moreover, the development of robust and conductive materials for membranes, electrodes, and bipolar plates is essential for improving the overall performance of fuel cells. Innovations in materials science, such as the use of high-entropy oxides and composite materials, have shown promise in enhancing the thermal stability and ionic conductivity of fuel cell components (Iwan *et al.*, 2015). Advancements in materials and catalysts not only improve the technical performance of fuel cells but also contribute to reducing manufacturing costs and increasing the scalability of production processes. These developments are critical for transitioning fuel cell technology from niche applications to mainstream energy solutions (Song, 2002).

3.2 OVERVIEW OF FUEL CELL TECHNOLOGIES

Fuel cells are electrochemical devices that convert chemical energy directly into electrical energy through redox reactions, offering high efficiency and low environmental impact. They are primarily classified based on the type of electrolyte used, which influences their operating temperature, fuel flexibility, and applications as shown in Table 3.2 (Kartha & Grimes, 1994). At the core of fuel cell operation lies the electrochemical conversion of fuel into electricity. A typical fuel cell comprises three primary components: the anode, cathode, and electrolyte. The anode serves as the site for fuel oxidation, commonly hydrogen, releasing electrons and protons. These electrons traverse an external circuit, generating an electric current, while protons migrate through the electrolyte to the cathode. At the cathode, oxygen molecules combine with the incoming electrons and protons to form water, completing the reaction cycle (Haile, 2003).

The electrolyte's role is pivotal, as it facilitates selective ion transport while preventing electron flow within the cell, ensuring the external circuit carries the electrons to perform useful work. Different fuel cell types employ various electrolytes, influencing their operating temperatures and suitable applications (Lai & Ellis, 2017). For instance, Proton Exchange Membrane Fuel Cells (PEMFCs) utilize a solid polymer electrolyte, operating efficiently at lower temperatures (~80°C), making them ideal for transportation and portable power applications (Wee, 2007). In contrast, Solid Oxide Fuel Cells (SOFCs) employ a solid ceramic electrolyte, functioning at higher temperatures (600–1000°C), suitable for stationary power generation (Singh & Minh, 2004).

The electrodes, typically composed of porous materials infused with catalysts like platinum, facilitate the redox reactions. Advancements in catalyst development aim to reduce reliance on

TABLE 3.2
Summary of Major Fuel Cell Types (Scofield *et al.*, 2015; Corigliano *et al.*, 2022; Raj, 2020; Jha, 2012; Frangini & Masi, 2016)

Fuel Cell Type	Electrolyte	Operating Temp.	Applications	Advantages	Challenges
PEMFC	Polymer Electrolyte Membrane	~80°C	Transportation, Portable Power	High power density, quick start-up	Sensitive to fuel impurities, costly catalysts
SOFC	Solid Oxide	600–1000°C	Stationary Power, Industrial Applications	Fuel flexibility, high efficiency	High operating temperature, material degradation
DMFC	Polymer Electrolyte Membrane	60–130°C	Portable Electronics, Military Applications	Simple fuel storage, compact design	Lower efficiency, methanol crossover
AFC	Aqueous Alkaline Solution	60–90°C	Space Applications, Military Systems	High efficiency, low cost	CO_2 sensitivity, electrolyte management
MCFC	Molten Carbonate Salt	600–700°C	Large-scale Power Generation	High efficiency, internal reforming capability	Corrosive electrolyte, durability issues

precious metals, thereby lowering costs and enhancing durability. Additionally, components such as gas diffusion layers and bipolar plates play crucial roles in distributing reactants evenly and managing water and heat within the cell (Dihrab *et al.*, 2009). Fuel cells exhibit higher energy conversion efficiencies compared to traditional combustion engines. PEMFCs, for example, achieve electrical efficiencies between 40% and 60%, while SOFCs can reach up to 65%. When integrated into combined heat and power (CHP) systems, overall efficiencies can soar to 85–90%, as waste heat is effectively utilized (Lindorfer *et al.*, 2020).

Power density is another metric that reflects the amount of power generated per unit volume or mass of the fuel cell. High power density is crucial for applications with space and weight constraints, such as in vehicles. PEMFCs are notable for their high power densities, with values reaching up to 2.5 kW/L, making them suitable for automotive applications (Chuahy & Kokjohn, 2019). The operational lifespan of fuel cells is a critical factor for commercial viability. Durability challenges stem from factors like catalyst degradation, membrane thinning, and material corrosion. For instance, PEMFCs have demonstrated lifespans of approximately 4,000 hours under automotive conditions, with ongoing research aiming to extend this to meet industry targets of 5,000 hours or more. SOFCs, due to their high operating temperatures, face challenges related to thermal cycling and material stability, impacting their longevity (Simons & Bauer, 2015).

3.3 KEY MATERIALS IN FUEL CELLS

The electrolyte membrane is central to fuel cell operation, facilitating ion transport while preventing electron flow within the cell. In Proton Exchange Membrane Fuel Cells (PEMFCs), the electrolyte is typically a perfluorosulfonic acid polymer, such as Nafion, which conducts protons and acts as a barrier to gases like hydrogen and oxygen (Maiti *et al.*, 2022). Nafion's high proton conductivity and chemical stability make it a preferred choice, although its performance can degrade at higher temperatures and low humidity. Alternative materials, including composite membranes and non-fluorinated polymers, are being explored to address these limitations (Maiti *et al.*, 2022).

In Solid Oxide Fuel Cells (SOFCs), the electrolyte is a dense ceramic material that conducts oxygen ions. Common materials include yttria-stabilized zirconia (YSZ) and gadolinium-doped ceria (GDC), which offer high ionic conductivity at elevated temperatures (Dwivedi, 2020). These materials enable SOFCs to operate efficiently at temperatures between 600°C and 1000°C, although efforts are ongoing to develop electrolytes that function effectively at lower temperatures to reduce system costs and material degradation (Dwivedi, 2020).

Catalysts are essential for facilitating the electrochemical reactions in fuel cells. In PEMFCs, platinum-based catalysts are commonly used at both the anode and cathode to promote hydrogen oxidation and oxygen reduction reactions, respectively. However, the high cost and scarcity of platinum have driven research into alternative catalysts, such as platinum alloys, non-precious metal catalysts, and carbon-based materials. For instance, nitrogen-doped carbon nanotubes have shown promise as cost-effective and durable catalysts, potentially reducing reliance on platinum (Huang *et al.*, 2021).

In SOFCs, the anode typically consists of a composite of nickel and YSZ, which provides both catalytic activity and electronic conductivity. The cathode often comprises materials like lanthanum strontium manganite (LSM), which facilitates oxygen reduction while maintaining compatibility with the electrolyte (Khan *et al.*, 2016). Advancements in catalyst materials aim to enhance performance, reduce costs, and improve durability under operating conditions.

Gas diffusion layers play a critical role in distributing reactant gases uniformly across the catalyst surface and facilitating the removal of water produced during the electrochemical reactions (Nguyen & Dinh, 2020). Typically made from porous carbon materials treated with hydrophobic agents like polytetrafluoroethylene (PTFE), GDLs ensure efficient gas transport and water management within the fuel cell. The structural and chemical properties of GDLs are crucial for maintaining optimal fuel cell performance, particularly under varying humidity and temperature conditions (Nguyen & Dinh, 2020).

Bipolar plates are integral components that separate individual cells within a fuel cell stack, distribute gases, and conduct electricity. They must exhibit high electrical and thermal conductivity, corrosion resistance, and mechanical strength. Materials used for bipolar plates include graphite, metal alloys, and composite materials (Tawfik *et al.*, 2007). Recent developments have focused on using high-performance thermoplastics and coated metals to reduce weight and manufacturing costs while maintaining performance and durability (Tawfik *et al.*, 2007).

Effective sealing is essential to prevent gas leakage and ensure the safe operation of fuel cells. Materials used for gaskets and seals must withstand the chemical and thermal environments within the cell. Common materials include silicone, fluoropolymers, and thermoplastics (Zhao *et al.*, 2023). Insulation materials are also employed to maintain temperature uniformity and protect components from thermal stresses, contributing to the overall efficiency and longevity of the fuel cell system (Zhao *et al.*, 2023).

3.4 ROLE OF CATALYSTS IN FUEL CELL REACTIONS

Catalysts are indispensable in fuel cells, facilitating the electrochemical reactions that convert chemical energy into electrical energy. They lower the activation energy required for these reactions, thereby enhancing the rate at which they occur. In hydrogen fuel cells, for instance, catalysts enable the efficient oxidation of hydrogen at the anode and the reduction of oxygen at the cathode. The effectiveness of these catalysts directly influences the overall performance, efficiency, and durability of the fuel cell (Peng & Wei, 2020).

Platinum (Pt) has long been the benchmark catalyst due to its exceptional activity and stability. However, its high cost and scarcity have spurred research into alternative materials, including non-precious metal catalysts and nanostructured materials. The development of these alternatives aims to reduce costs while maintaining or improving catalytic performance (Bhatt & Lee, 2020).

3.4.1 Anode and Cathode Reaction Mechanisms

At the anode of a hydrogen fuel cell, the hydrogen oxidation reaction (HOR) occurs. Hydrogen molecules (H_2) are adsorbed onto the catalyst surface, where they dissociate into protons (H^+) and electrons (e^-). The protons migrate through the electrolyte to the cathode, while the electrons travel through an external circuit, generating electrical power. The HOR on platinum surfaces typically follows two pathways: the Tafel–Volmer and the Heyrovsky–Volmer mechanisms (Sheng *et al.*, 2010).

Tafel–Volmer Mechanism:
Tafel Step:

$$H_2 \rightarrow 2H_{ads}$$

Volmer Step:

$$H_{ads} \rightarrow H^+ + e^-$$

Heyrovsky–Volmer Mechanism:
Heyrovsky Step:

$$H_2 + H_{ads} \rightarrow H^+ + e^- + H_{ads}$$

Volmer Step:

$$H_{ads} \rightarrow H^+ + e^-$$

These mechanisms involve the adsorption of hydrogen molecules onto the catalyst surface, followed by their dissociation and oxidation. The choice of pathway depends on various factors, including the nature of the catalyst and operating conditions.

At the cathode, the oxygen reduction reaction (ORR) takes place. Oxygen molecules (O_2) are reduced to water (H_2O) by gaining electrons and combining with protons. The ORR is more complex than the HOR and can proceed via two main pathways: the four-electron pathway and the two-electron pathway.

Four-Electron Pathway:

$$O_2 + 4H^+ + 4e^- \rightarrow 2H_2O$$

Two-Electron Pathway:

$$O_2 + 2H^+ + 2e^- \rightarrow H_2O_2$$

$$H_2O_2 + 2H^+ + 2e^- \rightarrow 2H_2O$$

The four-electron pathway is preferred due to its higher efficiency and the avoidance of hydrogen peroxide (H_2O_2) formation, which can be detrimental to fuel cell components. Platinum catalysts favor the four-electron pathway, but their high cost has led to the exploration of alternative catalysts that can also facilitate this pathway effectively.

3.4.2 Kinetics and Thermodynamic Considerations

The kinetics of the electrochemical reactions in fuel cells are critical for determining their performance. The rate at which these reactions occur is influenced by the activation energy, which catalysts aim to minimize. The exchange current density (j_0) is a key parameter representing the intrinsic rate of the electrochemical reaction at equilibrium (Divisek *et al.*, 1994). For the HOR, platinum exhibits a high exchange current density, indicating rapid kinetics. In contrast, the ORR has slower kinetics, even on platinum, necessitating higher catalyst loadings to achieve desired performance levels. This disparity underscores the importance of developing catalysts with enhanced ORR activity (Alonso-Vante, 2023).

The thermodynamics of fuel cell reactions are governed by the Gibbs free energy change (ΔG). For the overall reaction in a hydrogen fuel cell:

$$H_2 + \tfrac{1}{2}O_2 \rightarrow H_2O$$

The standard Gibbs free energy change (ΔG°) at 298 K is approximately −237.2 kJ/mol, corresponding to a theoretical cell voltage (E°) of 1.23 V. However, practical fuel cells operate at lower voltages due to various losses, including activation overpotentials, ohmic losses, and mass transport limitations. The activation overpotential is particularly significant for the ORR, as the reaction has a high activation energy barrier. Reducing this overpotential through the development of more active catalysts is essential for improving fuel cell efficiency.

3.4.3 Precious Metal Catalyst (Platinum and Platinum Alloys)

Platinum (Pt) remains the most effective and widely used precious metal catalyst in fuel cells, particularly proton exchange membrane fuel cells (PEMFCs), due to its remarkable catalytic activity for both the hydrogen oxidation reaction (HOR) at the anode and the oxygen reduction reaction (ORR) at the cathode. The kinetics of these reactions directly affect the overall efficiency and performance of fuel cells. Pt's ability to adsorb reactants and facilitate electron transfer with relatively low activation energy losses makes it the gold standard catalyst despite its high cost and limited availability (Gasteiger *et al.*, 2005). In PEMFCs, the sluggish kinetics of the ORR compared to the rapid HOR make the cathode catalyst a critical component in improving fuel cell efficiency. While Pt exhibits high intrinsic activity toward ORR, pure Pt catalysts still suffer from several drawbacks, such as poisoning by impurities like carbon monoxide (CO) and insufficient durability under prolonged operational conditions, motivating research into Pt alloys to enhance performance and reduce platinum loading (Debe, 2012).

The effectiveness of platinum catalysts depends heavily on their surface structure and the support materials used. Nanoparticulate Pt dispersed on carbon supports provides a high surface area to volume ratio, increasing the availability of active catalytic sites (Huang *et al.*, 2021). Typically, Vulcan carbon black is used as the support due to its excellent conductivity and surface area; however, carbon corrosion under fuel cell operating conditions can lead to catalyst degradation. To overcome these issues, alternative supports such as graphene, carbon nanotubes, and doped metal oxides have been developed to improve catalyst stability and prevent agglomeration of Pt nanoparticles (Chen *et al.*, 2011). The nanoscale dispersion of Pt enhances not only catalytic activity but also catalyst utilization efficiency, thereby reducing the amount of precious metal required. Although pure Pt is highly active, alloying platinum with other transition metals such as cobalt (Co), nickel (Ni), iron (Fe), copper (Cu), and manganese (Mn) has proven to be a powerful strategy for improving catalytic performance. These alloys exploit electronic (ligand) effects and geometric (strain) effects to modify the Pt surface properties, particularly the d-band electronic structure. This alteration optimizes the adsorption energies of key reaction intermediates involved in ORR and HOR, leading to enhanced reaction rates and lower overpotentials (Nørskov *et al.*, 2004). By tuning the binding energies of oxygen species, these alloys improve catalytic activity, reduce Pt loading, and can increase resistance to poisoning and degradation (Stamenkovic *et al.*, 2006).

Pt-Co alloys are among the most extensively studied and commercially promising Pt-based catalysts. The incorporation of Co atoms into the Pt lattice creates a compressive strain on the Pt surface, which weakens oxygen binding energy and thereby accelerates ORR kinetics by up to an order of magnitude relative to pure Pt (Stamenkovic *et al.*, 2006). However, the dissolution of Co under acidic fuel cell operating conditions can lead to stability issues and membrane contamination. To address this, controlled dealloying techniques are used to form Pt-rich shells around Pt-Co cores, preserving high activity while enhancing durability by shielding the less stable Co atoms beneath a protective Pt surface (Strasser *et al.*, 2010). This core-shell design maintains the beneficial strain and electronic effects while improving catalyst lifespan.

Similarly, Pt-Ni alloys have demonstrated extraordinary ORR activity, surpassing even Pt-Co alloys under certain conditions. The Pt_3Ni intermetallic phase, especially when exposing the (111) crystallographic facet, exhibits one of the highest specific activities for ORR reported in literature (Stamenkovic *et al.*, 2006). These facets provide optimal adsorption sites for oxygen reduction intermediates, reducing energy barriers for the reaction. Nonetheless, Ni is prone to dissolution under acidic environments, which limits long-term stability. Advanced catalyst designs incorporating Pt-Ni with protective Pt shells or doping with other metals have shown promising durability improvements (Li *et al.*, 2023). Furthermore, Pt-Ni nanostructures with tailored shapes, such as octahedra or nanoframes, maximize exposure of high-index facets, further boosting catalytic efficiency.

Pt-Fe alloys offer additional advantages through the formation of ordered intermetallic compounds such as Pt_3Fe, which provide enhanced stability and catalytic activity. Iron's presence adjusts the electronic structure of Pt, weakening oxygen binding and increasing ORR kinetics (Liu *et al.*, 2016). The ordered structure of Pt-Fe intermetallics increases thermodynamic stability, reducing the rate of metal dissolution compared to random alloys. Despite this, Fe dissolution remains a concern, and catalyst designs that stabilize intermetallic phases or incorporate protective layers are necessary for practical applications. Other platinum alloys such as Pt-Cu, Pt-Mn, and Pt-Sn have been explored for specific improvements. Pt-Cu alloys exhibit beneficial strain effects, but copper's higher dissolution rate reduces stability. Pt-Mn alloys show moderate ORR activity and better durability; however, manganese's susceptibility to oxidation and dissolution limits widespread application (Liu *et al.*, 2016). Pt-Sn alloys have been studied primarily for their ability to enhance CO tolerance in reformate hydrogen fuel cells and direct methanol fuel cells (DMFCs). Tin modifies the electronic structure of Pt and provides oxygen-containing species at low potentials, facilitating the oxidation of adsorbed CO and thus improving catalyst tolerance to poisoning (Chen *et al.*, 2011).

The catalytic performance of Pt and its alloys arises from intricate electronic and geometric effects. Alloying induces shifts in the Pt d-band center, which modulate the adsorption energies of ORR intermediates such as atomic oxygen (*O), hydroxyl (*OH), and peroxide (*OOH). The balance between adsorption and desorption of these intermediates determines the catalytic turnover frequency and pathway selectivity (Nørskov *et al.*, 2004). A fundamental principle, known as the Sabatier principle, states that optimal catalytic activity is achieved when the binding of intermediates is neither too strong nor too weak. Pt alloys shift the d-band center to reach this optimum, enhancing reaction kinetics. The ORR mechanism on Pt-based catalysts involves multiple proton-coupled electron transfer steps. Two major pathways exist: the preferred four-electron reduction producing water directly, and the less efficient two-electron pathway producing hydrogen peroxide, a reactive species that can degrade fuel cell components. Pt alloys promote the four-electron pathway by stabilizing ORR intermediates and lowering the formation energy of water, thereby improving efficiency and durability (Gasteiger *et al.*, 2005). Kinetic analyses reveal that alloying lowers the activation energy for ORR and increases the exchange current density, correlating with enhanced fuel cell performance.

Durability remains a major challenge for Pt and Pt alloy catalysts. Pt nanoparticles can dissolve and redeposit via Ostwald ripening, causing agglomeration and loss of active surface area. The carbon supports used to disperse Pt are vulnerable to oxidation at high potentials, leading to catalyst detachment and performance decay. Additionally, the dissolution of less noble alloying metals such as Co, Ni, and Fe leads to structural degradation and contamination of the proton exchange membrane, adversely affecting fuel cell longevity (Debe, 2012). To mitigate these issues, catalyst designs increasingly incorporate core-shell architectures with Pt-enriched surfaces that protect alloy cores, use corrosion-resistant supports such as doped carbons or metal oxides, and favor intermetallic phases with higher thermodynamic stability.

Various synthesis methods have been developed to produce Pt and Pt-alloy catalysts with controlled size, morphology, and composition. Chemical co-reduction in solution allows simultaneous reduction of Pt and alloy metals, yielding homogeneous alloys. Thermal annealing promotes alloying and ordering, enhancing catalytic activity and stability. Galvanic replacement and dealloying techniques create core-shell catalysts with a Pt-rich shell that improves durability. Atomic layer deposition (ALD) offers atomic-scale precision in depositing Pt shells on alloy cores, optimizing thickness and surface coverage for enhanced performance (Strasser *et al.*, 2010). Tailoring these parameters is crucial for balancing catalytic activity, stability, and cost.

3.4.4 Non-Precious Metal Catalysts

Non-precious metal catalysts (NPMCs) have emerged as promising alternatives to expensive platinum-based catalysts for oxygen reduction reaction (ORR) in fuel cells, aiming to reduce cost while maintaining high catalytic activity and durability. Among these, transition metal–nitrogen–carbon (M-N-C) catalysts, particularly iron (Fe-N-C) and cobalt (Co-N-C) based systems, have gained significant attention for proton exchange membrane fuel cells (PEMFCs) and alkaline fuel cells due to their remarkable ORR activity and stability under operating conditions (He *et al.*, 2019). M-N-C catalysts are typically prepared by pyrolyzing transition metal salts coordinated with nitrogen-containing organic precursors on high surface area carbon supports, creating atomically dispersed metal-nitrogen coordination sites embedded within a conductive carbon matrix (Zitolo *et al.*, 2015). These M-Nx sites are widely accepted as the primary active centers responsible for catalytic activity, where the metal atom coordinated by nitrogen atoms optimally adsorbs oxygen and facilitates its reduction (Chen *et al.*, 2022).

Fe-N-C catalysts are the most extensively studied M-N-C materials due to their superior ORR activity approaching that of platinum in acidic and alkaline media. The high activity is attributed to Fe-N4 moieties formed during high-temperature pyrolysis, which provide an optimal electronic environment for oxygen adsorption and reduction (Cheng *et al.*, 2019). The ORR catalyzed by Fe-N-C follows a four-electron transfer pathway directly producing water, which is preferred over the less efficient two-electron pathway generating hydrogen peroxide (H_2O_2), thereby increasing fuel cell efficiency and minimizing oxidative degradation (Guo *et al.*, 2016). These catalysts also display a significant tolerance toward CO and methanol poisoning, which commonly affect platinum catalysts in real fuel cell environments (Banham *et al.*, 2015). The tunability of Fe-N-C catalysts through controlled pyrolysis conditions, choice of nitrogen precursors (e.g., polyaniline, porphyrins), and metal loading enables optimization of active site density, porosity, and electrical conductivity (Zhang *et al.*, 2023).

Cobalt-based M-N-C catalysts (Co-N-C) similarly contain Co-N4 active centers embedded in carbon matrices and exhibit competitive ORR performance, particularly in alkaline media where the oxygen reduction kinetics are more favorable (Jaouen *et al.*, 2011). Co-N-C catalysts often demonstrate superior stability under fuel cell operating conditions compared to Fe-N-C counterparts due to the reduced susceptibility of cobalt to Fenton-type reactions that produce damaging reactive oxygen species (ROS) (Kiani *et al.*, 2021). Although Co-N-C typically exhibits slightly lower intrinsic activity than Fe-N-C in acidic environments, recent advances in catalyst synthesis have narrowed this gap by improving site density and electronic structure tuning (Kramm *et al.*, 2012). The stability advantage makes Co-N-C catalysts highly attractive for applications requiring prolonged operation, such as stationary power systems.

The catalytic activity of M-N-C materials largely stems from the unique electronic interactions between the transition metal center, nitrogen ligands, and the carbon support. The nitrogen dopants in the carbon lattice serve multiple roles: they stabilize the metal centers, modulate the electronic density of states near the Fermi level, and create favorable binding sites for oxygen molecules (Singh *et al.*, 2019). Pyridinic and graphitic nitrogen species are often found to enhance ORR activity by tuning charge distribution, while the embedded metal centers actively participate in breaking the O=O bond (Zitolo *et al.*, 2015). Advanced characterization techniques such as Mössbauer spectroscopy, X-ray absorption spectroscopy (XAS), and electron microscopy have confirmed the presence of atomically dispersed Fe-N4 and Co-N4 sites as dominant active species (Chen *et al.*, 2022). Furthermore, density functional theory (DFT) calculations have provided mechanistic insight into how these active sites optimize adsorption energies for key ORR intermediates like *O, *OH, and *OOH, following the Sabatier principle to achieve optimal catalytic performance (Hong *et al.*, 2017).

Despite their advantages, M-N-C catalysts face challenges related to durability and stability under fuel cell operating conditions. Fe-N-C catalysts can catalyze Fenton-like reactions generating hydroxyl radicals (•OH) from H2O2 intermediates, which attack the carbon matrix and cause structural degradation over time (Banham *et al.*, 2019). This oxidative degradation results in loss of active sites and performance decline. Strategies to mitigate this include improving the graphitization degree of the carbon support to enhance corrosion resistance, incorporating secondary heteroatoms such as sulfur or phosphorus to stabilize metal centers, and employing protective layers to limit radical formation (Hong *et al.*, 2017). Additionally, careful control of pyrolysis temperature and atmosphere optimizes the formation of stable M-N4 sites while reducing the presence of Fe nanoparticles or clusters that contribute to undesired side reactions (Zitolo *et al.*, 2015). Similarly, Co-N-C catalysts avoid some Fenton-related issues but still require optimization of metal dispersion and carbon support to maximize durability.

The preparation of M-N-C catalysts involves complex synthesis routes typically including the impregnation of nitrogen-rich ligands and metal salts onto carbon supports, followed by high-temperature pyrolysis under inert atmospheres. This process generates atomically dispersed metal-nitrogen moieties through thermally induced coordination and carbonization (Zitolo *et al.*, 2015). Post-pyrolysis acid leaching removes unstable metal nanoparticles and excess metal precursors, enriching the active site density. Recent advances include the use of metal-organic frameworks (MOFs) as precursors to create highly porous and uniform M-N-C catalysts with well-defined active sites and enhanced mass transport properties (Chen *et al.*, 2022). MOF-derived Fe-N-C catalysts show enhanced ORR activity due to improved site accessibility and tailored pore structures facilitating reactant diffusion.

Catalytic activity is strongly influenced by the surface area, porosity, and conductivity of the carbon support. Mesoporous and microporous carbon frameworks derived from graphene, carbon nanotubes, or biomass have been employed to enhance catalyst performance (Zitolo *et al.*, 2015). High surface areas increase active site exposure, while hierarchical pore structures improve oxygen and proton transport within the catalyst layer. Additionally, nitrogen doping not only creates active sites but also improves electron conductivity by altering the electronic structure of carbon materials (He *et al.*, 2019). Optimizing these physical properties is essential for maximizing the utilization of M-N-C active sites and improving overall fuel cell performance.

Performance benchmarks show that Fe-N-C catalysts can achieve ORR half-wave potentials approaching 0.85 V vs. RHE in acidic electrolytes, nearing those of commercial Pt/C catalysts (Banham *et al.*, 2015). In alkaline media, M-N-C catalysts often surpass Pt-based catalysts due to inherently faster ORR kinetics and reduced Pt poisoning (Jaouen *et al.*, 2011). However, durability tests reveal that Fe-N-C catalysts typically retain 60–80% of their initial activity after 30,000 voltage cycles, compared to over 90% retention for Pt catalysts (Banham *et al.*, 2019). Efforts to improve this include designing more robust carbon supports, stabilizing metal sites with stronger coordination, and alloying with secondary metals to inhibit degradation pathways.

Fe-N-C and Co-N-C catalysts have also demonstrated good performance in direct methanol fuel cells (DMFCs), showing enhanced tolerance to methanol crossover compared to Pt-based cathodes (Kiani *et al.*, 2021). This tolerance arises from the different adsorption mechanisms of methanol and oxygen on the active sites, reducing the parasitic oxidation of methanol and improving fuel utilization efficiency. This feature makes M-N-C catalysts highly attractive for portable power applications where fuel crossover is a critical issue.

The scalability and cost-effectiveness of M-N-C catalysts are additional advantages over platinum catalysts. Iron and cobalt are abundant and inexpensive metals, and the use of carbon and nitrogen-rich precursors from renewable or waste sources further reduces catalyst production costs (Zitolo *et al.*, 2015). The potential to integrate these materials into existing fuel cell manufacturing processes supports their commercial viability. Nonetheless, challenges remain in achieving consistent catalyst quality and performance in large-scale production due to the sensitivity of active site formation to synthesis conditions.

Emerging research directions include the development of single-atom catalysts (SACs) where isolated Fe or Co atoms are stabilized on nitrogen-doped carbon matrices, maximizing atom efficiency and catalytic activity (Cheng *et al.*, 2019). SACs have shown exceptional ORR performance with enhanced stability due to the absence of metal clusters prone to aggregation or dissolution. Moreover, dual-metal or bimetallic M-N-C catalysts combining Fe and Co or incorporating other transition metals offer synergistic effects, improving activity and durability beyond monometallic counterparts (He *et al.*, 2019). Machine learning and computational screening accelerate the discovery of optimal coordination environments and metal combinations for next-generation catalysts.

3.4.5 Nanostructured and Hybrid Catalysts

Nanostructured and hybrid catalysts have emerged as essential components in advancing fuel cell performance by optimizing catalytic activity, reducing precious metal usage, and enhancing durability. Core-shell catalysts are designed by encapsulating a core nanoparticle with a thin shell of catalytically active material, typically a noble metal such as platinum. This architecture capitalizes on the fact that the core can be made from less expensive or more abundant materials, while the shell provides the essential surface sites for electrochemical reactions (Zhang *et al.*, 2022). The interaction between the core and the shell leads to electronic and strain effects that modulate the surface properties of the shell atoms, thus enhancing catalytic activity and selectivity. In fuel cells, platinum-based core-shell catalysts have demonstrated substantial improvements in oxygen reduction reaction (ORR) kinetics. For example, platinum shells grown on palladium, nickel, or copper cores exhibit a shift in the d-band center of surface platinum atoms, optimizing the binding energies of oxygenated species and improving turnover frequency. Moreover, the shell thickness, often limited to just a few atomic layers, allows for maximal utilization of platinum atoms, a critical factor given the high cost of platinum. Such ultrathin shells ensure that nearly all platinum atoms participate in catalysis, markedly improving mass activity compared to bulk platinum catalysts (Zhang *et al.*, 2022).

Beyond enhancing activity, core-shell structures also contribute to enhanced durability. The core can impart structural stability to the shell, mitigating issues like platinum dissolution and particle aggregation that commonly plague fuel cell catalysts during prolonged operation. Innovative designs incorporating hollow or porous cores further increase accessible surface area and promote better mass transport of reactants and products within the catalyst layer, which is vital for sustaining high current densities under fuel cell operating conditions (Rojas-Carbonell *et al.*, 2017). Nevertheless, challenges persist, especially in achieving precise control over shell thickness, preventing core leaching, and maintaining structural integrity in the highly corrosive and dynamic electrochemical environment of a working fuel cell (Rojas-Carbonell *et al.*, 2017).

In parallel, bimetallic and trimetallic catalysts have gained attention due to their ability to harness synergistic effects arising from the intimate combination of different metals at the nanoscale. These catalysts often take the form of alloys, core-shell heterostructures, or phase-segregated nanoparticles, where the composition and atomic arrangement can be finely tuned to optimize catalytic properties (Zheng *et al.*, 2015). Platinum-based bimetallic alloys, such as Pt-Ni, Pt-Co, and Pt-Fe, are among the most extensively studied for ORR catalysis in proton exchange membrane fuel cells (PEMFCs). The incorporation of a second metal alters the electronic environment of platinum atoms, weakening the adsorption strength of oxygen intermediates and facilitating faster reaction kinetics. For instance, Pt-Ni alloys exhibit compressive strain on the platinum lattice and modified electronic states, leading to significant improvements in ORR activity over pure platinum (Zheng *et al.*, 2015). Similarly, Pt-Co and Pt-Fe alloys demonstrate enhanced catalytic activity alongside improved resistance to degradation caused by catalyst poisoning and corrosion.

Trimetallic catalysts add further complexity and opportunity by introducing a third metal, which can further tune the catalyst's electronic structure and geometric configuration. These multi-metal systems often exhibit improved performance compared to their bimetallic counterparts, as the third metal can provide additional active sites or influence stability (Banham *et al.*, 2015). For example, Pt-Ni-Fe and Pt-Co-Cu trimetallic catalysts have shown promising ORR activities and improved durability through optimized binding energies of reaction intermediates and enhanced resistance to oxidative degradation. The synthesis of such multi-metallic catalysts commonly involves methods like co-reduction, sequential deposition, or thermal annealing, allowing for controlled alloy formation or core-shell architectures. However, the increased compositional complexity demands rigorous characterization and mechanistic understanding to fully exploit their potential, as subtle variations in composition and structure can drastically affect catalytic behavior (Banham *et al.*, 2015).

Integral to the performance of these nanostructured catalysts is the nature of their support materials. Carbon-based supports remain the benchmark due to their high electrical conductivity, large surface area, and mechanical stability, which facilitate effective dispersion of catalyst nanoparticles and efficient electron transport within the fuel cell electrode. Conventional carbon black materials, such as Vulcan XC-72, have been widely used for catalyst support. However, these materials suffer from corrosion and oxidation during prolonged fuel cell operation, leading to catalyst detachment, loss of active surface area, and overall performance degradation. To address these issues, research has increasingly turned to advanced carbon nanomaterials such as graphene, carbon nanotubes (CNTs), and mesoporous carbons (Banham *et al.*, 2015).

Graphene and CNTs offer several advantages, including superior electrical conductivity, large surface area, and excellent chemical stability. Their two-dimensional and tubular structures, respectively, allow for high catalyst dispersion and strong metal-support interactions, which can inhibit nanoparticle aggregation and enhance durability (Guo *et al.*, 2016). Mesoporous carbons, characterized by their well-defined pore structures, further improve mass transport by facilitating the diffusion of reactants like oxygen and protons to the active sites and the removal of water and reaction products. Surface functionalization of these carbon supports, such as nitrogen doping or the introduction of oxygen-containing groups, has been shown to strengthen the interaction between metal nanoparticles and the support, thus enhancing catalyst stability and catalytic activity (Oh & Kim, 2012).

Hybrid support systems have also emerged, combining carbon materials with metal oxides or conducting polymers to leverage the strengths of each component. For example, carbon-TiO_2 composites can improve resistance to carbon corrosion and provide additional catalytic sites or promote water management in the catalyst layer (Kramm *et al.*, 2012). Such hybrid supports contribute not only to catalyst durability but also to enhanced electrochemical performance through improved charge transfer and active site accessibility. However, optimizing the pore structure, surface chemistry, and metal-support interaction in these complex hybrid systems remains a challenge requiring advanced synthetic and characterization techniques (Banham *et al.*, 2015).

Overall, the combination of nanostructured catalyst design and advanced carbon supports represents a powerful strategy to enhance fuel cell catalyst performance. Core-shell catalysts enable efficient utilization of precious metals and improved durability via tailored electronic and strain effects. Bimetallic and trimetallic alloys harness synergistic interactions between metals to optimize catalytic activity and stability (Wang *et al.*, 2014). Carbon-supported catalysts ensure high dispersion, electrical conductivity, and mechanical integrity while mitigating corrosion-related degradation through advanced nanomaterials and hybrid supports. Future progress in this field will depend on the continued refinement of synthesis methods to control catalyst nanostructure and composition at the atomic level, enabling precise tuning of surface properties. In situ and operando characterization techniques, alongside computational modeling, will deepen mechanistic understanding of catalytic processes and degradation pathways under realistic operating conditions. Additionally, scaling up the production of nanostructured catalysts and supports in a cost-effective and reproducible manner remains essential for their commercial deployment (Wang *et al.*, 2014).

Durability, particularly under start-stop cycling and high potential excursions, is a critical hurdle that must be overcome to extend the operational lifetime of fuel cells employing these advanced catalysts. Strategies such as stabilizing the catalyst-support interface, optimizing alloy compositions to resist dissolution, and designing robust core-shell architectures will be key to addressing this challenge (Wang *et al.*, 2014).

3.4.6 Catalyst Degradation Mechanisms

Despite significant advances in performance, one of the critical challenges hindering the widespread commercial adoption of fuel cells is catalyst degradation. The catalysts, typically noble metals such as platinum supported on carbon, facilitate key electrochemical reactions but are prone to deterioration under operational conditions. This degradation results in the loss of catalytic activity, reduced fuel cell efficiency, and ultimately limits the lifetime of the device. Among the various degradation pathways, sintering, corrosion, and poisoning are the predominant mechanisms by which catalysts deteriorate during fuel cell operation (Gasteiger *et al.*, 2005).

Sintering, a thermally activated process, involves the growth and agglomeration of catalyst nanoparticles, which reduces the electrochemically active surface area and hence diminishes catalytic performance. In proton exchange membrane fuel cells (PEMFCs), platinum nanoparticles are typically dispersed on carbon supports with an average size of a few nanometers. Under operational conditions, particularly at elevated temperatures and during potential cycling, these nanoparticles tend to migrate and coalesce. This process is driven by the minimization of surface energy, as smaller particles dissolve and redeposit onto larger particles—a phenomenon known as Ostwald ripening. The net result is an increase in average particle size and a corresponding decrease in the number of active sites accessible for the oxygen reduction reaction (ORR), which is the rate-limiting step in PEMFCs (Gasteiger *et al.*, 2005).

The sintering process is further exacerbated by instability in the catalyst support. Carbon black, commonly used as a support due to its high surface area and electrical conductivity, undergoes structural degradation under electrochemical conditions. As the carbon support deteriorates, it loses its ability to effectively anchor platinum nanoparticles, allowing them to migrate more freely and coalesce. Consequently, the combined effects of carbon support instability and particle sintering contribute significantly to performance degradation (Antolini, 2009). In solid oxide fuel cells (SOFCs), which operate at higher temperatures ranging from 700 to 1000 °C, sintering manifests primarily through the agglomeration of nickel particles in the anode cermet. The elevated temperatures accelerate particle growth and reduce nickel dispersion, impairing the catalytic activity for hydrogen oxidation and fuel reforming reactions (Khan *et al.*, 2016).

Alongside sintering, corrosion plays a critical role in catalyst degradation. In PEMFCs, the carbon support is thermodynamically unstable under the highly oxidative environment of the cathode, especially during transient operating conditions such as start-up and shutdown cycles. At potentials exceeding approximately 0.9 V versus the reversible hydrogen electrode (RHE), carbon oxidation to carbon dioxide becomes significant. This electrochemical carbon corrosion leads to the gradual loss of the support material, resulting in catalyst layer thinning, loss of electrical conductivity, and detachment of platinum nanoparticles. Such effects accelerate catalyst sintering and dissolution, culminating in severe performance losses.

Carbon corrosion is particularly pronounced during start-stop events when local cathode potentials can spike above 1.2 V. These transient high potentials promote rapid carbon oxidation, exacerbating catalyst layer degradation. The loss of support structure not only reduces the mechanical integrity of the catalyst layer but also facilitates platinum nanoparticle migration and aggregation, further diminishing the electrochemical surface area (Feng *et al.*, 2017). To address these challenges, research has focused on developing more corrosion-resistant supports, such as graphitized carbon, carbon nanotubes, graphene, and various metal oxides, including titanium dioxide and tin dioxide. These materials exhibit higher structural stability under fuel cell operating conditions and help suppress both carbon corrosion and platinum nanoparticle migration.

Corrosion is not limited to the support material; catalyst nanoparticles themselves are susceptible to degradation. Although platinum is relatively noble and corrosion-resistant, it can dissolve under high potentials experienced during fuel cell operation, particularly during start–stop cycles. Dissolved platinum ions may then migrate and redeposit onto existing particles, contributing to particle growth through Ostwald ripening, which again reduces the available active surface area (Mitsushima *et al.*, 2008). In SOFCs, non-noble metal catalysts such as nickel are vulnerable to oxidation and sintering. Nickel oxidation to inactive NiO phases under fluctuating fuel conditions or high oxygen partial pressures leads to loss of catalytic activity and mechanical failure of the anode.

In addition to sintering and corrosion, catalyst poisoning presents a formidable barrier to maintaining catalyst activity. Poisoning arises when impurities in the fuel or air streams adsorb onto active catalytic sites, blocking reactant access and hindering electrochemical reactions. Among poisons, carbon monoxide (CO) is one of the most well-studied, particularly in PEMFCs using hydrogen derived from reformate gases, which often contain trace amounts of CO (10–100 ppm). CO binds strongly to platinum surfaces via linear and bridged adsorption modes, effectively occupying sites needed for hydrogen oxidation and thus dramatically reducing reaction kinetics.

CO poisoning is highly dependent on operating temperature and catalyst structure. Higher temperatures facilitate CO desorption or oxidation to CO_2, helping to alleviate poisoning effects. Alloying platinum with ruthenium has been a successful strategy to improve CO tolerance; ruthenium facilitates the oxidative removal of CO at lower potentials by providing oxygen-containing species that react with adsorbed CO (Molochas & Tsiakaras, 2021). This synergy enhances the durability and efficiency of catalysts under reformate fuel conditions.

Sulfur-containing compounds such as hydrogen sulfide (H_2S) and sulfur dioxide (SO_2) pose even greater challenges. Sulfur species adsorb strongly and irreversibly on catalyst surfaces, forming sulfide layers that severely impair catalytic activity. Unlike CO, sulfur poisoning is more difficult to reverse and can cause rapid and permanent catalyst deactivation even at trace concentrations. Sulfur contamination is especially problematic in fuel cells fed by fossil fuel-derived gases, where purification is incomplete (Sethuraman & Weidner, 2010). Regeneration of sulfur-poisoned catalysts typically requires harsh conditions such as high temperature and potential cycling, which can themselves accelerate sintering and corrosion, leading to a complex trade-off in durability.

Other impurities, including ammonia, chlorine, and nitrogen oxides, have also been reported to adsorb on catalyst surfaces or degrade membrane materials, indirectly affecting catalyst performance. Although their poisoning effects are generally less severe than CO or sulfur, they contribute to the overall degradation of fuel cell components and reduce system reliability.

Efforts to mitigate catalyst degradation encompass material innovation and operational strategies. Developing catalysts with intrinsic resistance to sintering and dissolution is an active area of research. Advanced nanostructured catalysts featuring core-shell architectures, shape-controlled nanoparticles, or alloyed compositions have demonstrated enhanced stability by reducing surface energy and limiting atomic mobility (Zhang *et al.*, 2013). Likewise, engineering catalyst supports with enhanced corrosion resistance, including graphitized carbons and conductive metal oxides, provide physical and chemical stabilization to catalyst nanoparticles.

Operational strategies also play a critical role in extending catalyst lifetime. Avoiding high cathode potentials and rapid potential cycling through improved fuel cell system design reduces carbon corrosion and platinum dissolution. Careful control of start-up and shutdown protocols can minimize transient conditions that exacerbate degradation. Moreover, improving fuel purity through effective reformate gas cleaning reduces poisoning risks (Molochas & Tsiakaras, 2021).

Despite advances, there remains a delicate balance in mitigating one degradation mechanism without accelerating another. For instance, aggressive regeneration protocols designed to remove sulfur poisons can increase carbon corrosion and particle sintering. Thus, integrated approaches combining materials design, system engineering, and fuel processing are essential to achieve durable fuel cell catalysts capable of meeting commercial performance and lifetime requirements (Molochas & Tsiakaras, 2021).

3.5 RECENT TRENDS IN MATERIALS AND CATALYSTS DEVELOPMENT

Central to their performance and commercialization is the development of advanced materials and catalysts that can deliver high activity, stability, and cost-effectiveness under operating conditions. Recent years have witnessed the emergence of innovative material classes and catalyst architectures that promise to overcome the limitations of conventional systems. Among these, high-entropy alloys (HEAs), single-atom catalysts (SACs), doped carbon and two-dimensional (2D) materials, and functional composite membranes stand out as transformative advances in fuel cell technology.

High-entropy alloys represent a novel class of catalytic materials defined by the incorporation of multiple principal metallic elements, typically five or more, in near-equimolar ratios. This multi-elemental approach leads to highly disordered atomic structures and unique electronic properties that differ fundamentally from traditional binary or ternary alloys. The configurational entropy in HEAs stabilizes single-phase solid solutions with enhanced mechanical strength, corrosion resistance, and catalytic activity (Ren *et al.*, 2023). In the context of fuel cells, HEAs have attracted attention primarily as electrocatalysts for the oxygen reduction reaction (ORR), a key limiting step in proton exchange membrane fuel cells (PEMFCs).

The catalytic performance of HEAs arises from the synergistic interplay of their constituent metals, which tunes the adsorption energies of reactants and intermediates on the catalyst surface. This leads to optimized binding strengths that facilitate reaction kinetics beyond those achievable with pure platinum or traditional platinum-based alloys (Xin *et al.*, 2020). For instance, recent studies have demonstrated that FeCoNiCrMn HEAs exhibit enhanced ORR activity and stability compared to commercial Pt/C catalysts, owing to the presence of multiple active sites and improved resistance to sintering and dissolution (Yu *et al.*, 2024). Moreover, the inherent compositional flexibility of HEAs enables the rational design of catalysts tailored to specific fuel cell environments, including acidic and alkaline conditions, making them versatile candidates for next-generation devices.

Complementing HEAs, single-atom catalysts have emerged as a paradigm shift in catalyst design by maximizing atomic efficiency and enabling unique reaction pathways. SACs consist of isolated metal atoms anchored on suitable supports such as nitrogen-doped carbon or metal oxides, preventing aggregation into nanoparticles and exposing every atom as a potential active site (Qiao *et al.*, 2012). This atomic dispersion leads to exceptional catalytic activity, selectivity, and stability, particularly important for expensive noble metals like platinum, where maximizing utilization is critical for cost reduction.

In fuel cells, SACs have shown promising activity for both the ORR and hydrogen oxidation reaction (HOR). For example, single platinum atoms dispersed on nitrogen-doped graphene demonstrate enhanced ORR activity due to altered electronic structures and favorable adsorption of oxygen intermediates (Tang *et al.*, 2023). Additionally, SACs based on non-noble metals such as iron, cobalt, and nickel coordinated with nitrogen sites (M–N–C) have emerged as effective platinum group metal (PGM)-free catalysts. These materials exhibit remarkable ORR activity in alkaline and acidic media, challenging the dominance of platinum catalysts while significantly reducing material costs (Hong *et al.*, 2017). Furthermore, the strong metal-support interactions and site isolation in SACs impart superior stability against sintering and dissolution, addressing critical degradation pathways that limit conventional catalysts.

Another transformative trend is the development of doped carbon and two-dimensional materials as catalyst supports and active catalysts. Carbon materials are widely employed in fuel cells for their excellent electrical conductivity, high surface area, and chemical stability. Recent advances focus on heteroatom doping—incorporating elements such as nitrogen, sulfur, phosphorus, or boron into the carbon lattice—to modify electronic properties and create active sites for catalytic reactions (Zhang & Dai, 2015). Nitrogen-doped carbons, for example, generate defects and alter charge distributions that enhance oxygen adsorption and activation, improving ORR kinetics without requiring expensive noble metals (Asefa & Huang, 2017).

Two-dimensional materials, including graphene, transition metal dichalcogenides (TMDs), and MXenes, offer exceptional physical and chemical properties beneficial for fuel cell applications. Their atomic thickness, high surface area, tunable electronic structures, and mechanical flexibility make them ideal catalyst supports or active components. Graphene, with its high conductivity and chemical robustness, serves as an excellent support to anchor metal nanoparticles or single atoms, mitigating aggregation and enhancing electron transfer (Su & Hu, 2021). TMDs such as MoS_2 have shown intrinsic catalytic activity for hydrogen evolution and oxygen reduction when appropriately engineered or doped (Zhu *et al.*, 2019). MXenes, a family of 2D transition metal carbides and nitrides, combine metallic conductivity with hydrophilic surfaces and rich surface chemistry, enabling strong interactions with catalytic species and facilitating mass transport (Ravichandran *et al.*, 2024). These doped carbon and 2D materials thus form a versatile platform to engineer high-performance, durable catalysts with reduced precious metal content.

Alongside catalyst development, advances in functional composite membranes have been instrumental in enhancing fuel cell performance and durability. Proton exchange membranes (PEMs) are critical components that facilitate proton conduction while electrically isolating the electrodes. Conventional Nafion membranes, although widely used, suffer from limitations including high cost, limited operating temperature range, and susceptibility to chemical and mechanical degradation (Mauritz & Moore, 2004). To overcome these challenges, research has focused on composite membranes that incorporate inorganic fillers, polymer blends, or functionalized nanomaterials to improve proton conductivity, thermal stability, and mechanical strength.

Incorporating inorganic nanoparticles such as silica, titania, or zirconia into polymer matrices enhances water retention and thermal stability, thereby broadening the operational window of PEMs (Senthil *et al.*, 2023). Functionalized fillers bearing acidic groups or ionic liquids have also been employed to facilitate proton transport pathways independent of water content, enabling high conductivity under low humidity conditions (Escorihuela *et al.*, 2019). Moreover, composite membranes utilizing 2D materials like graphene oxide (GO) or layered silicates create tortuous pathways that improve gas barrier properties and reduce fuel crossover, enhancing fuel cell efficiency and lifetime (Pandey *et al.*, 2017).

Polymer blends and block copolymers with tailored microphase-separated morphologies have further advanced membrane technology by optimizing proton conduction channels and mechanical robustness simultaneously (Elabd & Hickner, 2011). Additionally, emerging functional membranes with self-healing capabilities, radical scavenging properties, or improved chemical stability against reactive oxygen species contribute to prolonged membrane and fuel cell durability (Zhu *et al.*, 2023).

These functional composite membranes not only address fundamental material limitations but also enable the design of fuel cells capable of operating under more demanding conditions, including elevated temperatures and dry atmospheres, which are crucial for automotive and portable applications.

The convergence of these material innovations—high-entropy alloys, single-atom catalysts, doped carbon and 2D materials, and functional composite membranes—represents a holistic approach to fuel cell advancement. By enhancing catalytic activity and durability while reducing precious metal usage and material costs, these trends collectively move fuel cells closer to commercial viability. Nevertheless, challenges remain in scaling synthesis methods, ensuring material stability under real operating conditions, and integrating new materials into practical fuel cell architectures.

High-entropy alloys, while promising, require further exploration to understand the complex structure-property relationships and to optimize compositions for specific electrochemical reactions. Precise control of atomic arrangements and defect engineering within HEAs could unlock even higher catalytic performance (Ren *et al.*, 2023). Single-atom catalysts face challenges related to atom stabilization and large-scale production, as isolated atoms can aggregate under operating conditions if not robustly anchored (Tang *et al.*, 2023). Advances in support design and synthesis techniques, such as atomic layer deposition and coordination chemistry, are crucial to overcoming these hurdles.

The functionalization and doping of carbon and 2D materials require fine-tuning to balance electrical conductivity, catalytic activity, and chemical stability. Achieving uniform doping and defect control on a scalable basis remains a significant barrier (Zhang & Dai, 2015). For functional composite membranes, challenges include achieving reproducible fabrication, mechanical flexibility, and long-term chemical stability under the harsh fuel cell environment. Additionally, integrating multiple functionalities such as proton conduction, gas barrier properties, and chemical resistance into a single membrane system necessitates careful material design and processing control (Pandey *et al.*, 2017).

3.6 CHALLENGES IN SCALING UP FUEL CELL MATERIALS AND CATALYSTS FOR INDUSTRIAL APPLICATIONS

Scaling up fuel cell materials and catalysts from laboratory-scale innovation to industrial-level production presents a multifaceted challenge that has impeded the widespread adoption of fuel cell technology despite its promising environmental and efficiency benefits. While fuel cells such as proton exchange membrane fuel cells (PEMFCs) and solid oxide fuel cells (SOFCs) have demonstrated excellent performance and low emissions in controlled environments, translating these successes into reliable, cost-effective industrial products remains a complex task due to intrinsic material, manufacturing, and economic barriers (Elabd & Hickner, 2011). One of the primary obstacles in this scaling process revolves around the catalysts, particularly those used at the cathode for the oxygen reduction reaction (ORR). Platinum, the benchmark catalyst due to its unmatched activity and stability, remains prohibitively expensive and scarce, creating a bottleneck in supply and inflating production costs (Elabd & Hickner, 2011). Although researchers have developed strategies to reduce platinum loading, such as dispersing platinum nanoparticles on high surface area supports or alloying platinum with cheaper metals, the challenge of maintaining catalytic performance while minimizing platinum usage becomes pronounced when transitioning to bulk manufacturing. Moreover, the volatility of platinum prices on the global market further complicates economic projections and supply chain security for large-scale producers (Tang *et al.*, 2023).

In addition to cost constraints, durability issues emerge as critical when catalysts are exposed to the harsh operating environments of real-world fuel cells. Catalysts synthesized at lab scale often degrade rapidly through processes like nanoparticle sintering, dissolution, or support corrosion when subjected to the temperature fluctuations, humidity cycles, and potential changes typical in industrial fuel cell applications (Yang *et al.*, 2023). Carbon supports, which have traditionally served as the backbone for dispersing platinum nanoparticles, are prone to oxidative corrosion at high potentials, leading to loss of catalyst surface area and subsequent performance decline (Yang *et al.*, 2023). Researchers have thus turned to alternative supports such as doped graphene, metal oxides, and recently, high-entropy alloys (HEAs), which show promising improvements in durability. However, these materials often require complex synthesis routes that are difficult to scale economically without sacrificing uniformity or performance (Yang *et al.*, 2023).

Membrane production also poses significant challenges in scaling fuel cells for industrial applications. Proton exchange membranes (PEMs) serve as the vital conduit for proton transport while blocking electron and gas crossover, and their fabrication must ensure excellent mechanical strength, chemical stability, and ion conductivity. Commercial membranes like Nafion are produced via intricate polymerization and casting processes that must be tightly controlled to avoid pinholes, thickness variations, and mechanical defects. Scaling these processes to produce larger membrane areas without compromising quality remains difficult because even minor defects can cause catastrophic fuel leakage or premature cell failure (Cai *et al.*, 2017). Additionally, newer applications targeting high-temperature or low-humidity environments require membranes that maintain conductivity and stability under more extreme conditions. Composite membranes incorporating inorganic fillers or ionic liquids enhance thermal and chemical resilience but add further complexity to manufacturing, often demanding specialized equipment and lengthy processing times that impede rapid scale-up (Yang *et al.*, 2023).

Beyond membranes and catalysts, the synthesis of nanomaterials for fuel cells introduces reproducibility and consistency challenges at scale. The catalytic performance of nanoparticles depends heavily on their size, morphology, and dispersion, attributes that are controllable in laboratory batch processes but become highly variable in industrial batch or continuous reactors (Wang *et al.*, 2023). The precise control over nucleation and growth kinetics, essential for producing uniform nanoparticles, is difficult to maintain during scale-up, leading to batch-to-batch variation that undermines fuel cell reliability. Furthermore, novel catalyst designs such as core-shell architectures and single-atom catalysts, where isolated metal atoms are anchored on supports for maximal efficiency, demand synthesis conditions that are extremely sensitive and often only reproducible at a small scale (Yang *et al.*, 2023). Scaling these methods requires innovative reactor designs and in situ monitoring technologies to ensure quality control, but such technologies are still in nascent stages of industrial adoption (Tang *et al.*, 2023).

Integrating these advanced materials into functional fuel cell electrodes and stacks further complicates the path to industrial deployment. Fuel cell electrodes are complex multilayered structures comprising catalysts, ionomers, and gas diffusion layers, where each layer's uniformity and microstructure critically affect overall performance (Zhang *et al.*, 2022). Achieving consistent catalyst loading and ionomer distribution over large electrode areas during roll-to-roll coating or spraying processes requires extensive optimization, especially when incorporating novel catalyst materials with different surface chemistries or particle sizes. Compatibility issues between new materials and existing fabrication processes can cause defects such as delamination or uneven porosity, which severely impair reactant transport and catalyst utilization. Furthermore, fuel cell stacks operate under dynamic thermal and mechanical stresses, and materials that perform well in isolated tests may degrade or fail prematurely when subjected to these complex, multiscale operational environments (Yang *et al.*, 2023).

Economic factors add another layer of complexity to scaling fuel cell materials. Industrial manufacturing demands large quantities of raw materials and well-established supply chains to reduce costs through economies of scale. The scarcity and cost of critical raw materials like platinum, rare earth metals, or specialized dopants create significant bottlenecks, especially when geopolitical or market fluctuations disrupt supply (Wang *et al.*, 2023). Additionally, many advanced catalyst and membrane synthesis techniques involve expensive precursors, energy-intensive steps, and hazardous solvents, all of which increase the environmental footprint and production cost (Cai *et al.*, 2017).

These issues call for the development of sustainable, green synthesis routes and circular economy approaches, including catalyst recycling and recovery technologies, which remain insufficiently developed for commercial fuel cell manufacturing (Yang *et al.*, 2023).

Despite these daunting challenges, ongoing research has produced promising strategies that could alleviate some of the barriers to scaling. The development of single-atom and high-entropy alloy catalysts aims to drastically reduce precious metal content while enhancing durability and activity, representing a paradigm shift in catalyst design (Yang *et al.*, 2023; Tang *et al.*, 2023). Continuous manufacturing methods such as roll-to-roll coating, spray pyrolysis, and atomic layer deposition are being adapted for large-scale production of membranes and electrodes with improved uniformity and throughput (Elabd & Hickner, 2011). Composite and hybrid membrane materials show enhanced thermal and chemical resilience, potentially enabling fuel cells to operate reliably under more severe conditions encountered in industrial settings (Yang *et al.*, 2023). Advances in real-time, in situ characterization techniques provide tools to monitor catalyst synthesis and fuel cell operation closely, improving quality control and early detection of degradation pathways that can be addressed proactively (Yang *et al.*, 2023). Moreover, efforts to implement recycling and material recovery protocols are gaining momentum to reduce raw material demand and environmental impact (Yang *et al.*, 2023).

3.7 EMERGING APPROACHES AND FUTURE RESEARCH DIRECTIONS

One of the most significant emerging trends is the green synthesis of catalysts, which addresses the environmental and economic issues associated with conventional catalyst fabrication. Traditional methods for catalyst synthesis often rely on hazardous solvents, high temperatures, and energy-intensive processes, leading to increased production costs and environmental burdens (Smith *et al.*, 2022). In contrast, green synthesis leverages benign solvents such as water or ethanol, low-temperature routes, and biogenic reducing agents derived from plants or microorganisms. For example, plant extracts rich in polyphenols and flavonoids can reduce metal salts to form nanoparticles with controlled morphology and size, without the need for toxic chemicals (Nas & Calimli, 2021). Such biosynthesis not only reduces environmental impact but also enhances catalyst biocompatibility and functionality through natural capping agents, which improve stability and prevent agglomeration. Additionally, green-synthesized catalysts have demonstrated comparable or even superior electrochemical performance to those produced by conventional means, especially in oxygen reduction and hydrogen evolution reactions critical for fuel cell operation (Muthukrishnan *et al.*, 2024). Further research is required to scale these bio-assisted processes, optimize reaction conditions, and understand the mechanistic pathways that govern particle nucleation and growth.

Parallel to green synthesis, bio-inspired and biomimetic materials are gaining traction as innovative fuel cell components. Nature's catalytic systems, such as enzymes and metal centers in proteins, operate under mild conditions with remarkable specificity and efficiency. Mimicking these biological catalysts offers a pathway to develop fuel cell catalysts that combine high activity with durability and low cost (Artero, 2017). For instance, hydrogenases and laccases provide models for designing molecular catalysts and electrode interfaces that efficiently catalyze hydrogen oxidation and oxygen reduction, respectively (Aralekallu *et al.*, 2023). Incorporation of such biomimetic catalysts into fuel cell electrodes has the potential to reduce or eliminate the dependence on precious metals like platinum, which is scarce and expensive. Advances in synthetic biology and protein engineering facilitate the tailoring of enzyme-like catalysts with enhanced stability against fuel cell operating conditions, such as temperature and pH variations. Moreover, biomimetic membranes inspired by cellular structures offer promising proton-conductive pathways with improved selectivity and mechanical strength, addressing key limitations of current polymer electrolyte membranes (Haider *et al.*, 2021). Challenges remain in integrating biological materials with inorganic components, ensuring long-term stability, and scaling production, but ongoing interdisciplinary research shows promise for hybrid bio-electrochemical systems.

Artificial intelligence (AI) and machine learning (ML) represent revolutionary tools that accelerate the discovery and optimization of fuel cell materials and catalysts. Traditionally, the development of catalysts involves laborious trial-and-error experimentation and computational simulations, often spanning years. AI algorithms, trained on vast datasets of catalyst properties and performance metrics, can identify hidden patterns and predict promising candidates with unprecedented speed and accuracy (Yin *et al.*, 2024). For example, ML models have been used to screen thousands of potential alloy compositions for oxygen reduction catalysts, predicting activity and stability with high confidence, thus narrowing the experimental workload significantly (Wang *et al.*, 2022). Deep learning techniques enable the design of complex catalyst structures, such as core-shell nanoparticles or single-atom catalysts, by optimizing atomic arrangements to enhance catalytic sites and durability (Park & Lee, 2024). Beyond catalyst design, AI aids in the modeling of fuel cell membranes and electrode architectures to optimize proton conductivity, gas transport, and mechanical integrity. Integration of AI into synthesis processes through real-time monitoring and feedback control can enhance reproducibility and scale-up efficiency. The challenge lies in curating comprehensive and high-quality datasets that capture the multi-scale phenomena in fuel cells and in bridging the gap between computational predictions and experimental validation. Nonetheless, AI-driven materials discovery is poised to transform fuel cell R&D by shortening innovation cycles and enabling rational design strategies.

In the broader energy landscape, integration of fuel cells with renewable energy systems represents a critical research direction for industrial applications. Fuel cells can complement intermittent renewable sources such as solar and wind by providing efficient energy storage and on-demand power generation through hydrogen production and utilization (Eriksson & Gray, 2017). Coupling water electrolysis powered by renewables with fuel cells creates a closed-loop green hydrogen economy, which requires materials that can withstand variable load conditions and operate efficiently at fluctuating temperatures and humidities. This integration necessitates the development of durable catalysts and membranes capable of rapid start-up, shut-down cycles, and load following without degradation (Wang, 2017). Furthermore, hybrid systems combining fuel cells with batteries or supercapacitors demand novel composite materials that facilitate synergistic energy management and thermal stability. Research on bifunctional catalysts that can efficiently catalyze both oxygen reduction and evolution reactions is critical for reversible fuel cells and electrolyzers in such integrated systems (Artero, 2017). The future of industrial-scale fuel cells lies in these versatile, adaptive materials that enable seamless integration into smart grids and decentralized energy networks.

The adoption of circular economy principles in the lifecycle of fuel cell materials and catalysts is an emerging imperative. Conventional manufacturing and disposal of fuel cell components contribute to resource depletion, waste generation, and environmental pollution. Circular economy strategies promote material recovery, recycling, and reuse to enhance sustainability and reduce costs (Pedram *et al.*, 2021). For catalysts, this involves developing processes to efficiently reclaim precious metals from spent electrodes, maintaining their activity through reprocessing, and minimizing losses. Advances in hydrometallurgical and pyrometallurgical recycling techniques, supported by green chemistry principles, are being explored to close the material loop (Centi & Perathoner, 2003). Additionally, designing catalysts and membranes with recyclability in mind—using easily separable and non-toxic components—facilitates end-of-life management. Life cycle assessment (LCA) tools integrated with AI can optimize production and recycling workflows to minimize carbon footprint and economic costs. Implementing circular approaches at an industrial scale requires collaboration across material scientists, process engineers, and policymakers to create supportive regulatory frameworks and market incentives.

3.8 CONCLUSION

The future of fuel cell materials and catalysts for industrial applications hinges on embracing green synthesis methods that reduce environmental impact and enhance functionality. Bio-inspired and biomimetic materials offer pathways to replace scarce precious metals and improve durability. AI and machine learning accelerate discovery and enable precision design of catalysts and membranes tailored for specific operating conditions. Integrating fuel cells with renewable energy systems demands adaptive materials capable of handling dynamic energy flows while maintaining efficiency. Adopting circular economy principles throughout the materials lifecycle ensures sustainability and economic viability. Together, these emerging approaches position fuel cells as vital components in the global transition to clean, reliable, and sustainable energy infrastructures. Continued interdisciplinary research and industrial collaboration will be essential to translate these promising directions into commercial realities.

REFERENCES

Abdin, Z. (2024). Shaping the stationary energy storage landscape with reversible fuel cells. *Journal of Energy Storage*, *86*, 111354.

Alonso-Vante, N. (2023). Parameters affecting the fuel cell reactions on platinum bimetallic nanostructures. *Electrochemical Energy Reviews*, *6*(1), 3.

Antolini, E. (2009). Carbon supports for low-temperature fuel cell catalysts. *Applied Catalysis B: Environmental*, *88*(1–2), 1–24.

Aralekallu, S., Sannegowda, L. K., & Singh, V. (2023). Developments in electrocatalysts for electrocatalytic hydrogen evolution reaction with reference to bio-inspired phthalocyanines. *International Journal of Hydrogen Energy*, *48*(44), 16569–16592.

Artero, V. (2017). Bioinspired catalytic materials for energy-relevant conversions. *Nature Energy*, *2*(9), 1–6.

Asefa, T., & Huang, X. (2017). Heteroatom-Doped carbon materials for electrocatalysis. *Chemistry–A European Journal*, *23*(45), 10703–10713.

Banham, D., Choi, J. Y., Kishimoto, T., & Ye, S. (2019). Integrating PGM-free catalysts into catalyst layers and proton exchange membrane fuel cell devices. *Advanced Materials*, *31*(31), 1804846.

Banham, D., Ye, S., Pei, K., Ozaki, J. I., Kishimoto, T., & Imashiro, Y. (2015). A review of the stability and durability of non-precious metal catalysts for the oxygen reduction reaction in proton exchange membrane fuel cells. *Journal of Power Sources*, *285*, 334–348.

Bhatt, M. D., & Lee, J. Y. (2020). Advancement of platinum (Pt)-free (non-Pt precious metals) and/or metal-free (non-precious-metals) electrocatalysts in energy applications: A review and perspectives. *Energy & Fuels*, *34*(6), 6634–6695.

Cai, Y., Yue, Z., & Xu, S. (2017). A novel polybenzimidazole composite modified by sulfonated graphene oxide for high temperature proton exchange membrane fuel cells in anhydrous atmosphere. *Journal of Applied Polymer Science*, *134*(25).

Centi, G., & Perathoner, S. (2003). Catalysis and sustainable (green) chemistry. *Catalysis Today*, *77*(4), 287–297.

Chen, B., Zhong, X., Zhou, G., Zhao, N., & Cheng, H. M. (2022). Graphene-supported atomically dispersed metals as bifunctional catalysts for next-generation batteries based on conversion reactions. *Advanced Materials*, *34*(5), 2105812.

Chen, Z., Higgins, D., Yu, A., Zhang, L., & Zhang, J. (2011). A review on non-precious metal electrocatalysts for PEM fuel cells. *Energy & Environmental Science*, *4*(9), 3167–3192.

Cheng, N., Zhang, L., Doyle-Davis, K., & Sun, X. (2019). Single-atom catalysts: From design to application. *Electrochemical Energy Reviews*, *2*, 539–573.

Chuahy, F. D., & Kokjohn, S. L. (2019). Solid oxide fuel cell and advanced combustion engine combined cycle: A pathway to 70% electrical efficiency. *Applied Energy*, *235*, 391–408.

Corigliano, O., Pagnotta, L., & Fragiacomo, P. (2022). On the technology of solid oxide fuel cell (SOFC) energy systems for stationary power generation: A review. *Sustainability*, *14*(22), 15276.

Debe, M. K. (2012). Electrocatalyst approaches and challenges for automotive fuel cells. *Nature*, *486*(7401), 43–51.

Dihrab, S. S., Sopian, K., Alghoul, M. A., & Sulaiman, M. Y. (2009). Review of the membrane and bipolar plates materials for conventional and unitized regenerative fuel cells. *Renewable and Sustainable Energy Reviews*, *13*(6–7), 1663–1668.

Divisek, J., De Haart, L. G. J., Holtappels, P., Lennartz, T., Mallener, W., Stimming, U., & Wippermann, K. (1994). The kinetics of electrochemical reactions on high temperature fuel cell electrodes. *Journal of Power Sources*, *49*(1–3), 257–270.

Dwivedi, S. (2020). Solid oxide fuel cell: Materials for anode, cathode and electrolyte. *International Journal of Hydrogen Energy*, *45*(44), 23988–24013.

Elabd, Y. A., & Hickner, M. A. (2011). Block copolymers for fuel cells. *Macromolecules*, *44*(1), 1–11.

Eriksson, E. L. V., & Gray, E. M. (2017). Optimization and integration of hybrid renewable energy hydrogen fuel cell energy systems–A critical review. *Applied Energy*, *202*, 348–364.

Escorihuela, J., García-Bernabé, A., Montero, Á., Sahuquillo, Ó., Giménez, E., & Compañ, V. (2019). Ionic liquid composite polybenzimidazol membranes for high temperature PEMFC applications. *Polymers*, *11*(4), 732.

Fakhreddine, O., Gharbia, Y., Derakhshandeh, J. F., & Amer, A. M. (2023). Challenges and solutions of hydrogen fuel cells in transportation systems: A review and prospects. *World Electric Vehicle Journal*, *14*(6), 156.

Feng, Q., Liu, G., Wei, B., Zhang, Z., Li, H., & Wang, H. (2017). A review of proton exchange membrane water electrolysis on degradation mechanisms and mitigation strategies. *Journal of Power Sources*, *366*, 33–55.

Frangini, S., & Masi, A. (2016). Molten carbonates for advanced and sustainable energy applications: Part II. Review of recent literature. *International Journal of Hydrogen Energy*, *41*(42), 18971–18994.

Gasteiger, H. A., Kocha, S. S., Sompalli, B., & Wagner, F. T. (2005). Activity benchmarks and requirements for Pt, Pt-alloy, and non-Pt oxygen reduction catalysts for PEMFCs. *Applied Catalysis B: Environmental*, *56*(1–2), 9–35.

Guo, D., Shibuya, R., Akiba, C., Saji, S., Kondo, T., & Nakamura, J. (2016). Active sites of nitrogen-doped carbon materials for oxygen reduction reaction clarified using model catalysts. *Science*, *351*(6271), 361–365.

Gupta, P., Toksha, B., & Rahaman, M. (2024). A critical review on hydrogen based fuel cell technology and applications. *The Chemical Record*, *24*(1), e202300295.

Haider, R., Wen, Y., Ma, Z. F., Wilkinson, D. P., Zhang, L., Yuan, X., ... & Zhang, J. (2021). High temperature proton exchange membrane fuel cells: progress in advanced materials and key technologies. *Chemical Society Reviews*, *50*(2), 1138–1187.

Haile, S. M. (2003). Fuel cell materials and components. *Acta materialia*, *51*(19), 5981–6000.

He, Y., Liu, S., Priest, C., Shi, Q., & Wu, G. (2020). Atomically dispersed metal–nitrogen–carbon catalysts for fuel cells: Advances in catalyst design, electrode performance, and durability improvement. *Chemical Society Reviews*, *49*(11), 3484–3524.

He, Y., Tan, Q., Lu, L., Sokolowski, J., & Wu, G. (2019). Metal-nitrogen-carbon catalysts for oxygen reduction in PEM fuel cells: Self-template synthesis approach to enhancing catalytic activity and stability. *Electrochemical Energy Reviews*, *2*, 231–251.

Hong, W. T., O'Brien, E. R., & Zhang, B. (2017). Recent advances in non-precious metal catalysts for oxygen reduction reaction. *Materials Today*, *20*, 380–387.

Huang, L., Zaman, S., Tian, X., Wang, Z., Fang, W., & Xia, B. Y. (2021). Advanced platinum-based oxygen reduction electrocatalysts for fuel cells. *Accounts of Chemical Research*, *54*(2), 311–322.

Iwan, A., Malinowski, M., & Pasciak, G. (2015). Polymer fuel cell components modified by graphene: Electrodes, electrolytes and bipolar plates. *Renewable and Sustainable Energy Reviews*, *49*, 954–967.

Jaouen, F., Proietti, E., Lefèvre, M., Chenitz, R., Dodelet, J. P., Wu, G., ... & Zelenay, P. (2011). Recent advances in non-precious metal catalysis for oxygen-reduction reaction in polymer electrolyte fuel cells. *Energy & Environmental Science*, *4*(1), 114–130.

Jha, A. R. (2012). *Next-generation batteries and fuel cells for commercial, military, and space applications* (pp. 312–315). Boca Raton, FL: CRC Press.

Kartha, S., & Grimes, P (1994). Fuel cells: Energy conversion for the next century. *Physics Today*, *47*(11), 54–61.

Khan, M. S., Lee, S. B., Song, R. H., Lee, J. W., Lim, T. H., & Park, S. J. (2016). Fundamental mechanisms involved in the degradation of nickel–yttria stabilized zirconia (Ni–YSZ) anode during solid oxide fuel cells operation: A review. *Ceramics International*, *42*(1), 35–48.

Kiani, M., Tian, X. Q., & Zhang, W. (2021). Non-precious metal electrocatalysts design for oxygen reduction reaction in polymer electrolyte membrane fuel cells: Recent advances, challenges and future perspectives. *Coordination Chemistry Reviews*, *441*, 213954.

Kramm, U. I., Herranz, J., Larouche, N., Arruda, T. M., Lefèvre, M., Jaouen, F., ... & Dodelet, J. P. (2012). Structure of the catalytic sites in Fe/N/C-catalysts for O 2-reduction in PEM fuel cells. *Physical Chemistry Chemical Physics*, *14*(33), 11673–11688.

Kubert, C. (2010). Stationary fuel cells and critical power applications. *Clean Energy States Alliance*.

Lai, J. S., & Ellis, M. W. (2017). Fuel cell power systems and applications. *Proceedings of the IEEE*, *105*(11), 2166–2190.

Li, C., Clament Sagaya Selvam, N., & Fang, J. (2023). Shape-controlled synthesis of platinum-based nanocrystals and their electrocatalytic applications in fuel cells. *Nano-Micro Letters*, *15*(1), 83.

Lindorfer, J., Rosenfeld, D. C., & Böhm, H. (2020). Fuel cells: Energy conversion technology. In *Future energy* (pp. 495–517). Elsevier.

Liu, J., Luo, W., Yang, X., & Wu, L. (2016). Robust model-based fault diagnosis for PEM fuel cell air-feed system. *IEEE Transactions on Industrial Electronics*, *63*(5), 3261–3270.

Maiti, T. K., Singh, J., Dixit, P., Majhi, J., Bhushan, S., Bandyopadhyay, A., & Chattopadhyay, S. (2022). Advances in perfluorosulfonic acid-based proton exchange membranes for fuel cell applications: A review. *Chemical Engineering Journal Advances*, *12*, 100372.

Mauritz, K. A., & Moore, R. B. (2004). State of understanding of Nafion. *Chemical reviews*, *104*(10), 4535–4586.

McNicol, B. D., Rand, D. A. J., & Williams, K. R. (2001). Fuel cells for road transportation purposes—yes or no?. *Journal of Power Sources*, *100*(1–2), 47–59.

Mitsushima, S., Koizumi, Y., Uzuka, S., & Ota, K. I. (2008). Dissolution of platinum in acidic media. *Electrochimica Acta*, *54*(2), 455–460.

Molochas, C., & Tsiakaras, P. (2021). Carbon monoxide tolerant Pt-based electrocatalysts for H2-PEMFC applications: Current progress and challenges. *Catalysts*, *11*(9), 1127.

Muthukrishnan, L., Kamaraj, S. K., Sánchez-Cárdenas, M., & Sánchez-Olmos, L. A. (2024). An overview of sustainable issues associated with bio-assisted power generation systems. *Photosynthesis-Assisted Energy Generation: From Fundamentals to Lab Scale and In-Field Applications*, *4*, 359–383.

Nas, M. S., & Calimli, M. H. (2021). Recent development of nanoparticle by green-conventional methods and applications for corrosion and fuel cells. *Current Nanoscience*, *17*(4), 525–539.

Nguyen, L. (2006). *A geographic market valuation for micro fuel cell technology in portable power applications*.

Nguyen, T. N., & Dinh, C. T. (2020). Gas diffusion electrode design for electrochemical carbon dioxide reduction. *Chemical Society Reviews*, *49*(21), 7488–7504.

Nørskov, J. K., Rossmeisl, J., Logadottir, A., Lindqvist, L. R. K. J., Kitchin, J. R., Bligaard, T., & Jonsson, H. (2004). Origin of the overpotential for oxygen reduction at a fuel-cell cathode. *The Journal of Physical Chemistry B*, *108*(46), 17886–17892.

Oh, H. S., & Kim, H. (2012). The role of transition metals in non-precious nitrogen-modified carbon-based electrocatalysts for oxygen reduction reaction. *Journal of Power Sources*, *212*, 220–225.

Pandey, R. P., Shukla, G., Manohar, M., & Shahi, V. K. (2017). Graphene oxide based nanohybrid proton exchange membranes for fuel cell applications: An overview. *Advances in Colloid and Interface Science*, *240*, 15–30.

Pardhi, S., Chakraborty, S., Tran, D. D., El Baghdadi, M., Wilkins, S., & Hegazy, O. (2022). A review of fuel cell powertrains for long-haul heavy-duty vehicles: Technology, hydrogen, energy and thermal management solutions. *Energies*, *15*(24), 9557.

Park, J., & Lee, J. (2024). Electrochemical energy conversion and storage processes with machine learning. *Trends in Chemistry*, *6*(6), 302–313.

Pedram, S., Batool, M., Yapp, K., Bonville, L., & Jankovic, J. (2021). A review on bioinspired proton exchange membrane fuel cell: Design and materials. *Advanced Energy and Sustainability Research*, *2*(7), 2000092.

Peng, L., & Wei, Z. (2020). Catalyst engineering for electrochemical energy conversion from water to water: Water electrolysis and the hydrogen fuel cell. *Engineering*, *6*(6), 653–679.

Qasem, N. A., & Abdulrahman, G. A. (2024). A recent comprehensive review of fuel cells: History, types, and applications. *International Journal of Energy Research*, *2024*(1), 7271748.

Qiao, B., Liu, J., Allard, L., Wang, A., Cui, Y., Zhang, T., … & Jiang, Z. (2012). Single-atom catalysis: Pt1/FeOx for CO oxidation and preferential oxidation of CO in H2. *Microscopy and Microanalysis*, *18*(S2), 350–351.

Raj, V. (2020). Direct methanol fuel cells in portable applications: Materials, designs, operating parameters, and practical steps toward commercialization. In *Direct methanol fuel cell technology*, Kingshuk Dutta (ed.), (pp. 495–525). Elsevier.

Ravichandran, B., Ravichandran, S., Liu, H., Zhang, W., Narayanan, N., Xu, Q., & Su, H. (2024). Promising application of MXene-based materials in direct methanol fuel cells: A review. *Journal of Alloys and Compounds*, *993*, 174674.

Ren, J. T., Chen, L., Wang, H. Y., & Yuan, Z. Y. (2023). High-entropy alloys in electrocatalysis: From fundamentals to applications. *Chemical Society Reviews*, *52*(23), 8319–8373.

Rojas-Carbonell, S., Santoro, C., Serov, A., & Atanassov, P. (2017). Transition metal-nitrogen-carbon catalysts for oxygen reduction reaction in neutral electrolyte. *Electrochemistry Communications*, *75*, 38–42.

Scofield, M. E., Liu, H., & Wong, S. S. (2015). A concise guide to sustainable PEMFCs: Recent advances in improving both oxygen reduction catalysts and proton exchange membranes. *Chemical Society Reviews*, *44*(16), 5836–5860.

Senthil, T., Prabukanthan, P., Paradesi, D., & Dinakaran, K. (2023). TiO2 nanoparticle enhanced high temperature proton conductivity in hyperbranched sulfonated polyarylene aliphatic ketones for proton exchange membrane fuel cell applications. *Journal of Applied Polymer Science*, *140*(15), e53737.

Sethuraman, V. A., & Weidner, J. W. (2010). Analysis of sulfur poisoning on a PEM fuel cell electrode. *Electrochimica Acta*, *55*(20), 5683–5694.

Sheng, W., Gasteiger, H. A., & Shao-Horn, Y. (2010). Hydrogen oxidation and evolution reaction kinetics on platinum: Acid vs alkaline electrolytes. *Journal of The Electrochemical Society*, *157*(11), B1529.

Simons, A., & Bauer, C. (2015). A life-cycle perspective on automotive fuel cells. *Applied Energy*, *157*, 884–896.

Singh, P., & Minh, N. Q. (2004). Solid oxide fuel cells: Technology status. *International Journal of Applied Ceramic Technology*, *1*(1), 5–15.

Singh, S. (2021). Energy crisis and climate change: Global concerns and their solutions. *Energy: Crises, Challenges and Solutions*, *25*, 1–17.

Singh, S., Rao, D. G., & Dixit, M. (2023). Future of transportation-a comparison between internal combustion engine, electric vehicles and fuel cell vehicles. *Universal Journal of Mechanical Engineering*, *11*(1), 13–23.

Singh, S. K., Takeyasu, K., & Nakamura, J. (2019). Active sites and mechanism of oxygen reduction reaction electrocatalysis on nitrogen-doped carbon materials. *Advanced Materials*, *31*(13), 1804297.

Smith, B. J., Graziano, D. J., Riddle, M. E., Liu, D. J., Sun, P., Iloeje, C., ... & Diamond, D. (2022). *Platinum Group Metal Catalysts: Supply Chain Deep Dive Assessment. USDOE Office of Policy (PO)*, Washington DC (United States).

Song, C. (2002). Fuel processing for low-temperature and high-temperature fuel cells: Challenges, and opportunities for sustainable development in the 21st century. *Catalysis today*, *77*(1–2), 17–49.

Stambouli, A. B. (2011). Fuel cells: The expectations for an environmental-friendly and sustainable source of energy. *Renewable and Sustainable Energy Reviews*, *15*(9), 4507–4520.

Stamenkovic, V., Mun, B. S., Mayrhofer, K. J., Ross, P. N., Markovic, N. M., Rossmeisl, J., ... & Nørskov, J. K. (2006). Changing the activity of electrocatalysts for oxygen reduction by tuning the surface electronic structure. *Angewandte Chemie International Edition*, *45*(18), 289–901.

Strasser, P., Koh, S., Anniyev, T., Greeley, J., More, K., Yu, C., ... & Nilsson, A. (2010). Lattice-strain control of the activity in dealloyed core–shell fuel cell catalysts. *Nature Chemistry*, *2*(6), 454–460.

Su, H., & Hu, Y. H. (2021). Recent advances in graphene-based materials for fuel cell applications. *Energy Science & Engineering*, *9*(7), 958–983.

Tang, M., Yang, T., Yang, X., Li, Y., Shi, Z., Wang, X., ... & Ge, J. (2023). Single-atom catalysts for proton exchange membrane fuel cell: Anode anti-poisoning & characterization technology. *Electrochimica Acta*, *446*, 142120.

Tawfik, H., Hung, Y., & Mahajan, D. (2007). Metal bipolar plates for PEM fuel cell—A review. *Journal of Power Sources*, *163*(2), 755–767.

Wang, J. (2017). System integration, durability and reliability of fuel cells: Challenges and solutions. *Applied Energy*, *189*, 460–479.

Wang, X., He, B., Hu, Z., Zeng, Z., & Han, S. (2014). Current advances in precious metal core–shell catalyst design. *Science and Technology of Advanced Materials*, *15*(4), 043502.

Wang, Y., Pang, Y., Xu, H., Martinez, A., & Chen, K. S. (2022). PEM Fuel cell and electrolysis cell technologies and hydrogen infrastructure development–a review. *Energy & Environmental Science*, *15*(6), 2288–2328.

Wang, Z., Liu, Z., Fan, L., Du, Q., & Jiao, K. (2023). Application progress of small-scale proton exchange membrane fuel cell. *Energy Reviews*, *2*(2), 100017.

Wee, J. H. (2007). Applications of proton exchange membrane fuel cell systems. *Renewable and Sustainable Energy Reviews*, *11*(8), 1720–1738.

Xin, Y., Li, S., Qian, Y., Zhu, W., Yuan, H., Jiang, P., ... & Wang, L. (2020). High-entropy alloys as a platform for catalysis: Progress, challenges, and opportunities. *Acs Catalysis*, *10*(19), 11280–11306.

Yang, P., Jiang, Z., Shi, Y., Ren, X., Liang, L., Shao, Q., & Zhu, K. (2023). Enhancement of oxygen evolution reaction performance of FeCoNiCrMn high entropy alloy thin film electrodes through in-situ reconstruction. *Journal of Alloys and Compounds*, *947*, 169699.

Yin, P., Niu, X., Li, S. B., Chen, K., Zhang, X., Zuo, M., ... & Liang, H. W. (2024). Machine-learning-accelerated design of high-performance platinum intermetallic nanoparticle fuel cell catalysts. *Nature Communications*, *15*(1), 415.

Younas, M., Shafique, S., Hafeez, A., Javed, F., & Rehman, F. (2022). An overview of hydrogen production: Current status, potential, and challenges. *Fuel*, *316*, 123317.
Yu, Z., Boukhvalov, D. W., Tan, H., Xiong, D., Feng, C., Wang, J., … & Liu, L. (2024). Sulfur and phosphorus co-doped FeCoNiCrMn high-entropy alloys as efficient sulfion oxidation reaction catalysts enabling self-powered asymmetric seawater electrolysis. *Chemical Engineering Journal*, *494*, 153094.
Zhang, J., & Dai, L. (2015). Heteroatom-doped graphitic carbon catalysts for efficient electrocatalysis of oxygen reduction reaction. *Acs Catalysis*, *5*(12), 7244–7253.
Zhang, M., Lai, C., Xu, F., Huang, D., Liu, S., Fu, Y., … & Chen, L. (2022). Atomically dispersed metal catalysts confined by covalent organic frameworks and their derivatives for electrochemical energy conversion and storage. *Coordination Chemistry Reviews*, *466*, 214592.
Zhang, S., Shao, Y., Yin, G., & Lin, Y. (2013). Recent progress in nanostructured electrocatalysts for PEM fuel cells. *Journal of Materials Chemistry A*, *1*(15), 4631–4641.
Zhang, Y., Wang, J., & Yao, Z. (2023). Recent development of fuel cell core components and key materials: a review. *Energies*, *16*(5), 2099.
Zhao, J., Guo, H., Xing, Y., Ping, S., Lin, W., Yang, Y., … & Ma, T. (2023). A review on the sealing structure and materials of fuel-cell stacks. *Clean Energy*, *7*(1), 59–69.
Zheng, Y., Jiao, Y., Jaroniec, M., & Qiao, S. Z. (2015). Advancing the electrochemistry of the hydrogen-evolution reaction through combining experiment and theory. *Angewandte Chemie International Edition*, *54*(1), 52–65.
Zhu, K., Li, C., Jing, Z., Liu, X., He, Y., Lv, X., … & Liu, K. (2019). Two-dimensional transition-metal dichalcogenides for electrochemical hydrogen evolution reaction. *FlatChem*, *18*, 100140.
Zhu, L., Li, Y., Ye, P., Zhao, J., Liu, J., Lei, J., … & Xue, R. (2023). Ultra-stable, highly proton conductive, and self-healing proton exchange membranes based on molecule intercalation technique and noncovalent assembly nanostructure. *Advanced Functional Materials*, *33*(10), 2210453.
Zitolo, A., Goellner, V., Armel, V., Sougrati, M. T., Mineva, T., Stievano, L., … & Jaouen, F. (2015). Identification of catalytic sites for oxygen reduction in iron-and nitrogen-doped graphene materials. *Nature Materials*, *14*(9), 937–942.

4 Hydrogen Production and Its Environmental Implications

Olusola T. Lawal

4.1 INTRODUCTION

A major worldwide concern is climate change, which is mostly caused by human activities such as the burning of fossil fuels, deforestation, and industrial pollutants (Filonchyk *et al.*, 2024). The fact that global temperatures have increased by about 1°C over pre-industrial levels and are expected to rise by another 1.5°C to 2°C this century unless significant mitigation measures are taken highlights the urgent need to switch from fossil fuels to cleaner energy sources (Matthews & Wynes, 2022). One possible approach to decarbonizing energy-intensive industry and transportation is hydrogen, especially green hydrogen.

Energy storage, industrial uses, and clean power generation are made possible by hydrogen, a clean and low-carbon energy carrier that opens the door to a low-carbon economy (Zhang *et al.*, 2024). Its potential is particularly noteworthy in industries that are difficult to electrify, such as heavy transportation, steel production, and chemical manufacture, where hydrogen can be a key component in lowering greenhouse gas emissions. Carbon emissions are eliminated during the production and consumption phases of green hydrogen, which is created through electrolysis with renewable energy. Hydrogen is becoming increasingly recognized as a crucial element of future energy systems due to improvements in electrolyzer performance and falling costs for renewable electricity (Sadeq et al., 2024).

Energy production is undergoing a paradigm shift, and green hydrogen is showing promise as a substitute. Green hydrogen generation provides a greener future by removing carbon emissions through the use of renewable energy sources. Interestingly, hydrogen fuel cells are more efficient than conventional combustion engines, resulting in notable decreases in pollutants and fuel usage (Saha *et al.*, 2024).

Various obstacles impede the creation of hydrogen. A switch from hydrogen's requirement for high energy to renewable sources would reduce its emissions (Whitehead *et al.*, 2023). The major worries with hydrogen production are limited water supplies, as the process is known to be water-intensive electrolysis (Rödl, 2025). The significant amounts of CO_2 are released during complex techniques such as steam methane reforming without carbon capture (Whitehead *et al.*, 2023). This necessitates the need for specific, safety-focused infrastructure due to the flammability of hydrogen (Cavaliere, 2023). The great roles of hydrogen in the global transition to cleaner and sustainable energy are multifaceted. Hence, there is a high need for climate-favourable and sustainable sources of hydrogen production, which can be scaled up to tackle the environmental and economic issues (Jeje *et al.*, 2024). Investment in renewable-based production and enhancing infrastructure would enable hydrogen to emerge as a low-carbon energy source and thereby lower foreseeable environmental implications (Swadi *et al.*, 2024).

DOI: 10.1201/9781003645405-4

4.2 ENVIRONMENTAL IMPACT OF DIFFERENT HYDROGEN PRODUCTION METHODS

Transitioning to a low-carbon economy through hydrogen energy is vital for sustainable development. Yet, the success of this transition depends on multiple production routes and environmental determinants. While hydrogen produces zero emissions at the point of use, its overall environmental impact varies greatly, from highly polluting to nearly carbon-neutral—depending on the energy source and production method employed (Whitehead *et al.*, 2023).

The environmental influence and the sources of energy greatly affect the production of hydrogen. Large quantities of carbon dioxide are released via the production of grey hydrogen by steam methane reforming (SMR) from natural gas. Similar steps are taken with blue hydrogen; however, carbon capture and storage (CCS) is used to cut emissions. Green hydrogen, on the other hand, is the most environmentally friendly choice because it is produced by electrolysis with renewable energy sources and has almost no carbon footprint. The environmental assessment is made more difficult by other cutting-edge techniques, including pink hydrogen from nuclear-powered electrolysis and turquoise hydrogen from methane pyrolysis (Saeidi *et al.*, 2023). Figure 4.1 depicts the atmosphere and hydrogen generation.

4.3 FOSSIL FUEL-BASED HYDROGEN (GREY, BROWN, BLACK, AND BLUE)

Although hydrogen is frequently seen as a clean energy source, the production process has an impact on the environment. Over 95% of the world's hydrogen production now comes from fossil fuels, mostly from coal gasification and steam methane reforming (SMR), which results in large emissions of carbon dioxide (CO_2) and other greenhouse gases (GHGs) (Younas *et al.*, 2022). Grey, brown, black, and blue hydrogen are the primary forms of hydrogen produced from fossil fuels; each has its own environmental issues (Saha *et al.*, 2024). Figure 4.1 compares hydrogen derived from fossil fuels and their consequences on the environment.

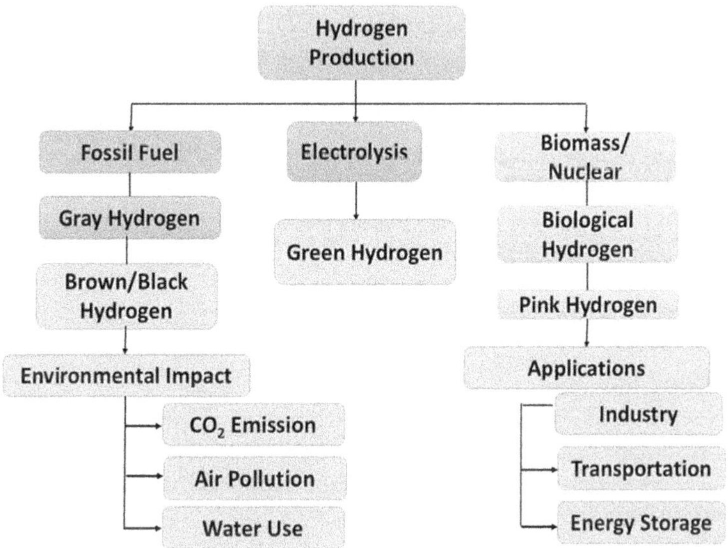

FIGURE 4.1 Production of hydrogen and its effects on the environment

4.3.1 Grey Hydrogen: Steam Methane Reforming (SMR)

$$CO + H_2O \rightarrow CO_2 + H_2$$

This method's low cost and high efficiency (around 70–85%) make it useful.

4.3.1.1 Environmental Impact

There are several environmental problems with using Steam Methane Reforming (SMR) to produce hydrogen. An estimated 830 million tons of CO_2 are liberated annually, equivalent to the entire emissions in Germany. This contributes enormously to changes in climatic conditions with the release of 9–12 of CO_2 per kilogram of hydrogen (Jahami *et al.*, 2025). Methane (a more powerful gas with 80 times the warming potential than CO_2) is also released during the production and transportation of natural gas. The high SMR has a large carbon footprint owing to its high energy needs, which are often produced from fossil fuels (Jahami *et al.*, 2025).

4.3.2 Brown and Black Hydrogen: Coal Gasification

4.3.2.1 Production Process

The syngas, such as carbon monoxide, is produced by reacting coal, oxygen, and steam at high temperature by the process, coal gasification. Brown and black hydrogen are also produced by this process using lignite and hard coal, respectively (Kumar et al., 2024):

$$C + H_2O \rightarrow CO + H_2$$

$$CO + H_2O \rightarrow CO_2 + H_2$$

Black and brown are historically and richly available in countries like India, South Africa, and China.

4.3.2.2 Environmental Impact

The quantities of CO_2 produced by blue hydrogen have been estimated as 18–20 kg, which is twice that of grey hydrogen, contributing 5 million tons of CO_2 emissions per year (Tetteh and Salehi, 2023). The offensive substances, such as nitrogen oxides, sulfur dioxide, nitrogen oxide, and particulate matter, are released during coal gasification. These pollutants can lead to respiratory illnesses, smog, and acid rain. This method of hydrogen production is not sustainable in an environment where there is a limited source of water, as it requires a lot of water, and as much as 7–10 liters per kilogram of hydrogen are needed. The practices such as deforestation, groundwater contamination, and land degradation further worsen the environmental effects of extensive coal mining activities (Hannun & Razzaq, 2022).

4.3.3 Blue Hydrogen: Fossil Fuels with Carbon Capture and Storage (CCS)

4.3.3.1 Production Process

Blue hydrogen is identical to grey (SMR or coal gasification) as both have a similar phase where carbon capture and storage (CCS) is needed to reduce CO_2 emissions. Blue hydrogen naturally takes up 50–90% of CO_2 emissions, incorporating with saline aquifers, depleted oil fields, or deep geological formations (AlHumaidan *et al.*, 2023).

4.3.3.2 Environmental Impact

Blue hydrogen made from natural gas, using carbon capture and storage (CCS), has several restrictions and difficulties. Blue hydrogen is less comparable to green hydrogen as its production depends on natural gas, which suffers limitations of methane leakage and indirect emissions (10–50% of CO_2 emissions), despite that it is connected with CCS (AlHumaidan *et al.*, 2023). There is a 15–25% decrease in the efficiency of blue hydrogen compared to grey hydrogen because of an increase in the requirement of CCS for energy, produced from fossil fuels, negating some of the climatic benefits.

Carbon leakage may also arise owing to geological storage issues, including seismic activity and groundwater contamination, and improper management of storage sites (Gholami *et al.*, 2021). The blue hydrogen is limited due to scaling up owing to the high cost of developing CCS infrastructure, comprising CO_2 injection and transport, and the small number of CCS projects worldwide (Massarweh *et al.*, 2023).

4.3.4 Other Challenges Encountered during Hydrogen Production

4.3.4.1 Air Pollution

The production of hydrogen from fossil fuels leads to the emission of several harmful pollutants:

- Nitrogen Oxides (NOx): Produced from high-temperature combustion in SMR and gasification processes, leading to acid rain, respiratory illnesses, and ground-level ozone generation (Dubey and Dubey 2024).
- Sulfur Oxides (SOx): Mainly discharged during the synthesis of brown and black hydrogen from coal, these pollutants cause acid rain and adverse environmental effects (Mehmood *et al.*, 2025).
- Particulate Matter: Fine particles produced by coal gasification deteriorate air quality, endangering human health and causing climate change by changing atmospheric radiation (Yang *et al.*, 2025).

4.3.5 Water Consumption and Pollution in Hydrogen Production

There are major water-related issues with the production of hydrogen using coal gasification and steam methane reforming (SMR). The process of extracting hydrogen uses water as a feedstock, which puts more strain on freshwater supplies. Furthermore, the operations produce hazardous effluent that can contaminate water bodies since it contains residual hydrocarbons, heavy metals, and thermal pollutants. Sustainability issues are raised by the fact that even blue hydrogen, which lowers CO_2 emissions through carbon capture and storage (CCS), uses a substantial amount more water during the capture process (Zoback and Smit, 2023).

4.4 RENEWABLE AND LOW-CARBON HYDROGEN

Low-carbon and renewable methods can be used to manufacture hydrogen, greatly minimising the impact on the environment. Nuclear and green hydrogen are the two primary varieties in this group (Table 4.1).

4.4.1 Green Hydrogen

Green hydrogen is one of the products of hydrogen generated by the traditional method. It is a low-carbon, sustainable, environmentally friendly, and global energy source. The production of green hydrogen occurs by splitting of water molecule into hydrogen and oxygen by electrolysis. Green hydrogen produced by electrolysis is void of greenhouse gases as it is manufactured from renewable sources such as solar, wind, and hydro. Green hydrogen plays an important role as clean energy in decarbonization, and industries including transportation and power generation (Hassan *et al.*, 2024).

TABLE 4.1
Comparative Summary of Fossil Fuel-Based Hydrogen Environmental Impact

Hydrogen Type	Production Method	Energy Source	CO_2 Emissions (kg/kg H_2)	Efficiency (%)	Key Environmental Concerns	Sustainability Rating	References
Grey Hydrogen	Steam Methane Reforming (SMR)	Natural Gas	9–12	70–85%	High CO_2, methane leakage, and high water use	Low	Kumar *et al.* (2023); Shahulhameed *et al.* (2025)
Brown Hydrogen	Lignite Coal Gasification	Coal	18–20	45–60%	High CO_2, SOx, NOx, water use, deforestation	Low	Dufour *et al.* (2010)
Black Hydrogen	Hard Coal Gasification	Coal	18–20	45–60%	Similar to brown hydrogen, with added land degradation	Low	IEA (2022)
Blue Hydrogen	SMR + Carbon Capture	Natural Gas	1–5 (with CCS)	60–70%	Methane leaks, CCS inefficiency, energy loss	Medium	IEA (2022); Howarth & Jacobson (2021)
Green Hydrogen	Electrolysis	Renewable Electricity	~0	60–90%	High water demand, intermittent energy input	High	IRENA (2020); Kumar *et al.* (2023); Dutta (2014)
Pink Hydrogen	Electrolysis (Nuclear Power)	Nuclear Energy	~0	60–90%	Nuclear waste, safety risks, high capex	Medium	Uddin *et al.* (2021)
Biological Hydrogen	Fermentation, photobiological	Organic Waste / Biomass	~0	Variable (low–moderate)	Land use, competition with food, low scalability	Medium	Das & Veziroglu (2008); Das & Peu (2022)
Turquoise Hydrogen	Methane Pyrolysis	Natural Gas	Low	~70%	Solid carbon waste, tech not fully mature	Medium	Dutta (2014); IEA (2022)

4.4.1.1 Production Process of Green Hydrogen

The main types of electrolysis technology by which green hydrogen is produced are:

1. **Alkaline Electrolysis (AEL)**: This method produces green hydrogen using potassium hydroxide and sodium hydroxide. The process is known as alkaline electrolysis. Alkaline electrolysis has low comparative efficiency (60–80%) but is highly economical (Sabahi & Razfar, 2018).
   The equation for Alkaline Electrolysis (AEL) is: $2H_2O \rightarrow 2H_2 + O_2$
   At the cathode (negative electrode) $2H_2O + 2e^- \rightarrow H_2 + 2OH^-$
   At the anode (positive electrode: $2OH^- \rightarrow 1/2O_2 + H_2O + 2e^-$
2. **Proton Exchange Membrane (PEM) Electrolysis**: Nafion and Flemion are typical polymer membranes and electrolytes used in PEM electrolysis. These electrolytes react very fast to fluctuating renewable energy sources and run with high efficiency. The hydrogen produced by PEM has higher purity than alkaline electrolysis, although it is more expensive.

PEM electrolysis has an efficiency of 70–90% for producing hydrogen (Mirsalami & Mirsalami, 2024).

3. **Solid Oxide Electrolysis (SOEC)**: An electrolyte referred to as ceramic electrolyte, which operates at high temperature, 700–1000°C is utilized in solid oxide electrolysis (SOEC). Despite being the most efficient technology, SOEC is still in its early stages of commercial development. Notably, SOEC achieves a remarkable hydrogen production rate, boasting an efficiency of 85–90% (Hauch *et al.*, 2020).

4.4.2 Renewable Energy Sources for Green Hydrogen

The sustainability of green hydrogen depends on its electricity source, which must come from renewable energy:

- **Solar Energy**: Photovoltaic (PV) solar panels generate electricity for electrolysis.
- **Wind Energy**: Wind turbines provide clean power to hydrogen production plants.
- **Hydropower**: Large-scale hydroelectric dams or small-scale hydro projects offer a steady electricity supply.
- **Geothermal and Biomass Energy**: Green energy is usually produced from geothermal and biomass. The energy from the Earth's core is utilized by geothermal, and the heat generated is used to power electrolysis. Organic matter in biomass is converted into hydrogen through gasification or biological processes. Both geothermal and biomass energy can provide constant and reliable energy for hydrogen production (Sayed *et al.*, 2021).

4.5 CHALLENGES AND LIMITATIONS OF GREEN HYDROGEN

The adoption of green hydrogen faces several challenges. High production costs are currently a significant barrier, as green hydrogen is more expensive than grey and blue hydrogen due to high electrolysis costs and renewable energy intermittency. However, costs are projected to decrease by 50% by 2030 with technological advancements and economies of scale. Additionally, infrastructure development is required, including new hydrogen pipelines, storage facilities, and refueling stations, as current distribution networks are limited. Furthermore, water resource concerns are also a challenge, as electrolysis requires significant water input, particularly in arid regions, although solutions such as seawater electrolysis and using wastewater for hydrogen production are being explored (Ueckerdt *et al.*, 2024).

4.5.1 Environmental Benefits of Green Hydrogen

Green hydrogen offers numerous environmental benefits, particularly in the emissions and air pollution. Notably, it boasts zero carbon emissions, producing no CO_2 emissions during either production or use, unlike its grey or blue hydrogen counterparts (Saha *et al.*, 2024). The reduction in emissions helps decrease dependence on fossil fuels and contributes significantly to global decarbonization efforts. Furthermore, green hydrogen production does not release harmful pollutants like NOx, SOx, or particulate matter, which are commonly associated with fossil fuel-based hydrogen production and contribute to smog and respiratory diseases (Amin *et al.*, 2022).

4.6 BIOLOGICAL HYDROGEN PRODUCTION

Microorganisms such as bacteria, algae, and fungi are utilized in the production of biological hydrogen. This process has no harmful impact on the environment. The production of this type of hydrogen is facilitated by the metabolic activities of microorganisms under ambient conditions, unlike fossil fuels and electrolysis used in traditional methods (Jeje et al., 2024). This process provides a feasible, low-carbon, and energy-efficient alternative to conventional techniques.

The renewable resources, such as organic waste, wastewater, and sunlight, are used to manufacture biological hydrogen production. These materials offer a sustainable and energy-efficient method with minor greenhouse gas emissions and air pollution (Akram *et al.*, 2024). The hydrogen produced by this method is affordable compared to electrolysis and does not require an external energy source. This approach permits the utilization of renewable resources by converting waste products, such as biomass and industrial waste, into useful hydrogen fuel, advancing sustainability and the circular economy (Sharma *et al.*, 2020).

Photobiological water splitting, dark fermentation, and photofermentation are techniques used in the production of biological hydrogen (Hitam & Jalil, 2023). Dark fermentation uses anaerobic bacteria in the absence of light, at ambient temperatures, with the help of organic waste and wastewater to degrade organic matter and produce hydrogen. Photo fermentation uses purple non-sulfur bacteria to produce hydrogen. This method is more energy-efficient; however, some microbes are sensitive to certain light conditions. Photobiological water splitting uses solar energy produced by microalgae and cyanobacteria to divide water into hydrogen and oxygen, resulting in simulating photosynthesis and providing a renewable, carbon-free technique. These biological methods yield viable alternatives to the production of hydrogen in a sustainable approach (Hitam & Jalil, 2023).

$$C_6H_{12}O_6\,(\text{glucose}) \rightarrow 2CH_3CH_2OH\,(\text{ethanol}) + 2CO_2 + 2H_2 - \text{Dark fermentation}$$

$$CH_3COOH\,(\text{acetate}) + 2H_2O + \text{light energy} \rightarrow 2CO_2 + 4H_2 - \text{Photo fermentation}$$

$$2H_2O + \text{light energy} \rightarrow 2H_2 + O_2 - \text{Photobiological water splitting}$$

4.7 NUCLEAR HYDROGEN PRODUCTION

Nuclear hydrogen is emerging as a vital component of the global transition to a low-carbon energy system, offering several key advantages. Since nuclear power doesn't emit CO_2, it's a cleaner option than grey and blue hydrogen, which is one of its most significant benefits (Addo *et al.*, 2023). A steady and extensive power supply is another benefit of nuclear energy over sporadic renewable energy sources like wind and solar. The production of thermochemical hydrogen is made possible by the ability of advanced nuclear reactors to function at high temperatures (Hercog *et al.*, 2024). Because nuclear power lessens dependency on imported fossil fuels, it promotes energy independence and improves energy security.

4.7.1 Electrolysis Using Nuclear Power

The nuclear power technique of hydrogen production is a cleaner method to produce electricity using alkaline electrolysis (AEC) to power different electrolysis processes (Sebbahi *et al.*, 2024). Compared to AEC, proton exchange membrane electrolysis (PEM), which makes use of a polymer electrolyte membrane, offers higher efficiency and quicker reaction times (Mao *et al.*, 2024). Finally, when combined with nuclear heat, Solid Oxide Electrolysis (SOEC) greatly increases efficiency due to its high operating temperatures, which range from 700 to 1,000°C (Liu *et al.*, 2024).

4.7.2 Thermochemical Water Splitting

Nuclear reactors, such as Generation IV reactors, provide the heat needed for thermochemical water splitting, which produces hydrogen without the use of energy, at extremely high temperatures (Hercog *et al.*, 2024). The two primary techniques in this process are the Sulfur-Iodine (S-I) Cycle, which uses nuclear reactor heat (~800°C) to initiate chemical reactions that split water into hydrogen and oxygen, and the Hybrid Sulfur Process, which combines thermal and electrochemical steps and improves efficiency by using nuclear heat (Zeng *et al.*, 2024). These thermochemical methods offer significant efficiency improvements over conventional electrolysis while maintaining nearly zero carbon emissions (Xue *et al.*, 2025).

4.7.3 Steam Methane Reforming with Nuclear Heat

Natural gas is usually converted to hydrogen using steam methane reforming (SMR), although this process has a major negative impact on the environment due to CO_2 emissions. However, the ecological impact of the SMR process can be reduced by partially decarbonizing it by using nuclear heat instead of fossil fuels. This novel hybrid strategy has been dubbed "pink hydrogen," indicating its transitional role toward cleaner hydrogen generation.

4.7.4 Challenges of Nuclear Hydrogen Production

There are lots of obstacles to overcome before nuclear energy can be used to produce hydrogen.

High initial costs and protracted construction timetables are the challenges of nuclear reactors, since they require large capital investments. The development of sophisticated reactors, which include high-temperature gas-cooled reactors, is ongoing (Stewart et al., 2021). Public opinions and safety concerns are other obstacles. Nuclear energy often encounters resistance because of radioactive waste, mishaps, and proliferation threats (Aydın, 2020). Political and regulatory obstacles further worsen the process as they cause delays due to stringent clearances and expensive compliance. Nuclear energy is already phased out in many nations, and therefore limits the potential for producing hydrogen. Managing the radioactive wastes produced by nuclear power is still difficult and calls for long-term fixes (Drace et al., 2022).

4.8 LAND USE AND BIOMASS SOURCING CONCERNS

Biological hydrogen production is a viable substitute for hydrogen generated from fossil fuels. This process uses microorganisms to produce hydrogen from organic materials. However, there are serious issues with biomass sourcing and land use when this technique is scaled up. It is imperative to address many critical issues, which include the availability of adequate land for biomass cultivation, the rivalry with food production, and the sustainability of biomass feedstocks to guarantee the long-term sustainability of biological hydrogen as a clean energy source (Obiora et al., 2024).

4.8.1 Land Use Concerns

4.8.1.1 Competition with Agriculture

The competition for fertile land for agriculture is a major obstacle to biomass-based hydrogen. Reduced land availability for food crops due to large-scale biomass for hydrogen production may result in food shortages and price rises. It might also promote habitat degradation and deforestation, which could affect ecosystem stability and biodiversity. Furthermore, growing biomass frequently necessitates large water inputs, and this puts strain on freshwater supplies and exacerbates already-existing problems with water management (Obiora et al., 2024).

4.8.1.2 Deforestation and Environmental Impacts

The ecology is seriously threatened by the increase in land use for biomass sources, especially in tropical areas where forests are frequently cut down to make way for more agriculture. Widespread deforestation may lead to the loss of essential sinks that absorb CO_2 emissions (Raihan et al., 2022).

Long-term productivity and land fertility may also be destroyed by soil erosion and degradation. The goal of switching to a more sustainable energy source is ultimately undermined by the disastrous effects of ecosystem disruption, which also include the loss of wildlife habitats and a fall in biodiversity (Kolawole & Iyiola, 2023).

4.8.1.3 Urban and Industrial Land Constraint

Absence of land for large-scale biomass production in heavily industrialized or crowded areas calls for creative solutions. Alternative strategies that can be applied include vertical ferming and controlled-environment agriculture, which maximize land utilization for bioenergy crops, and urban waste-to-hydrogen projects, which turn organic waste into hydrogen (Cholomskė, 2023). Also, a workable way to get around land restrictions and support sustainable energy production is to incorporate biomass production with current agricultural systems in addition to food and fiber crops (Saleem, 2022)

4.8.2 Biomass Sourcing Concerns

For biological hydrogen generation to be feasible in the long run, biomass feedstock sustainability is essential. While seasonal variations in biomass availability might result in irregular hydrogen generation, excessive harvesting of biomass can cause the depletion of natural resources. The necessity for careful control of biomass supply is further highlighted by the fact that energy-intensive harvesting and processing can raise the overall carbon footprint (Bhatia *et al.*, 2021).

A resource with many uses, biomass is utilized in the production of chemicals, paper and pulp, animal feed, and biofuels. Redirecting biomass to produce hydrogen might make it economically unfeasible in some areas due to rising costs and demand. The significance of creating effective and sustainable biomass procurement strategies is highlighted by this rivalry for biomass resources (Obiora *et al.*, 2024).

Utilizing organic waste materials to produce hydrogen is one possible way to address the problems associated with biomass sources. It is possible to make use of industrial by-products such as food processing waste and paper mill sludge, as well as agricultural waste like crop residues and manure. Municipal solid waste can also be used, especially organic fractions from commercial and residential garbage. In addition to lowering waste disposal problems, this strategy offers a reliable and accessible source of biomass for the production of hydrogen (Divyabharathi *et al.*, 2024).

4.8.2.1 Competition with Food Supply for Bio-based Feedstocks

Biomass, waste materials (feedstocks), and agricultural wastes are examples of organic resources that are used as feedstocks for biological hydrogen generation. However, worries about bio-based feedstocks competing with food production are growing as demand for them rises. Crops used for energy instead of food can increase land-use conflicts, affect global food security, and raise food costs (Zhuang *et al.*, 2022). To guarantee that biological hydrogen production continues to be a viable and moral energy source, several issues must be resolved.

4.8.2.2 Use of Edible Crops for Hydrogen Production

The utilization of edible crops to produce hydrogen has sparked concerns about the global food supply. Certain biological hydrogen production systems require starch-based and sugar-rich crops, such as corn, sugarcane, wheat, cassava, potatoes, soybeans, and oilseed crops. If these food crops are widely diverted for energy generation, the world's food supply might be significantly diminished, which would disproportionately damage low-income people who rely on staple crops for their survival (Younas *et al.*, 2022).

4.8.2.3 Increased Food Price

As more and more agricultural land and resources are allocated to the production of bio-based hydrogen, food crops may become less available, perhaps leading to a cascading impact. This reduction in the supply of food crops may make it more difficult for many people to obtain staple meals, particularly low-income populations in emerging nations who will be most negatively impacted monetarily. Furthermore, price fluctuations that could have a detrimental impact on both farmers and consumers could be indicative of potential market instability (Ghaedi *et al.*, 2024).

4.8.2.4 Land and Water Use Competition

Biomass generation for hydrogen synthesis requires a large amount of land and water, which could divert resources from food production. This raises a number of difficulties, including the potential reduction in arable land for food crops as a result of increasing the production of bioenergy crops. Due to the high watering requirements of certain bio-feedstocks, water scarcity issues may intensify in regions that are prone to drought. The risk of deforestation, which can lead to biodiversity loss, soil degradation, and other environmental problems, also rises when farmland is expanded for bio-feedstocks (Osman *et al.*, 2022).

4.8.2.5 Impact on the Agricultural System

The transition to bio-based hydrogen production may have a significant influence on conventional farming practices and could have a cascading effect on agricultural systems. Large-scale production of hydrogen feedstocks could encourage monoculture farming, which would accelerate soil degradation and reduce biodiversity (Rasul *et al.*, 2022). Land and water contamination may result from increased use of fertilizers and pesticides brought on by enhanced bio-feedstock agriculture. Farmers may choose energy crops over food crops if bio-feedstocks become more feasible, which might upset regional food systems and alter rural economies (De Vrieze *et al.*, 2020).

4.9 ENVIRONMENTAL RISK OF HYDROGEN STORAGE DISTRIBUTIONS

4.9.1 Hydrogen Leakage and Indirect Greenhouse Effects

The potential to lower carbon emissions is drawing attention to the switch to hydrogen as an alternative energy source. But leakage is a major environmental issue related to the distribution and storage of hydrogen. Being the lightest and smallest molecule, hydrogen is very prone to leakage from pipelines, storage tanks, and transit networks. It is still difficult to completely stop hydrogen leaks, even with the most sophisticated containment devices (Hashemi *et al.*, 2025).

4.9.2 Mechanisms of Hydrogen Leakage

At several points throughout the manufacturing, delivery, and storage of hydrogen, leaks may occur. Storage tanks are one of the main places where leaks occur because hydrogen's small molecular size causes it to gradually diffuse through materials over time (Magliano *et al.*, 2024). Another susceptible area is pipelines, where conventional materials like steel may undergo hydrogen embrittlement, leading to leaks and fissures (Shehata & El-Shamy, 2023). If valves and fittings are not properly maintained, they could become leaky. If safety precautions are not properly followed, hydrogen-powered cars and refueling facilities can also leak, especially during refueling operations (Genovese *et al.*, 2024).

4.9.2.1 Indirect Greenhouse Effects of Hydrogen Leakage

Hydrogen is not a greenhouse gas in and of itself, but when it is released into the atmosphere, it can alter atmospheric chemistry and cause indirect climate consequences. The following are the main ways that hydrogen leaks affect the environment.

4.9.2.2 Reduction in Hydroxyl Radicals (OH)

In atmospheric chemistry, hydroxyl radicals are essential because they degrade methane (CH_4), a powerful greenhouse gas. Hydroxyl radicals are less available to break down methane when hydrogen escapes into the atmosphere and combines with them. Methane's atmospheric lifespan increases as a result of this process, increasing the gas' potential to contribute to global warming (Wang *et al.*, 2022). Since methane absorbs heat significantly more effectively than carbon dioxide, the indirect impacts of hydrogen leakage could offset the environmental benefits of hydrogen energy (Goita *et al.*, 2025).

4.9.2.3 Increased Stratospheric and Tropospheric Water Vapor

Hydrogen leaks can potentially cause increases in water vapor in the stratosphere. More water vapor is created when hydrogen oxidizes as it enters the upper atmosphere (Lakshmanan & Bhati, 2024). Water vapor in the stratosphere primarily regulates Earth's radiative balance, and an excess of it can have further warming effects. Changes in cloud formation and atmospheric radiation dynamics brought on by an increase in water vapor further increase climate uncertainty (Voigt *et al.*, 2021).

4.9.2.4 Formation of Ozone and Other Pollutants

Emissions of hydrogen can affect ground-level ozone, one of the primary components of smog. Excess hydrogen in the atmosphere can affect the oxidation balance, which can lead to an increase in tropospheric ozone (Archibald *et al.*, 2020). While ozone in the stratosphere helps shield the body from harmful UV radiation, its accumulation at lower altitudes causes air pollution and respiratory health issues.

4.9.2.5 Impact on Nitrous Oxide and Other Greenhouse Gases

Other greenhouse gases like nitrous oxide (N_2O) may be indirectly impacted by the changed atmospheric composition brought on by hydrogen leakage. The natural balance of different gases may be altered by changes in chemical processes brought on by hydrogen's interaction with atmospheric constituents, which would further contribute to climate change (Von Schneidemesser *et al.*, 2015).

4.10 SAFETY RISKS (FLAMMABILITY, EXPLOSION HAZARD)

Hydrogen has been embraced more as a sustainable energy source because of its potential to reduce carbon emissions and dependency on fossil fuels. Despite its benefits, hydrogen's tendency toward flammability and explosion makes it a major safety concern during transport and storage. Strict safety measures are required to prevent accidents due to the unique properties of hydrogen, including its low ignition energy, broad range of combustible and detonation potential. Resolving these risks is essential to the safe integration of hydrogen into energy systems (Hosseini & Wahid, 2016).

4.10.1 Properties of Hydrogen that Increase Safety Risk

Handling of hydrogen by humans is deleterious owing to its dangerous characteristics, such as fires and explosions. Low ignition point of hydrogen (0.02 millijoules) makes it highly dangerous and produces harmful signals such as sparks, friction, and electrical discharges (Cavaliere, 2023). Its flammability properties of 4–75% could cause danger in environments and result in accidental leaks. It possesses hard-to-detect, faint blue, and rapid flame fire speed, which are more difficult to handle (Proud *et al.*, 2023). Leaving the hydrogen in a confined enclosure could result in its buildup and high ignition as it diffuses and disperses easily. Therefore, a need for careful observation of handling, storage, and safety measures of hydrogen (Yang *et al.*, 2021).

4.10.2 Explosion Hazard of Hydrogen

There is a high susceptibility of hydrogen to a pronounced danger owing to its high risk of explosion and wide-ranging boundaries of destruction. Hydrogen in confined enclosures tends to explode owing to its highly explosive mixture, with an estimated 18–59% (Li *et al.*, 2023). The shockwave emanating from the detonation of hydrogen could cause major damage to buildings. Hydrogen embrittlement could result in corrosion of metals, leading to catastrophic failure and explosion. There is a high risk of explosion if there is limited air in a confined space where hydrogen is stored. Therefore, handling hydrogen in a confined container with strict safety precautions and adequate ventilation is crucial (Sobola and Dallaev, 2024).

4.10.2.1 Potential Accident Scenario in Hydrogen Storage and Distribution

Many practical examples around the globe have been highlighted, showing the dangers of hydrogen to ignite and its ability to explode. These are seen in degradation or embrittlement as a result of an unexpected explosion or rupture, and pipeline explosions and leaks (Cavaliere, 2023). Accidental leaks at hydrogen filling stations can also cause flames, rupture, and explosion of hydrogen-powered cars (Genovese *et al.*, 2024). Strong safety procedures and emergency preparedness are crucial since accidents at hydrogen production facilities and industrial storage tank failures might result in significant fires or explosions (Cavaliere, 2023).

4.11 HYDROGEN LIQUEFACTION AND STORAGE WATER FOOTPRINT

The environmental consequences of hydrogen production, storage, and distribution are becoming increasingly significant as it emerges as a significant future energy source. The water footprint of hydrogen liquefaction and storage is a major worry (Osman *et al.*, 2022). The energy-intensive process of hydrogen liquefaction necessitates extremely low temperatures (−253°C or 20K), and the cooling systems employed frequently depend heavily on water (Al Ghafri *et al.*, 2022). Significant water use can also result from hydrogen storage, especially in high-pressure tanks or subterranean reservoirs. The increasing demand for hydrogen, notably in industrial and transportation sectors, needs an evaluation of the water-intensive processes involved to enable sustainable hydrogen deployment (Liu *et al.*, 2023).

4.11.1 Understanding the Water Footprint in Hydrogen Liquefaction

Since hydrogen has a low energy density per unit volume and exists as a gas under normal air conditions, liquefaction is required to make transportation and storage easier. Long-distance transportation, improved storage efficiency, and support for applications requiring liquid hydrogen are all made possible by cooling hydrogen to cryogenic temperatures (−253°C) and liquefying it (Xie *et al.*, 2024).

However, a lot of cooling is needed in hydrogen liquefaction plants, which frequently use water-based cooling systems. Water-cooled heat exchangers are required to dissipate the significant heat produced by the use of water for cooling and refrigeration. Water-intensive energy production is also indirectly facilitated by the fact that power stations that use water for cooling frequently provide the electricity needed for liquefaction. The necessity of effective water management in hydrogen liquefaction is further highlighted by the direct water losses that can result from purge gas and evaporative cooling systems (He *et al.*, 2023).

4.11.2 Water Footprint in Hydrogen Storage

Hydrogen liquefaction is important for effective transport and storage, as it has a poor energy density as a gas. Cooling it to −253°C improves long-distance transit, boosts storage efficiency, and supports applications needing liquid hydrogen. Nevertheless, liquefaction facilities need a lot of cooling, frequently with water-based systems, which can result in water loss and consumption for refrigeration, cooling, and energy production. Efficient water management is necessary to mitigate these impacts (Wang *et al.*, 2024).

4.11.2.1 Cryogenic Storage of Liquid Hydrogen (LH_2)

The ultra-cold temperatures of liquid hydrogen (LH_2) require storage in insulated cryogenic tanks. This storage technique does have certain drawbacks, though, such as the need for constant cooling, which causes cooling towers to use water for heat dispersion. Moreover, boil-off losses happen when the liquid hydrogen progressively warms up and evaporates, requiring more cooling cycles that use even more water (Xu *et al.*, 2023).

4.11.2.2 Underground Hydrogen Storage

Significant water-related effects may result from the large-scale storage of hydrogen in subterranean facilities like aquifers, salt caverns, or depleted gas fields. For example, water injection may be necessary to maintain pressure stability in these storage locations, which may change the chemistry and availability of subsurface water. Furthermore, the construction of salt caverns requires water-intensive leaching procedures that dissolve rock salt and produce subterranean caverns using millions of liters of water (Gupta & Yadav, 2020).

4.11.2.3 High-pressure Hydrogen Gas Storage

At very high pressures, hydrogen is frequently stored as a compressed gas. However, during production procedures such as resin curing and material treatment, a significant amount of water is needed to produce these tanks, particularly those built from fiber-reinforced composites (Bouhala *et al.*, 2024).

4.11.2.4 Impacts of the Water Footprint in Hydrogen Liquefaction and Storage

In areas where water scarcity is already a problem, the extensive installation of hydrogen infrastructure may make it worse. Conflicts over resource distribution may result from competing needs for water from different sectors (Kountouris *et al.*, 2024).

Because the cooling water released by hydrogen liquefaction plants can damage aquatic life and disturb aquatic ecosystems, these facilities can also contribute to thermal pollution. Furthermore, there is a chance that groundwater contamination from subterranean hydrogen storage will degrade the quality of the local water supply (Wei *et al.*, 2024).

4.12 HYDROGEN PLAYS A ROLE IN DECARBONIZATION AND SUSTAINABILITY

Utilizing hydrogen as a clean substitute for fossil fuels considerably lowers CO_2 emissions in a variety of businesses. According to Griffiths *et al.* (2021), the effects of hydrogen are most noticeable in industries like steel manufacturing, transportation, and chemicals that have historically been difficult to decarbonize.

4.12.1 Hydrogen Impacts on Reducing Emissions in Industries

4.12.1.1 Steel Industry

Green hydrogen technology has helped the steel sector transition to more sustainable production methods. In Direct Reduced Iron processes, green hydrogen can be used in place of coking coal to produce water rather than CO_2, which is a substantial CO_2 emission from traditional coal-powered blast furnaces (Zang *et al.*, 2023). To lessen their environmental impact, major corporations, including SSAB, ArcelorMittal, and Thyssenkrupp, are investing in hydrogen-based steelmaking (Ledari *et al.*, 2023)

4.12.1.2 Heavy Transport (Trucks, Ships, Trains, and Aviation)

As a zero-emission energy source for heavy transportation, such as trucks, ships, trains, and airplanes, hydrogen is being researched. Long-haul trucks are powered by hydrogen fuel cells, which reduces the need for diesel (Pardhi *et al.*, 2022). To lessen carbon footprints, synthetic fuels and hydrogen-based ammonia are also being developed for use in transport and aviation. As a greener substitute for diesel locomotives, nations like Germany and Japan are also investing in hydrogen-powered trains (Alavi-Borazjani *et al.*, 2025).

4.12.1.3 Chemicals and Refining

Hydrogen, which is usually obtained from fossil fuels (grey hydrogen), is essential to the chemical industry's production of ammonia and methanol. Nonetheless, switching to green hydrogen, which is created by electrolysis with renewable energy, can significantly lower CO_2 emissions. To further reduce emissions and encourage a cleaner production process, refineries can also use hydrogen to desulfurize fuels (de la Cruz-Soto *et al.*, 2025).

4.12.1.4 Power Generation and Energy Storage

A possible alternative for energy storage and power generation is hydrogen. Gas turbines can burn it to generate energy with reduced carbon emissions. A clean and effective method of storing energy for later use, such as during times of high demand or when renewable sources are unavailable, is to electrolyze excess renewable energy from solar or wind power to produce hydrogen (Hosseini & Wahid, 2020).

4.13 ROLES IN HEAVY TRANSPORT AVIATION AND DECARBONIZATION

4.13.1 Heavy Transport

When it comes to decarbonization, heavy-duty transportation, such as long-haul trucks, buses, trains, and ships, presents considerable obstacles. Often, battery-electric systems are inadequate due to their limitations in weight, range, and recharge times. In contrast, hydrogen combustion and fuel cell electric vehicles help hydrogen to produce clean and efficient energy (Waseem *et al.*, 2023). For heavy-duty transportation, hydrogen fuel cell technology is especially well-suited. Fuel cell vehicles release only water vapor while converting hydrogen into electricity to run electric motors. Like diesel trucks, hydrogen vehicles can also be refueled in a matter of minutes, which makes them perfect for long-distance driving. Hydrogen also provides higher energy density than batteries, making it suitable for heavy loads and long distances (Dash *et al.*, 2022).

4.13.2 Current Development

Several companies, including Toyota, Hyundai, and Nikola, are actively developing hydrogen-powered trucks, marking a significant step toward cleaner transportation (Soleimani *et al.*, 2024). In the rail sector, hydrogen trains, such as Alstom's Coradia iLint, are being introduced in Europe as a replacement for diesel-powered locomotives on non-electrified rail lines (Gurrì *et al.*, 2023).

The maritime industry is also turning to hydrogen and ammonia as potential marine fuels to meet the International Maritime Organization's (IMO) carbon reduction targets for 2030 and 2050 (Louizou, 2024). This shift is expected to play a crucial role in minimizing the environmental impact of maritime transport.

4.13.3 Hydrogen-Powered Fuel Cells and Their Emissions Profile

An electrochemical process between hydrogen and oxygen produces electricity in hydrogen-powered fuel cells, a clean energy source that only produces heat and water vapor as byproducts (Singla *et al.*, 2021). They are therefore ideal for use in transportation, industrial equipment, and backup power systems—particularly in regions where air quality and emissions are significant concerns, because they release no CO_2, particulate matter, or hazardous gases when in use.

However, the method used to manufacture the hydrogen affects the hydrogen fuel cells' total emissions profile. While green hydrogen, which is created by electrolysis using renewable energy, is almost emission-free, grey hydrogen, which is made from fossil fuels, produces considerable carbon emissions (Hammi *et al.*, 2024). Thus, even if fuel cells are clean in and of themselves, the climatic impact is dependent on the carbon intensity of hydrogen production, underscoring the significance of increasing the use of hydrogen derived from renewable sources (Akpasi *et al.*, 2025).

Hydrogen fuel cells provide major environmental advantages, such as silent operation and no local emissions, in contrast to diesel engines and even some electric vehicles (depending on the electricity mix) (Togun *et al.*, 2024). Hydrogen fuel cells are positioned to play a significant role in global decarbonization as green hydrogen becomes more accessible and reasonably priced, particularly in industries that are difficult to electrify, such as heavy transportation, shipping, and aviation (Ali *et al.*, 2025).

4.14 STRATEGIES FOR SUSTAINABLE HYDROGEN PRODUCTION AND USE

4.14.1 Improving the Effectiveness of Hydrogen Production from Renewable Resources

Green hydrogen is a viable and adjustable approach to global decarbonization, but it requires improving the efficiency of renewable-based hydrogen production (Zhang *et al.*, 2024; Hassan *et al.*, 2024). The most common method, water electrolysis, splits water into hydrogen and oxygen using energy from renewable sources such as solar, wind, or hydro power (Akpasi *et al.*, 2025). Improving the efficiency of this process will reduce energy consumption and the overall cost of producing hydrogen (Younas *et al.*, 2022).

The development of advanced electrolyzer technologies, such as solid oxide electrolyzers (SOEC), alkaline, and proton exchange membranes (PEM), is one of the best tactics (Sebbahi *et al.*, 2022; 2024). These devices aim to improve endurance, reduce reliance on expensive or rare components such as platinum or iridium, and increase conversion efficiency. In order to optimize uptime and maintain flexibility even in the case of variations in the energy supply, research is also being conducted on the direct integration of electrolyzers with intermittent renewable sources (Wang *et al.*, 2024).

Utilizing waste heat from industrial processes or merging hydrogen production facilities with renewable energy plants are examples of system-level integration that may lead to efficiency benefits in addition to technological breakthroughs (Kukula *et al.*, 2024). The widespread use of clean hydrogen solutions is made possible by advancements in smart grid technology, real-time demand management, and energy storage, which all contribute to the reliable and efficient production of hydrogen from renewable sources (Abdin, 2024).

4.14.2 Carbon Capture Requirement for Elevated Capacity of Hydrogen and Emission Reduction

When combined with steam methane reforming (SMR) or autothermal reforming (ATR), carbon capture, utilization, and storage (CCUS) produces blue hydrogen from natural gas and reduces associated CO_2 emissions (Ishaq & Crawford, 2025). Although blue hydrogen is not as emission-free as green hydrogen, it is a good interim solution, particularly in places where natural gas infrastructure already exists (Pathak *et al.*, 2023). To maximize the climatic benefits of blue hydrogen, carbon capture technology must be improved.

Raising the amount of captured carbon from 60% percent to more than 90% or even 95% is the main strategy for improving sustainability (Wilberforce *et al.*, 2019). This can be accomplished by enhancing the efficiency and reducing the energy needs of the capture process through advancements in membrane separation, absorption materials, and cryogenic technologies (Dubey and Arora, 2022).

Coupling reforming units with pre- or post-combustion capture devices increases the possibility for overall emissions reduction (Law *et al.*, 2024).

The feasibility of blue hydrogen production is associated with the maintenance of captured carbon. By investing in geological storage sites like depleted oil and gas sources and developing carbon reuse uses like fuels or construction materials, blue hydrogen's environmental profile can be enhanced (Al Osman *et al.*, 2022). With improved collection rates and careful storage, blue hydrogen might be a lower-carbon link in the shift to a fully renewable hydrogen economy (Boretti and Pollet, 2024).

4.14.3 Low-carbon Hydrogen Distribution System and Enhancement

Low-carbon hydrogen production and distribution system is crucial to scale hydrogen as a clean energy carrier and guarantee its sustainability across production, storage, transportation, and end use (Ibrahim and Al-Mohannadi, 2023). Even if hydrogen is created with little or no emissions, its climate benefits may not be wholesome due to supply chain inefficiencies or carbon-intensive methods. Making supply chains strong, effective, and decarbonized is, therefore, a top strategic goal (Carlson and Trencher, 2024).

Thus, there will be a need to establish regional centers for hydrogen production, storage, and consumption to alleviate emissions and transportation requirements (Rattle *et al.*, 2024). To provide safe and energy-efficient logistics when long-distance transport is needed, cutting-edge technologies like liquid hydrogen, ammonia conversion, or solid-state carriers are being researched (Alavi-Borazjani *et al.* 2025). Furthermore, upgrading existing natural gas pipelines to accommodate hydrogen blending or constructing dedicated hydrogen pipelines can substantially lower the carbon footprint associated with energy distribution (Topolski et al., 2022).

The hydrogen supply chains are increasingly integrating digital technologies and smart monitoring systems to track emissions, optimize energy use, and ensure reliability and safety (Feng et al., 2024). When combined with global standards and certification schemes, such as those distinguishing green and or blue hydrogen. These approaches not only reduce carbon emissions but also enhance transparency and trust in the hydrogen market (Gallegos, 2024). Ultimately, developing a low-carbon hydrogen supply chain is vital for positioning hydrogen as a practical and scalable component of the global clean energy transition (Boretti and Pollet, 2024).

4.14.4 Global Collaborative Hydrogen Production Initiative

Collaborative efforts in hydrogen production are crucial for building a practical and adaptable global partnership, given the uneven distribution of hydrogen resources across countries. Hydrogen is available in abundance in many countries because of their access to renewable resources that could generate at a cheaper price; the reverse is the case for others owing to a limited renewable capacity (Boretti and Pollet, 2024). Global partnership permits access to effective trading in hydrogen, sustainable energy, and transfer of technology (Abdin, 2024).

Establishing cross-border pipelines and international hydrogen corridors, signing bilateral and multilateral agreements, and developing uniform certification procedures for low-carbon hydrogen are important tactics (Gallegos, 2024; Pinto, 2024). By preventing market or regulatory fragmentation, these initiatives guarantee that hydrogen produced sustainably in one area may be consistently exported and accepted in the importing zone (Ansari and Pepe, 2023). Such international collaboration is promoted by programs such as the International Partnership for Hydrogen and Fuel Cells in the Economy (IPHE), Mission Innovation, and the Hydrogen Council (Delaval *et al.*, 2023; Alam *et al.*, 2024).

Across the hydrogen value chain, collaborative investments in workforce development, infrastructure, and research can spur innovation and cut costs (Harichandan and Kar, 2023). Coordinating efforts on legislative frameworks, safety standards, and finance sources can reduce project risk and attract the private sector (Scholvin *et al.*, 2024). In essence, international cooperation supports global net-zero goals by transforming hydrogen from a localized potential into a globally integrated clean energy solution (Kumari, 2024).

4.15 POLICY RECOMMENDATIONS AND FUTURE DIRECTIONS

4.15.1 The Requirement for Stringent Carbon Accounting in the Manufacture of Hydrogen

As a crucial element of global decarbonization strategies, strict and transparent carbon accounting in hydrogen production is becoming increasingly significant. In the lack of clear restrictions, greenwashing, the marketing of hydrogen as "clean" even though it is produced using high-emission technologies, is a possibility. Accurate carbon accounting ensures that hydrogen's environmental claims are trustworthy, comparable, and consistent with global climate goals (Di Vaio and Ali, 25).

Policymakers should consider the full emissions impact of hydrogen production, including feedstock sourcing, energy inputs, and downstream consumption, using standardized lifecycle assessment (LCA) approaches (Osman *et al.*, 2024). Goita *et al.* (2025) state that these assessments must include measures for indirect emissions, such as methane leakage in natural gas supply chains, and account for differences in production methods (such as green, blue, and grey hydrogen). The implementation of emissions thresholds and certification mechanisms for low-carbon hydrogen will enable fair market competition and investment guidance toward truly sustainable alternatives (Gale *et al.*, 2024).

For the purpose of harmonizing carbon accounting systems internationally, governments and regulatory bodies should work together in the future. This will guarantee that climate funding, subsidies, and policy incentives are focused on genuinely low-emission hydrogen solutions in addition to promoting global hydrogen trade (Mahto *et al.*, 2023). In the end, rigorous carbon accounting guarantees accountability, fosters confidence in the hydrogen economy, and speeds up the transition to a net-zero future (Di Vaio and Ali, 2025).

4.15.2 Investments in Clean Hydrogen Technologies

Unlocking hydrogen's full potential as a pillar of the global clean energy transition requires significant and ongoing investment in clean hydrogen technology (Abdin, 2024). Governments, corporations, and international organizations must prioritize financing and incentives to enhance hydrogen production, improve efficiency, and reduce costs, especially for green hydrogen produced from renewable resources (Alam *et al.*, 2024).

Next-generation electrolyzers, sophisticated carbon capture technologies for blue hydrogen, and novel storage and transportation options like liquid hydrogen, ammonia, and solid-state carriers should all be encouraged by policy frameworks (Islam *et al.*, 2024). Public–private partnerships can promote commercialization and reduce the risk of large-scale projects, while public investments in research and development (R&D) can hasten innovations in cost-effective production and assist in overcoming technical obstacles (Favour, 2025).

Concurrently, infrastructure investments, such as pipelines, integrated hydrogen hubs, and hydrogen filling stations, are needed to enable widespread adoption in industries, transportation, and power (Kim *et al.*, 2023). By providing subsidies, tax breaks, loan guarantees, and market-based instruments like contracts for difference (CfDs), which guarantee a stable price for low-carbon hydrogen, governments can stimulate private investment (Guha, 2025). The investment would ensure the world climate benchmark as it encourages innovation and economic growth (Zhang *et al.*, 2022).

4.15.3 Potential Study Focus: Cost Reductions, Storage Inventions, and Productivity Improvements

Hydrogen needs to be improved, especially in research and innovation, and other crucial areas, if hydrogen is to achieve its potential and sustainability as an energy source. The hydrogen production efficiency, especially in electrolysis and thermochemical processes, should be the focus of future studies (Jeje *et al.*, 2024). The developments should also include strong and efficient membranes and electrocatalysts that offer high performance and decrease energy input and overhead cost (Jin *et al.*, 2023).

Proper research on reducing the hydrogen value chain is germane. This can only be achieved by replacing the costly and unusual materials such as catalysts, iridium, or platinum, including enhancing production bulk and minimizing the costs of electrolyzers (Mathur, 2025). Integrating a renewable energy system alongside a hybrid energy system, forming system-level innovation, may help minimize hydrogen cost (Sahraie *et al.*, 2024).

Advanced hydrogen storage and transportation are required to increase the utility of hydrogen in many industries (Cheekatamarla, 2024). Improved compressed and cryogenic hydrogen that brings about non-violent, more transportable, and better, inexpensive storage choices could be developed through innovative research (Kumar *et al.*, 2024). Advancing areas such as long-distance transportation, network balancing, and seasonal energy storage of hydrogen will make it more flexible and robust for future applications (Kamran & Turzyński, 2024).

4.16 CONCLUSION

Multiple and open-ended total zero solution is offered by hydrogen to the topmost countries at the helm of the energy transition. This chapter displays the incredible environmental footmark of hydrogen production, in the order of high-emitting fossil fuel-based systems, green and biological systems with close to zero emissions. Although green hydrogen offers the most climate-friendly option in the long run, a lot of challenges and issues are left unsolved. Its expensive nature of production, the absence of a sufficient infrastructure, the requirement for water, and the technological issues associated with the distribution and storage have to be addressed. Air pollution, land competition, and hydrogen leakage are also other necessities for a comprehensive strategy that addresses both environmental and socioeconomic effects.

Multi-faceted approaches such as enhancing low-carbon supply chains, raising carbon capture for bridging blue hydrogen, developing electrolyzer technologies, and facilitating international collaboration and policy harmony are required to unveil the full hydrogen potential. Early investment in clean infrastructure, research, and development can be used to reduce costs and increase production on a large scale in a sustainable way. With the right legislation, technology, and global collaboration, hydrogen has the potential to entirely revolutionize the way we transport energy, industrial power, and store renewable energy for future generations.

REFERENCES

Abdin, Z. (2024). Empowering the hydrogen economy: The transformative potential of blockchain technology. *Renewable and Sustainable Energy Reviews*, *200*, 114572.

Addo, E. K., Kabo-Bah, A. T., Diawuo, F. A., & Debrah, S. K. (2023). The role of nuclear energy in reducing greenhouse gas (GHG) emissions and energy security: A systematic review. *International Journal of Energy Research*, *2023*(1), 8823507.

Akpasi, S. O., Smarte Anekwe, I. M., Tetteh, E. K., Amune, U. O., Mustapha, S. I., & Kiambi, S. L. (2025). Hydrogen as a clean energy carrier: Advancements, challenges, and its role in a sustainable energy future. *Clean Energy*, *9*(1), 52–88.

Akram, F., Fatima, T., Ibrar, R., & ul Haq, I. (2024). Biohydrogen: Production, promising progressions and challenges of a green carbon-free energy. *Sustainable Energy Technologies and Assessments*, *69*, 103893.

Al Ghafri, S. Z., Munro, S., Cardella, U., Funke, T., Notardonato, W., Trusler, J. M., … & May, E. F. (2022). Hydrogen liquefaction: A review of the fundamental physics, engineering practice and future opportunities. *Energy & environmental science*, *15*(7), 2690–2731.

Alam, S. N., Khalid, Z., Singh, B., & Guldhe, A. (2024). Integration of government policies on the global level for green hydrogen production. In *Green hydrogen economy for environmental sustainability. Volume 1: Fundamentals and feedstocks* (pp. 1–28). American Chemical Society.

Alavi-Borazjani, S. A., Adeel, S., & Chkoniya, V. (2025). Hydrogen as a sustainable fuel: Transforming maritime logistics. *Energies*, *18*(5), 1231.

AlHumaidan, F. S., Halabi, M. A., Rana, M. S., & Vinoba, M. (2023). Blue hydrogen: Current status and future technologies. *Energy Conversion and Management*, *283*, 116840.

Ali, S. A., Bangash, I. A., Sajjad, H., Karim, M. A., Ahmad, F., Ahmad, M., … & Qudoos, A. (2025). Review on the role of electrofuels in decarbonizing hard-to-abate transportation sectors: Advances, challenges, and future directions. *Energy & Fuels*.

Amin, M., Shah, H. H., Fareed, A. G., Khan, W. U., Chung, E., Zia, A., … & Lee, C. (2022). Hydrogen production through renewable and non-renewable energy processes and their impact on climate change. *International Journal of Hydrogen Energy*, *47*(77), 33112–33134.

Ansari, D., & Pepe, J. M. (2023). Toward a hydrogen import strategy for Germany and the EU. *Priorities, countries, and multilateral frameworks. SWP (Research Division Global Issues, 1).[online]*https://www.swp-berlin.org/publications/products/arbeitspapiere/Ansari_Pepe_2023_Hydrogen_Import_Strategy_WP.pdf, accessed, 4, 2024.

Archibald, A. T., Neu, J. L., Elshorbany, Y. F., Cooper, O. R., Young, P. J., Akiyoshi, H., … & Zeng, G. (2020). tropospheric ozone assessment report a critical review of changes in the tropospheric ozone burden and budget from 1850 to 2100. *Elementa: Science of the Anthropocene*, *8*(1).

Aydın, C. İ. (2020). Nuclear energy debate in Turkey: Stakeholders, policy alternatives, and governance issues. *Energy Policy*, *136*, 111041.

Bhatia, S. K., Mehariya, S., Bhatia, R. K., Kumar, M., Pugazhendhi, A., Awasthi, M. K., … & Yang, Y. H. (2021). Wastewater based microalgal biorefinery for bioenergy production: Progress and challenges. *Science of the Total Environment*, *751*, 141599.

Boretti, A., & Pollet, B. G. (2024). Hydrogen economy: Paving the path to a sustainable, low-carbon future. *International Journal of Hydrogen Energy*, *93*, 307–319.

Bouhala, L., Karatrantos, A., Reinhardt, H., Schramm, N., Akin, B., Rauscher, A., … & Tanoglu, M. (2024). Advancement in the Modeling and design of composite pressure vessels for hydrogen storage: A Comprehensive Review. *Journal of Composites Science*, *8*(9), 339.

Carlson, J. T., & Trencher, G. (2024). A framework for considering decarbonisation risks emerging from low-carbon hydrogen supply chains. *Energy Research & Social Science*, *116*, 103685.

Cavaliere, P. (2023). Safety issues and regulations. In *Water electrolysis for hydrogen production* (pp. 729–791). Cham: Springer International Publishing.

Cheekatamarla, P. (2024). Hydrogen and the global energy transition—path to sustainability and adoption across all economic sectors. *Energies*, *17*(4), 807.

Cholomskė, V. (2023). Potential of an artificial intelligence application in buildings sector: A systematic literature review. In *CYSENI 2023 19th international conference of young scientists on energy and natural sciences issues* 23–26 May 2023, Kaunas, Lithuania, Lietuvos energetikos institutas.

Das, A., & Peu, S. D. (2022). A comprehensive review on recent advancements in thermochemical processes for clean hydrogen production to decarbonize the energy sector. *Sustainability*, *14*(18), 11206.

Das, D., & Veziroglu, T. N. (2008). Advances in biological hydrogen production processes. *International Journal of Hydrogen Energy*, *33*(21), 6046–6057.

Dash, S. K., Chakraborty, S., Roccotelli, M., & Sahu, U. K. (2022). Hydrogen fuel for future mobility: Challenges and future aspects. *Sustainability*, *14*(14), 8285.

de la Cruz-Soto, J., Azkona-Bedia, I., Cornejo-Jimenez, C., & Romero-Castanon, T. (2025). Assessment of levelized costs for green hydrogen production for the national refineries system in Mexico. *International Journal of Hydrogen Energy*, *108*, 121–132.

De Vrieze, J., Verbeeck, K., Pikaar, I., Boere, J., Van Wijk, A., Rabaey, K., & Verstraete, W. (2020). The hydrogen gas bio-based economy and the production of renewable building block chemicals, food and energy. *New Biotechnology*, *55*, 12–18.

Delaval, B., Rapson, T., Sharma, R., Hugh-Jones, W., McClure, E., Temminghoff, M., & Srinivasan, V. (2023). Hydrogen RD&D collaboration opportunities.

Di Vaio, A., & Ali, T. (2025). *Accounting for carbon neutrality: corporate accountability in the hydrogen economy*. Taylor & Francis.

Divyabharathi, R., Kalidasan, B., Sakthi Suriya Raj, J. S. & Chinnasamy, S. (2024). Recent advances in sustainable agro residue utilisation, barriers and remediation for environmental management: Present insights and future challenges. *Industrial Crops and Products*, *216*, 118790.
Drace, Z., Ojovan, M. I., & Samanta, S. K. (2022). Challenges in planning of integrated nuclear waste management. *Sustainability*, *14*(21), 14204.
Dubey, A., & Arora, A. (2022). Advancements in carbon capture technologies: A review. *Journal of Cleaner Production*, *373*, 133932.
Dubey, R., & Dubey, S. S. (2024). Air Pollution: Health Impact and Control. *Asian Journal of Research in Chemistry*, *17*(5), 271–277.
Dufour, P., Kilic, M., Fontaine, G., Bergeron, P., Lachapelle, F. R., Kleinman, S. J., & Leggett, S. K. (2010). The discovery of the most metal-rich white dwarf: Composition of a tidally disrupted extrasolar dwarf planet. *The Astrophysical Journal*, *719*(1), 803.
Dutta, S. (2014). A review on production, storage of hydrogen and its utilization as an energy resource. *Journal of Industrial and Engineering Chemistry*, *20*(4), 1148–1156.
Favour, A. A. (2025). The role of public-private partnerships in scaling graphene CO_2 capture.
Feng, Z., Eiubovi, I., Shao, Y., Fan, Z., & Tan, R. (2024). Review of digital twin technology applications in hydrogen energy. *Chain*, *1*(1), 54–74.
Filonchyk, M., Peterson, M. P., Zhang, L., Hurynovich, V., & He, Y. (2024). Greenhouse gases emissions and global climate change: Examining the influence of CO_2, CH_4, and N_2O. *Science of The Total Environment*, 173359.
Gale, F., Goodwin, D., Lovell, H., Murphy-Gregory, H., Beasy, K., & Schoen, M. (2024). Renewable hydrogen standards, certifications, and labels: A state-of-the-art review from a sustainability systems governance perspective. *International Journal of Hydrogen Energy*, *59*, 654–667.
Gallegos, F. (2024). Bridging the Gap: Certification Schemes for Sustainable Hydrogen in Global Trade. *Journal for European Environmental & Planning Law*, *21*(3–4), 239–273.
Genovese, M., Blekhman, D., & Fragiacomo, P. (2024). An exploration of safety measures in hydrogen refueling stations: Delving into hydrogen equipment and technical performance. *Hydrogen*, *5*(1), 102–122.
Ghaedi, M., Foukolaei, P. Z., Asari, F. A., Khazaei, M., Gholian-Jouybari, F., & Hajiaghaei-Keshteli, M. (2024). Pricing electricity from blue hydrogen to mitigate the energy rebound effect: A case study in agriculture and livestock. *International Journal of Hydrogen Energy*, *84*, 993–1003.
Gholami, R., Raza, A., & Iglauer, S. (2021). Leakage risk assessment of a CO2 storage site: A review. *Earth-Science Reviews*, *223*, 103849.
Goita, E. G., Beagle, E. A., Nasta, A. N., Wissmiller, D. L., Ravikumar, A., & Webber, M. E. (2025). Effect of hydrogen leakage on the life cycle climate impacts of hydrogen supply chains. *Communications Earth & Environment*, *6*(1), 160.
Griffiths, S., Sovacool, B. K., Kim, J., Bazilian, M., & Uratani, J. M. (2021). Industrial decarbonization via hydrogen: A critical and systematic review of developments, socio-technical systems and policy options. *Energy Research & Social Science*, *80*, 102208.
Guha, M. (2025). Role of the inflation reduction acts in financing hydrogen energy projects: Opportunities and challenges. *The Role of Green and Transition Finance in Achieving Carbon Neutrality and Sustainable Energy*, 207–236.
Gupta, P. K., & Yadav, B. (2020). Leakage of CO_2 from geological storage and its impacts on fresh soil–water systems: A review. *Environmental Science and Pollution Research*, *27*(12), 12995–13018.
Gurrì, S., Santacaterina, E., Guarrera, M., & Dalla Chiara, B. (2023). Driving modal shift on low-traffic railway lines through technological innovation: A case study in Piedmont (Italy) including hydrogen fuel-cells as an alternative. *Transportation Research Procedia*, *69*, 99–106.
Hammi, Z., Labjar, N., Dalimi, M., El Hamdouni, Y., Lotfi, E. M., & El Hajjaji, S. (2024). Green hydrogen: A holistic review covering life cycle assessment, environmental impacts, and color analysis. *International Journal of Hydrogen Energy*, *80*, 1030–1045.
Hannun, R. M., & Razzaq, A. H. A. (2022, March). Air pollution resulted from coal, oil and gas firing in thermal power plants and treatment: A review. In *IOP conference series: Earth and environmental science* (Vol. 1002, No. 1, p. 012008). IOP Publishing.
Harichandan, S., & Kar, S. K. (2023). Financing the hydrogen industry: Exploring demand and supply chain dynamics. *Environmental Science and Pollution Research*, *32*, 13281–13298.
Hashemi, M., Sedaee, B., & Fathi, Y. (2025). Assessing hydrogen leakage in underground hydrogen storage: insights from parametric analysis. *Energy & Fuels*. *39*, 3668–3682.
Hassan, N. S., Jalil, A. A., Rajendran, S., Khusnun, N. F., Bahari, M. B., Johari, A., … & Ismail, M. (2024). Recent review and evaluation of green hydrogen production via water electrolysis for a sustainable and clean energy society. *International Journal of Hydrogen Energy*, *52*, 420–441.

Hauch, A., Küngas, R., Blennow, P., Hansen, A. B., Hansen, J. B., Mathiesen, B. V., & Mogensen, M. B. (2020). Recent advances in solid oxide cell technology for electrolysis. *Science*, *370*(6513), eaba6118.

He, W., Zhang, J., Li, H., Guo, R., Liu, S., Wu, X., … & Wang, Y. (2023). Effects of different water-cooled heat sinks on the cooling system performance in a data center. *Energy and Buildings*, *292*, 113162.

Hercog, J., Kupecki, J., Świątkowski, B., Kowalik, P., Boettcher, A., Malesa, J., … & Tchorek, G. (2024). Advancing production of hydrogen using nuclear cycles-integration of high temperature gas-cooled reactors with thermochemical water splitting cycles. *International Journal of Hydrogen Energy*, *52*, 1070–1083.

Hitam, C. N. C., & Jalil, A. A. (2023). A review on biohydrogen production through photo-fermentation of lignocellulosic biomass. *Biomass Conversion and Biorefinery*, *13*(10), 8465–8483.

Hosseini, S. E., & Wahid, M. A. (2016). Hydrogen production from renewable and sustainable energy resources: Promising green energy carrier for clean development. *Renewable and Sustainable Energy Reviews*, *57*, 850–866.

Hosseini, S. E., & Wahid, M. A. (2020). Hydrogen from solar energy, a clean energy carrier from a sustainable source of energy. *International Journal of Energy Research*, *44*(6), 4110–4131.

Howarth, R. W., & Jacobson, M. Z. (2021). How green is blue hydrogen?. *Energy Science & Engineering*, *9*(10), 1676–1687.

Ibrahim, Y., & Al-Mohannadi, D. M. (2023). Optimization of low-carbon hydrogen supply chain networks in industrial clusters. *International Journal of Hydrogen Energy*, *48*(36), 13325–13342.

International Energy Agency. (2022, January 1). *Electricity – Fuels & technologies*. IEA. https://www.iea.org/fuels-and-technologies/electricity

IRENA (2020). Reaching zero with renewables: Eliminating CO_2 emissions from industry and transport in line with the 1.5°C climate goal, International Renewable Energy Agency, Abu Dhabi, www.irena.org/publications/2020/Sep/Reaching-Zero-withRenewables

Ishaq, H., & Crawford, C. (2025). Towards less carbon-intensive blue hydrogen: Integrated natural gas reforming and CO2 capture approach. *Journal of Environmental Chemical Engineering*, *13*(2), 115043.

Islam, A., Islam, T., Mahmud, H., Raihan, O., Islam, M. S., Marwani, H. M., … & Awual, M. R. (2024). Accelerating the green hydrogen revolution: A comprehensive analysis of technological advancements and policy interventions. *International Journal of Hydrogen Energy*, *67*, 458–486.

Jahami, M., Singh, P., & Khandelwal, B. (2025). Life cycle assessment of SMR and electrified-SMR with renewable energy systems: Projecting emissions and optimizing hydrogen production for california's 2035 goals. *Renewable Energy*, *243*, 122501.

Jeje, S. O., Marazani, T., Obiko, J. O., & Shongwe, M. B. (2024). Advancing the hydrogen production economy: A comprehensive review of technologies, sustainability, and future prospects. *International Journal of Hydrogen Energy*, *78*, 642–661.

Jin, H., Xu, J., Liu, H., Shen, H., Yu, H., Jaroniec, M., … & Qiao, S. Z. (2023). Emerging materials and technologies for electrocatalytic seawater splitting. *Science Advances*, *9*(42), eadi7755.

Kamran, M., & Turzyński, M. (2024). Exploring hydrogen energy systems: A comprehensive review of technologies, applications, prevailing trends, and associated challenges. *Journal of Energy Storage*, *96*, 112601.

Kim, C., Cho, S. H., Cho, S. M., Na, Y., Kim, S., & Kim, D. K. (2023). Review of hydrogen infrastructure: The current status and roll-out strategy. *International Journal of Hydrogen Energy*, *48*(5), 1701–1716.

Kolawole, A. S., & Iyiola, A. O. (2023). Environmental pollution: Threats, impact on biodiversity, and protection strategies. In *Sustainable utilization and conservation of Africa's biological resources and environment* (pp. 377–409). Singapore: Springer Nature Singapore.

Kountouris, I., Bramstoft, R., Madsen, T., Gea-Bermúdez, J., Münster, M., & Keles, D. (2024). A unified European hydrogen infrastructure planning to support the rapid scale-up of hydrogen production. *Nature Communications*, *15*(1), 5517.

Kukula, I., Ermanoski, I., & Stechel, E. B. (2024). Clean hydrogen potential for carbon-neutral copper mining. *Energy & Environmental Science*, *17*(23), 9164–9184.

Kumar, D., Adsul, T., Ghosh, S., & Varma, A. K. (2024). Coal for hydrogen production and storage. *Solid-Gaseous Biofuels Production*, 487–524.

Kumar, S. S., Ni, A., Himabindu, V., & Lim, H. (2023). Experimental and simulation of PEM water electrolyser with Pd/PN-CNPs electrodes for hydrogen evolution reaction: Performance assessment and validation. *Applied Energy*, *348*, 121565.

Kumari, P. (2024). Towards net-zero: A framework for sustainable energy transition and global cooperation. *Legal Spectrum Journal*, *4*, 1.

Lakshmanan, S., & Bhati, M. (2024). Unravelling the atmospheric and climate implications of hydrogen leakage. *International Journal of Hydrogen Energy*, *53*, 807–815.

Law, L. C., Gkantonas, S., Mengoni, A., & Mastorakos, E. (2024). Onboard pre-combustion carbon capture with combined-cycle gas turbine power plant architectures for LNG-fuelled ship propulsion. *Applied Thermal Engineering*, *248*, 123294.

Ledari, M. B., Khajehpour, H., Akbarnavasi, H., & Edalati, S. (2023). Greening steel industry by hydrogen: Lessons learned for the developing world. *International Journal of Hydrogen Energy*, *48*(94), 36623–36649.

Li, X., Zhang, H., Yang, C., Liu, Y., Dong, C., Ye, X., & Jia, S. (2023). Effects of confined space flow fields on explosion and hazard analysis. *Combustion Science and Technology*, *195*(5), 923–957.

Liu, H., Ampah, J. D., Afrane, S., Adun, H., Jin, C., & Yao, M. (2023). Deployment of hydrogen in hard-to-abate transport sectors under limited carbon dioxide removal (CDR): Implications on global energy-land-water system. *Renewable and Sustainable Energy Reviews*, *184*, 113578.

Liu, H., Yu, M., Tong, X., Wang, Q., & Chen, M. (2024). High temperature solid oxide electrolysis for green hydrogen production. *Chemical Reviews*, *124*(18), 10509–10576.

Louizou, D. (2024). *Shipping in energy transition pathways to meet IMO 2050 greenhouse gas reduction targets* (Master's thesis, Πανεπιστήμιο Πειραιώς).

Magliano, A., Perez Carrera, C., Pappalardo, C. M., Guida, D., & Berardi, V. P. (2024). A comprehensive literature review on hydrogen tanks: Storage, safety, and structural integrity. *Applied Sciences*, *14*(20), 9348.

Mahto, R. K., Saxena, S., & Kaushal Mahan, V. P. P. P. (2023). Global Standards for Carbon Accounting: An Agenda for G20.

Mao, J., Li, Z., Xuan, J., Du, X., Ni, M., & Xing, L. (2024). A review of control strategies for proton exchange membrane (PEM) fuel cells and water electrolyser: From automation to autonomy. *Energy and AI*, *17*, 100406.

Massarweh, O., Al-Khuzaei, M., Al-Shafi, M., Bicer, Y., & Abushaikha, A. S. (2023). Blue hydrogen production from natural gas reservoirs: A review of application and feasibility. *Journal of CO2 Utilization*, *70*, 102438.

Mathur, J. (2025). A global assessment of green hydrogen value chain: Synthesizing policy learnings. In *FDI, MSMEs, digitalization, and green industrialization: challenges, opportunities and policy lessons for india* (pp. 371–393). Singapore: Springer Nature Singapore.

Matthews, H. D., & Wynes, S. (2022). Current global efforts are insufficient to limit warming to 1.5 C. *Science*, *376*(6600), 1404–1409.

Mehmood, H., Akbar, H., Nilsalab, P., & Gheewala, S. H. (2025). Exploring the spectrum: An environmental examination of hydrogen's diverse colors. *Energy Advances*, *4*, 224–238.

Mirsalami, S. M., & Mirsalami, M. (2024). Comparative study of random and block SPEEK copolymers for high-temperature proton exchange membrane electrolysis. *Colloids and Surfaces A: Physicochemical and Engineering Aspects*, *702*, 134982.

Obiora, N. K., Ujah, C. O., Asadu, C. O., Kolawole, F. O., & Ekwueme, B. N. (2024). Production of hydrogen energy from biomass: Prospects and challenges. *Green Technologies and Sustainability*, *2*, 100100.

Osman, A. I., Mehta, N., Elgarahy, A. M., Hefny, M., Al-Hinai, A., Al-Muhtaseb, A. A. H., & Rooney, D. W. (2022). Hydrogen production, storage, utilisation and environmental impacts: A review. *Environmental Chemistry Letters*, *20*, 1–36.

Osman, A. I., Nasr, M., Mohamed, A. R., Abdelhaleem, A., Ayati, A., Farghali, M., ... & Rooney, D. W. (2024). Life cycle assessment of hydrogen production, storage, and utilization toward sustainability. *Wiley Interdisciplinary Reviews: Energy and Environment*, *13*(3), e526.

Pardhi, S., Chakraborty, S., Tran, D. D., El Baghdadi, M., Wilkins, S., & Hegazy, O. (2022). A review of fuel cell powertrains for long-haul heavy-duty vehicles: Technology, hydrogen, energy and thermal management solutions. *Energies*, *15*(24), 9557.

Pathak, P. K., Yadav, A. K., & Padmanaban, S. (2023). Transition toward emission-free energy systems by 2050: Potential role of hydrogen. *International Journal of Hydrogen Energy*, *48*(26), 9921–9927.

Pinto, M. C. (2024). Opportunities and challenges of green hydrogen in the energy transition framework: Analysis of potential cross-border cooperations through a multi-dimensional approach.

Proud, D., Gee, A., Evans, M., Smith, N., & Medwell, P. (2023). Flame visibility in hydrogen appliances.

Raihan, A., Begum, R. A., Nizam, M., Said, M., & Pereira, J. J. (2022). Dynamic impacts of energy use, agricultural land expansion, and deforestation on CO2 emissions in Malaysia. *Environmental and Ecological Statistics*, *29*(3), 477–507.

Rasul, M. G., Hazrat, M. A., Sattar, M. A., Jahirul, M. I., & Shearer, M. J. (2022). The future of hydrogen: Challenges on production, storage and applications. *Energy Conversion and Management*, *272*, 116326.

Rattle, I., Gailani, A., & Taylor, P. G. (2024). Decarbonisation strategies in industry: Going beyond clusters. *Sustainability Science*, *19*(1), 105–123.

Rödl, A. (2025). Freshwater demand and availability for green hydrogen production. In *Powerfuels* (pp. 155–177). Springer, Cham.

Sabahi, N., & Razfar, M. R. (2018). Investigating the effect of mixed alkaline electrolyte (NaOH+ KOH) on the improvement of machining efficiency in 2D electrochemical discharge machining (ECDM). *The International Journal of Advanced Manufacturing Technology*, *95*, 643–657.

Sadeq, A. M., Homod, R. Z., Hussein, A. K., Togun, H., Mahmoodi, A., Isleem, H. F., … & Moghaddam, A. H. (2024). Hydrogen energy systems: Technologies, trends, and future prospects. *Science of The Total Environment*, *939*, 173622.

Saeidi, S., Sápi, A., Khoja, A. H., Najari, S., Ayesha, M., Kónya, Z., … & Rodrigues, A. E. (2023). Evolution paths from grey to turquoise hydrogen via catalytic steam methane reforming: Current challenges and future developments. *Renewable and Sustainable Energy Reviews*, *183*, 113392.

Saha, P., Akash, F. A., Shovon, S. M., Monir, M. U., Ahmed, M. T., Khan, M. F. H., … & Akter, R. (2024). Grey, blue, and green hydrogen: A comprehensive review of production methods and prospects for zero-emission energy. *International Journal of Green Energy*, *21*(6), 1383–1397.

Sahraie, E., Kamwa, I., Moeini, A., & Mohseni-Bonab, S. M. (2024). Component and system levels limitations in power-hydrogen systems: Analytical review. *Energy Strategy Reviews*, *54*, 101476.

Saleem, M. (2022). Possibility of utilizing agriculture biomass as a renewable and sustainable future energy source. *Heliyon*, *8*(2).

Sayed, E. T., Wilberforce, T., Elsaid, K., Rabaia, M. K. H., Abdelkareem, M. A., Chae, K. J., & Olabi, A. G. (2021). A critical review on environmental impacts of renewable energy systems and mitigation strategies: Wind, hydro, biomass, and geothermal. *Science of the Total Environment*, *766*, 144505.

Scholvin, S., Black, A., & Robbins, G. (2024). Green Hydrogen as a driver of development? De-risking and production linkages with new value chains in South Africa and Chile.

Sebbahi, S., Assila, A., Belghiti, A. A., Laasri, S., Kaya, S., Hlil, E. K., … & Hajjaji, A. (2024). A comprehensive review of recent advances in alkaline water electrolysis for hydrogen production. *International Journal of Hydrogen Energy*, *82*, 583–599.

Sebbahi, S., Nabil, N., Alaoui-Belghiti, A., Laasri, S., Rachidi, S., & Hajjaji, A. (2022). Assessment of the three most developed water electrolysis technologies: Alkaline water electrolysis, proton exchange membrane and solid-oxide electrolysis. *Materials Today: Proceedings*, *66*, 140–145.

Shahulhameed, B., Gopinath, A. S., Jishor, E. K., Thangavelu, R., Koyyala, K. K., & Dilba, R. K. (2025). Feasibility Analysis of Green Hydrogen Generation from Offshore Energy Resources. *Journal of Coastal Research*, *113*(SI), 971–975.

Sharma, S., Basu, S., Shetti, N. P., & Aminabhavi, T. M. (2020). Waste-to-energy nexus for circular economy and environmental protection: Recent trends in hydrogen energy. *Science of the Total Environment*, *713*, 136633.

Shehata, M. F., & El-Shamy, A. M. (2023). Hydrogen-based failure in oil and gas pipelines a review. *Gas Science and Engineering*, *115*, 204994.

Singla, M. K., Nijhawan, P., & Oberoi, A. S. (2021). Hydrogen fuel and fuel cell technology for cleaner future: A review. *Environmental Science and Pollution Research*, *28*(13), 15607–15626.

Sobola, D., & Dallaev, R. (2024). Exploring hydrogen embrittlement: Mechanisms, consequences, and advances in metal science. *Energies*, *17*(12), 2972.

Soleimani, A., Hosseini Dolatabadi, S. H., Heidari, M., Pinnarelli, A., Mehdizadeh Khorrami, B., Luo, Y., … & Brusco, G. (2024). Progress in hydrogen fuel cell vehicles and up-and-coming technologies for eco-friendly transportation: An international assessment. *Multiscale and Multidisciplinary Modeling, Experiments and Design*, *7*(4), 3153–3172.

Stewart, W. R., Velez-Lopez, E., Wiser, R., & Shirvan, K. (2021). Economic solution for low carbon process heat: A horizontal, compact high temperature gas reactor. *Applied Energy*, *304*, 117650.

Swadi, Mahmood, et al. "Investigating and predicting the role of photovoltaic, wind, and hydrogen energies in sustainable global energy evolution." *Global Energy Interconnection* 7.4 (2024): 429–445.

Tetteh, D. A., & Salehi, S. (2023). The blue hydrogen economy: A promising option for the near-to-mid-term energy transition. *Journal of Energy Resources Technology*, *145*(4), 042701.

Togun, H., Aljibori, H. S. S., Abed, A. M., Biswas, N., Alshamkhani, M. T., Niyas, H., … & Paul, D. (2024). A review on recent advances on improving fuel economy and performance of a fuel cell hybrid electric vehicle. *International Journal of Hydrogen Energy*, *89*, 22–47.

Topolski, K., Reznicek, E. P., Erdener, B. C., San Marchi, C. W., Ronevich, J. A., Fring, L., … & Chung, M. (2022). Hydrogen blending into natural gas pipeline infrastructure: Review of the state of technology.

Uddin, N., Langley, J., Zhang, C., Fung, A. K., Lu, H., Yin, X., … & Yin, Z. (2021). Zero-emission multivalorization of light alcohols with self-separable pure H_2 fuel. *Applied Catalysis B: Environmental*, *292*, 120212.

Ueckerdt, F., Verpoort, P. C., Anantharaman, R., Bauer, C., Beck, F., Longden, T., & Roussanaly, S. (2024). On the cost competitiveness of blue and green hydrogen. *Joule*, *8*(1), 104–128.

Voigt, A., Albern, N., Ceppi, P., Grise, K., Li, Y., & Medeiros, B. (2021). Clouds, radiation, and atmospheric circulation in the present-day climate and under climate change. *Wiley Interdisciplinary Reviews: Climate Change*, *12*(2), e694.

Von Schneidemesser, E., Monks, P. S., Allan, J. D., Bruhwiler, L., Forster, P., Fowler, D., ... & Sutton, M. A. (2015). Chemistry and the linkages between air quality and climate change. *Chemical reviews*, *115*(10), 3856–3897.

Wang, L., He, L., & He, Y. (2024). Review on absorption refrigeration technology and its potential in energy-saving and carbon emission reduction in natural gas and hydrogen liquefaction. *Energies*, *17*(14), 3427.

Wang, Y., Ming, T., Li, W., Yuan, Q., De Richter, R., Davies, P., & Caillol, S. (2022). Atmospheric removal of methane by enhancing the natural hydroxyl radical sink. *Greenhouse Gases: Science and Technology*, *12*(6), 784–795.

Waseem, M., Amir, M., Lakshmi, G. S., Harivardhagini, S., & Ahmad, M. (2023). Fuel cell-based hybrid electric vehicles: An integrated review of current status, key challenges, recommended policies, and future prospects. *Green Energy and Intelligent Transportation*, *2*(6), 100121.

Wei, H., Hasanuzzaman, M., & Selvaraj, J. (2024). Energy/freshwater sustainability in urban areas through a novel solar-driven system with H_2 production/liquefaction: Techno-economic evaluation and multi-objective optimization. *Process Safety and Environmental Protection*, *188*, 317–335.

Whitehead, J., Newman, P., Whitehead, J., & Lim, K. L. (2023). Striking the right balance: Understanding the strategic applications of hydrogen in transitioning to a net zero emissions economy. *Sustainable Earth Reviews*, *6*(1), 1.

Wilberforce, T., Baroutaji, A., Soudan, B., Al-Alami, A. H., & Olabi, A. G. (2019). Outlook of carbon capture technology and challenges. *Science of the total environment*, *657*, 56–72.

Xie, Z., Jin, Q., Su, G., & Lu, W. (2024). A review of hydrogen storage and transportation: Progress and challenges. *Energies*, *17*(16), 4070.

Xu, S., Zhang, H., & Wang, Z. (2023). Thermal management and energy consumption in air, liquid, and free cooling systems for data centers: A review. *Energies*, *16*(3), 1279.

Xue, X., Li, G., Wang, Y., Han, W., Liu, C., & Jiao, F. (2025). Proposal and evaluation of a near-zero carbon emissions hydrogen production system coupled with photovoltaic, photothermal and coal gasification. *Applied Energy*, *377*, 124400.

Yang, F., Wang, T., Deng, X., Dang, J., Huang, Z., Hu, S., ... & Ouyang, M. (2021). Review on hydrogen safety issues: Incident statistics, hydrogen diffusion, and detonation process. *International journal of hydrogen energy*, *46*(61), 31467–31488.

Yang, W., Wang, B., & Gupta, R. (2025). Coal and climate change. In *Energy and climate change* (pp. 153–194). Academic Press.

Younas, M., Shafique, S., Hafeez, A., Javed, F., & Rehman, F. (2022). An overview of hydrogen production: Current status, potential, and challenges. *Fuel*, *316*, 123317.

Zang, G., Sun, P., Elgowainy, A., Bobba, P., McMillan, C., Ma, O., ... & Koleva, M. (2023). Cost and life cycle analysis for deep CO_2 emissions reduction for steel making: Direct reduced iron technologies. *Steel Research International*, *94*(6), 2200297.

Zeng, J., Zhang, J., Ling, B., He, Y., Weng, W., & Wang, Z. (2024). Hydrogen production by sulfur-iodine thermochemical cycle—Current status and recent advances. *International Journal of Hydrogen Energy*, *86*, 677–702.

Zhang, L., Jia, C., Bai, F., Wang, W., An, S., Zhao, K., ... & Sun, H. (2024). A comprehensive review of the promising clean energy carrier: Hydrogen production, transportation, storage, and utilization (HPTSU) technologies. *Fuel*, *355*, 129455.

Zhang, S., Luo, S., & Afshan, S. (2022). Role of climate technologies, financial development, and renewable energy in the facilitation of social, economic, and environmental goals. *Renewable Energy*, *199*, 169–178.

Zhuang, D., Abbas, J., Al-Sulaiti, K., Fahlevi, M., Aljuaid, M., & Saniuk, S. (2022). Land-use and food security in energy transition: Role of food supply. *Frontiers in Sustainable Food Systems*, *6*, 1053031.

Zoback, M., & Smit, D. (2023). Meeting the challenges of large-scale carbon storage and hydrogen production. *Proceedings of the National Academy of Sciences*, *120*(11), e2202397120.

5 Integration of Fuel Cells with Renewable Energy Sources

Olusola D. Ogundele, Oluwatosin D. Owolala, and Johnson Jayeola

5.1 INTRODUCTION

The 21st century has witnessed an unprecedented surge in global energy consumption, driven by rapid industrialization, urbanization, and population growth. According to the International Energy Agency (IEA), global energy demand rose by 2.2% in 2024, significantly higher than the average annual increase of 1.3% between 2013 and 2023, with emerging and developing economies accounting for over 80% of this growth (Asif & Muneer, 2007). This escalating energy demand has significant implications for climate change. The combustion of fossil fuels for energy production is the primary source of greenhouse gas emissions, contributing to global warming and associated environmental challenges. The Intergovernmental Panel on Climate Change (IPCC) emphasizes the urgent need to limit global temperature rise to 1.5°C above pre-industrial levels to mitigate the adverse effects of climate change (Masson-Delmotte *et al.*, 2018).

In response, there is a global push toward transitioning to low-carbon and renewable energy sources. Renewable energy technologies, such as solar and wind, have gained prominence due to their potential to reduce carbon emissions and enhance energy security. However, the integration of these intermittent energy sources into existing energy systems presents several challenges that need to be addressed to achieve a sustainable energy future. While renewable energy sources offer numerous environmental benefits, their inherent intermittency and variability pose significant challenges to energy reliability and grid stability (Widén *et al.*, 2015; Appleby, 1988). Solar energy generation is contingent on sunlight availability, leading to fluctuations in power output due to diurnal and seasonal variations, as well as weather conditions. Similarly, wind energy is subject to variability in wind speeds, resulting in inconsistent power generation (Graabak & Korpås, 2016; Alamri & Alamri, 2009).

These fluctuations can lead to mismatches between energy supply and demand, necessitating the need for energy storage solutions or backup power systems to ensure a stable and reliable energy supply. Moreover, the integration of high levels of intermittent renewable energy into the grid can cause issues related to voltage regulation, frequency stability, and power quality (Weschenfelder *et al.*, 2020; Jerez *et al.*, 2013). To mitigate these challenges, energy storage technologies such as batteries and pumped hydro storage have been employed. However, these solutions have limitations in terms of scalability, cost, and environmental impact. Therefore, there is a growing interest in exploring alternative solutions that can complement renewable energy sources and enhance the resilience and reliability of energy systems.

Fuel cells have emerged as a promising technology to address the limitations of standalone renewable energy systems. A fuel cell is an electrochemical device that converts the chemical energy of a fuel, typically hydrogen, directly into electricity, with water and heat as the only byproducts (Sharaf & Orhan, 2014). This clean energy conversion process offers several advantages, including high efficiency, scalability, and environmental friendliness. Integrating fuel cells with renewable energy sources can provide a reliable and continuous power supply by compensating for the

DOI: 10.1201/9781003645405-5

intermittency of renewables. For instance, excess electricity generated from solar or wind can be used to produce hydrogen through electrolysis, which can then be stored and utilized in fuel cells during periods of low renewable generation (Abdelkareem *et al.*, 2021). This approach not only enhances energy storage capabilities but also facilitates the decarbonization of various sectors, including transportation, industry, and residential energy use.

Fuel cells can be deployed in microgrids and distributed energy systems, providing localized and resilient power solutions. Their modular nature allows for scalability to meet varying energy demands, making them suitable for a wide range of applications. Additionally, fuel cells operate quietly and have a small physical footprint, which is advantageous in urban and space-constrained environments (Sharaf & Orhan, 2014).

5.2 FUNDAMENTALS OF FUEL CELL TECHNOLOGIES

Fuel cells are electrochemical devices that convert the chemical energy of a fuel directly into electrical energy through redox reactions, with water and heat as byproducts. Unlike conventional combustion-based power generation, fuel cells offer higher efficiencies and lower emissions, making them a promising technology for sustainable energy solutions (Qasem & Abdulrahman, 2024). The concept of fuel cells was first demonstrated by Sir William Grove in 1839 using a hydrogen-oxygen cell (Brusso, 2021). However, practical applications only emerged in the 20th century with NASA's use of alkaline fuel cells (AFCs) in space missions (Carrette *et al.*, 2000). Today, fuel cells power vehicles, portable electronics, and stationary power plants (Mench, 2008).

At the core of a fuel cell's operation is the electrochemical reaction between a fuel, typically hydrogen, and an oxidant, usually oxygen. This reaction occurs across two electrodes separated by an electrolyte. At the anode, hydrogen molecules are oxidized, releasing electrons and protons. The electrons travel through an external circuit, providing electrical power, while the protons migrate through the electrolyte to the cathode, where they combine with oxygen and electrons to form water (Srinivasan, 2006).

The overall reaction in a hydrogen fuel cell can be represented as:

Anode Reaction: $H_2 \rightarrow 2H^+ + 2e^-$
Cathode Reaction: $\frac{1}{2}O_2 + 2H^+ + 2e^- \rightarrow H_2O$
Overall Reaction: $H_2 + \frac{1}{2}O_2 \rightarrow H_2O + \text{Electrical Energy} + \text{Heat}$

This process is highly efficient and environmentally friendly, as it produces only water and heat as byproducts. Figure 5.1 shows the types of fuel cells.

5.2.1 PROTON EXCHANGE MEMBRANE FUEL CELLS (PEMFC)

Proton Exchange Membrane Fuel Cells (PEMFCs) are a pivotal technology in the realm of clean energy conversion, offering efficient, low-emission power generation from hydrogen fuel. These fuel cells operate at relatively low temperatures (~60–80°C) and have quick start-up capabilities, making them highly attractive for transportation, portable power, and stationary applications (Barbir, 2012). PEMFCs utilize a solid polymer electrolyte membrane that conducts protons while acting as an electronic insulator and gas barrier, enabling electrochemical reactions that convert hydrogen and

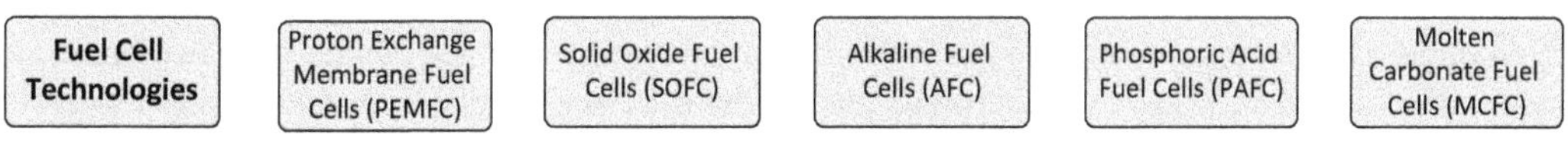

FIGURE 5.1 Types of fuel cells.

oxygen into electricity and water (Haider *et al.*, 2021). Due to their compact size, high power density, and environmental friendliness, PEMFCs have attracted extensive research and development attention over the past decades (Wang *et al.*, 2011).

The fundamental operation of PEMFCs involves the electrochemical conversion of hydrogen fuel into electricity through redox reactions at the anode and cathode. Hydrogen gas (H_2) is supplied to the anode, where a platinum catalyst facilitates its oxidation, splitting hydrogen molecules into protons (H^+) and electrons (e^-) (Barbir, 2012). The protons migrate through the proton exchange membrane to the cathode side, while electrons travel through an external circuit, generating electrical power (Haider *et al.*, 2021).

At the cathode, oxygen gas (O_2) from air reacts with the incoming protons and electrons to form water (H_2O), the only byproduct of the reaction (Wang *et al.*, 2011). This electrochemical process can be summarized by the overall reaction:

$$2H_2 + O_2 \rightarrow 2H_2O + \text{electricity} + \text{heat}$$

The proton exchange membrane, typically composed of a perfluorosulfonic acid polymer for selectively allowing proton transport while preventing electron and gas crossover (Kreuer, 2001). Maintaining membrane hydration is critical since proton conductivity drastically decreases when the membrane dries out (Zawodzinski *et al.*, 1993).

A typical PEMFC consists of several key components working synergistically to optimize performance:

5.2.1.1 Proton Exchange Membrane

The polymer electrolyte membrane is the heart of PEMFCs. Nafion is widely used because of its high proton conductivity, mechanical robustness, and chemical stability under fuel cell conditions (Kreuer, 2001). Its hydrophilic sulfonic acid groups enable proton hopping via vehicle and Grotthuss mechanisms (Zawodzinski *et al.*, 1993). However, Nafion membranes require hydration to maintain conductivity and are limited to temperatures below ~80°C due to dehydration issues (Mauritz & Moore, 2004).

5.2.1.2 Electrodes and Catalysts

The anode and cathode consist of porous carbon supports impregnated with platinum nanoparticles, which serve as catalysts for hydrogen oxidation and oxygen reduction reactions, respectively (Debe, 2012). The oxygen reduction reaction (ORR) at the cathode is kinetically slower and accounts for most efficiency losses (Zhang *et al.*, 2009). Hence, extensive research focuses on enhancing catalyst activity and reducing platinum loading to cut costs and improve durability (Debe, 2012).

5.2.1.3 Gas Diffusion Layers (GDL)

The GDL facilitates uniform distribution of reactant gases to the catalyst layers, aids water management by removing excess water, and provides electrical conductivity between electrodes and bipolar plates (Wang *et al.*, 2011). The GDL typically consists of carbon fiber paper or cloth with hydrophobic treatments to balance water retention and removal (Litster & McLean, 2004).

PEMFC technology offers several advantages over other fuel cell types and conventional power generation methods. Operating at low temperatures (~60–80°C) allows PEMFCs to start quickly and operate under mild conditions, suitable for mobile and portable applications (Barbir, 2006). Their high volumetric power density enables compact fuel cell stacks, advantageous in space-constrained uses like vehicles (Wang *et al.*, 2011). Additionally, PEMFCs generate only water and heat as byproducts, eliminating greenhouse gas emissions at the point of use and contributing to cleaner air quality (Haider *et al.*, 2021). Furthermore, the modular design of PEMFC stacks allows flexible scaling of power output, from a few watts for portable devices to hundreds of kilowatts for automotive and stationary power plants (Barbir, 2006). These properties, combined with quiet operation and high efficiency (40–60% electrical efficiency), position PEMFCs as a promising technology for sustainable energy systems (Zhang *et al.*, 2009).

Despite their advantages, PEMFCs face several technical and economic challenges limiting widespread commercialization. One significant issue is the high cost and limited durability of platinum catalysts. Platinum scarcity drives up system costs, while catalyst degradation from carbon corrosion, sintering, and poisoning reduces fuel cell lifespan (Debe, 2012). Membrane durability is another concern. Nafion membranes are susceptible to chemical attack by reactive oxygen species formed during fuel cell operation and mechanical degradation caused by swelling and drying cycles (Mauritz & Moore, 2004). Additionally, membrane dehydration leads to proton conductivity loss, while excess water flooding in the catalyst layer impedes gas diffusion, requiring precise water management strategies (Litster & McLean, 2004).

Operating temperature limitations (below 80°C) restrict the ability to tolerate fuel impurities such as carbon monoxide (CO), which poison the platinum catalyst and reduce performance (Barbir, 2006). High-temperature PEMFCs using alternative membranes are under development to address these issues but face material stability challenges (Haider *et al.*, 2021). Hydrogen storage and infrastructure remain bottlenecks for PEMFC adoption, particularly in transportation, where a safe, compact, and affordable hydrogen supply is needed (Wang *et al.*, 2011).

To overcome these challenges, significant research efforts are dedicated to improving catalyst performance, membrane materials, and system design.

Research into non-platinum group metal catalysts (NPMCs), platinum alloys, and core-shell structures aims to reduce precious metal loadings and improve durability without compromising activity (Debe, 2012). Advances in nanoparticle synthesis and catalyst support materials also contribute to enhanced stability and resistance to degradation. Development of alternative membranes with higher thermal stability and conductivity at low humidity, such as phosphoric acid-doped polybenzimidazole (PBI) membranes, enables high-temperature PEMFC operation (Haider *et al.*, 2021). Composite membranes incorporating inorganic fillers enhance mechanical strength and water retention, improving overall membrane longevity (Mauritz & Moore, 2004).

Improved gas diffusion layer designs and flow field configurations enhance water removal and membrane hydration, preventing flooding and dehydration simultaneously (Litster & McLean, 2004). Advanced modeling and diagnostics facilitate real-time monitoring and control of water distribution within fuel cells (Wang *et al.*, 2011). Integration of PEMFCs with renewable hydrogen production (via electrolysis powered by solar or wind) enables sustainable, zero-emission power generation (Barbir, 2012). Hybrid systems combining PEMFCs with batteries or supercapacitors improve transient response and overall system efficiency (Haider *et al.*, 2021).

PEMFCs have been implemented across diverse sectors. In the automotive industry, fuel cell vehicles (FCVs) leverage PEMFCs for zero-emission transportation with fast refueling and long driving ranges, demonstrated by commercial models from Toyota, Hyundai, and Honda (Wang *et al.*, 2011). Portable power applications include backup generators, military equipment, and remote power supplies, where PEMFCs provide quiet, reliable energy (Barbir, 2012). Stationary power generation uses PEMFCs for combined heat and power (CHP) systems in residential and commercial buildings, integrating with renewable energy sources to provide continuous, clean electricity (Haider *et al.*, 2021). These applications highlight PEMFCs' versatility and potential for broad impact in clean energy transitions.

5.2.2 Solid Oxide Fuel Cells (SOFC)

Solid Oxide Fuel Cells (SOFCs) represent a promising class of high-temperature fuel cells that convert chemical energy directly into electrical energy with high efficiency and fuel flexibility. Operating typically between 600°C and 1000°C, SOFCs utilize a solid ceramic electrolyte that conducts oxygen ions from the cathode to the anode, enabling electrochemical oxidation of fuels (Singhal & Kendall, 2003). SOFCs stand out among fuel cell technologies due to their high efficiency, fuel versatility, and potential for combined heat and power (CHP) applications, making them suitable for stationary power generation, auxiliary power units, and industrial use (Goodenough & Huang, 2007).

The operation of SOFCs is based on the electrochemical oxidation of fuel, typically hydrogen or hydrocarbon gases, facilitated by oxygen ion transport through a dense ceramic electrolyte. The SOFC comprises three main components: the anode (fuel electrode), the cathode (oxygen electrode), and the electrolyte (Singhal & Kendall, 2003). At the cathode side, molecular oxygen from air is reduced by gaining electrons from the external circuit to form oxygen ions (O^{2-}). These oxygen ions migrate through the solid oxide electrolyte to the anode, where they react with the fuel (commonly H_2 or carbon monoxide) to produce water, carbon dioxide (in case of hydrocarbons), and release electrons that flow back through the external circuit, generating an electric current (Goodenough & Huang, 2007). The overall reaction for hydrogen fuel is:

$$\text{Anode: } H_2 + O^{2-} \rightarrow H_2O + 2e^-$$
$$\text{Cathode: } 12O_2 + 2e^- \rightarrow O^{2-}$$
$$\text{Overall: } H_2 + 12O_2 \rightarrow H_2O + \text{electricity} + \text{heat}$$

Unlike low-temperature fuel cells such as PEMFCs, SOFCs conduct oxygen ions rather than protons through their electrolyte, which is typically a ceramic oxide material like yttria-stabilized zirconia (YSZ) (Singhal & Kendall, 2003). This ionic conduction at high temperature enables direct utilization of various fuels without extensive external reforming, including natural gas, biogas, and even liquid hydrocarbons (Steele & Heinzel, 2001).

A typical SOFC consists of several key components working synergistically to optimize performance.

5.2.2.1 Electrolyte

The electrolyte is a dense, gas-impermeable ceramic that selectively conducts oxygen ions while blocking electrons, ensuring electrochemical reactions occur only at the electrodes. Yttria-stabilized zirconia (YSZ) is the most commonly used electrolyte material due to its high oxygen ion conductivity and chemical stability at elevated temperatures (Steele & Heinzel, 2001). Other materials like gadolinium-doped ceria (GDC) and lanthanum gallate-based ceramics have been explored for lower operating temperatures and improved ionic conductivity (Sun *et al.*, 2010).

5.2.2.2 Anode

The anode is typically a porous composite of nickel and YSZ (Ni-YSZ cermet), which provides catalytic activity for fuel oxidation and electronic conductivity (Singhal & Kendall, 2003). The porosity allows fuel diffusion to the reactive sites and facilitates the removal of reaction products. Ni catalyzes hydrogen oxidation and is tolerant to carbon monoxide, allowing use of reformed hydrocarbon fuels (Goodenough & Huang, 2007).

5.2.2.3 Cathode

The cathode catalyzes the reduction of oxygen molecules to oxygen ions. Materials such as lanthanum strontium manganite (LSM) and lanthanum strontium cobalt ferrite (LSCF) are common cathodes, chosen for their mixed ionic-electronic conductivity and compatibility with the electrolyte (Singhal & Kendall, 2003). Cathode performance is critical for overall cell efficiency, and improving oxygen reduction reaction kinetics is an active research area (Sun *et al.*, 2010).

SOFCs offer several notable advantages compared to other fuel cell technologies. First, their high operating temperatures (600–1000°C) allow for internal reforming of hydrocarbon fuels, reducing the need for complex external fuel processing and increasing system efficiency (Steele & Heinzel, 2001). This capability enables SOFCs to use readily available fuels like natural gas, biogas, and even liquid hydrocarbons, increasing fuel flexibility (Goodenough & Huang, 2007). SOFCs achieve electrical efficiencies up to 60% when operated in standalone mode and can exceed 85% efficiency when used in combined heat and power (CHP) systems, capturing waste heat for heating or additional power generation (Singhal & Kendall, 2003). This high efficiency contributes to lower greenhouse gas emissions and reduced fuel consumption compared to conventional combustion-based power generation (Steele & Heinzel, 2001).

SOFCs produce negligible pollutant emissions, as their electrochemical conversion avoids combustion and associated nitrogen oxides (NOx) and particulate emissions, making them environmentally friendly (Goodenough & Huang, 2007). Additionally, their solid-state construction with no liquid electrolytes simplifies sealing and reduces corrosion issues typical of other fuel cell types (Singhal & Kendall, 2003).

SOFCs face several challenges hindering widespread commercialization. The high operating temperatures cause thermal stresses and material degradation, limiting the lifespan and reliability of fuel cell stacks (Steele & Heinzel, 2001). Long start-up and cool-down times due to thermal inertia restrict their use in applications requiring rapid power modulation or intermittent operation (Singhal & Kendall, 2003). Material compatibility at elevated temperatures also poses challenges. Differences in thermal expansion coefficients among cell components can induce mechanical failure or delamination (Sun *et al.*, 2010). Moreover, redox cycling during operation may degrade the Ni-YSZ anode microstructure, reducing catalytic activity and electrical conductivity over time (Goodenough & Huang, 2007). Cost is another barrier, as SOFC materials and manufacturing techniques remain expensive relative to other power generation technologies (Singhal & Kendall, 2003). Developing cost-effective fabrication methods and durable materials remains a focus of ongoing research (Sun *et al.*, 2010).

Advances in materials science and cell design are addressing many SOFC limitations. Efforts to lower the operating temperature to the intermediate range (500–700°C) using alternative electrolytes like ceria-based oxides reduce material degradation and allow the use of cheaper metallic interconnects (Sun *et al.*, 2010). This temperature reduction also decreases start-up times and improves system durability. Nanostructured electrodes with enhanced surface area and improved catalytic properties have demonstrated better performance and tolerance to carbon deposition during hydrocarbon fuel use (Goodenough & Huang, 2007). Composite cathodes with mixed ionic-electronic conductivity reduce polarization losses, enhancing oxygen reduction kinetics and overall efficiency (Singhal & Kendall, 2003).

Manufacturing innovations such as tape casting, screen printing, and infiltration techniques improve microstructure control and scalability of SOFC stacks, driving down costs (Sun *et al.*, 2010). System integration with renewable fuels, such as biomass-derived syngas or hydrogen from electrolysis, further enhances environmental benefits (Steele & Heinzel, 2001).

SOFCs are primarily suited for stationary power generation, especially in distributed generation and combined heat and power (CHP) systems. Their high efficiency and fuel flexibility make them ideal for industrial facilities, commercial buildings, and microgrids requiring reliable, clean power (Singhal & Kendall, 2003). SOFCs have been demonstrated in backup power units and remote locations where grid access is limited (Goodenough & Huang, 2007). Additionally, SOFCs are under development for auxiliary power units (APUs) in transportation and aerospace sectors, where efficient, quiet, and low-emission power sources are desirable (Steele & Heinzel, 2001). Hybrid systems coupling SOFCs with gas turbines or batteries provide high power densities and improved dynamic response (Sun *et al.*, 2010).

5.2.3 Alkaline Fuel Cells (AFC)

Alkaline Fuel Cells (AFCs) are one of the earliest developed and most mature types of fuel cell technologies. They operate on the principle of converting chemical energy from hydrogen and oxygen into electrical energy through electrochemical reactions in an alkaline electrolyte (Appleby & Foulkes, 1988). AFCs utilize an aqueous alkaline solution, commonly potassium hydroxide (KOH), as the electrolyte to facilitate ionic conduction, specifically the transport of hydroxide ions (OH^-)

from the cathode to the anode (Ferriday & Middleton, 2021). This unique operating chemistry offers AFCs high electrical efficiencies and rapid electrode kinetics compared to acidic electrolyte fuel cells. Despite some challenges related to electrolyte management and CO_2 sensitivity, AFCs continue to be attractive for applications requiring high efficiency and low temperature operation, such as space missions and portable power systems (Barbir, 2006).

Alkaline Fuel Cells operate on the principle of electrochemical oxidation of hydrogen at the anode and reduction of oxygen at the cathode in an alkaline medium (Barbir, 2006). The alkaline electrolyte facilitates the conduction of hydroxide ions (OH^-) from the cathode side to the anode side, which differentiates AFCs fundamentally from proton exchange membrane fuel cells (PEMFCs), where proton conduction occurs.

Anode reaction: $2H_2 + 4OH^- \rightarrow 4H_2O + 4e^-$
Cathode reaction: $O_2 + 2H_2O + 4e^- \rightarrow 4OH^-$
Overall reaction: $2H_2 + O_2 \rightarrow 2H_2O + \text{electricity} + \text{heat}$

The hydroxide ions generated at the cathode migrate through the alkaline electrolyte to the anode, completing the circuit and enabling continuous electricity generation (Appleby & Foulkes, 1988). The electrons flow externally from the anode to the cathode, powering an electrical load. The alkaline environment in AFCs allows the use of non-precious metal catalysts such as nickel and silver, lowering material costs compared to PEMFCs, which require expensive platinum catalysts (Ferriday & Middleton, 2021). Moreover, the alkaline medium enhances the oxygen reduction reaction (ORR) kinetics, improving overall cell efficiency (Barbir, 2006).

A typical AFC consists of several key components working synergistically to optimize performance.

5.2.3.1 Electrolyte

The electrolyte in AFCs is typically a concentrated aqueous solution of potassium hydroxide (KOH) with concentrations ranging from 20% to 40% by weight (Appleby & Foulkes, 1988). This alkaline solution provides high ionic conductivity for hydroxide ions, facilitating fast ion transport between electrodes. The electrolyte may be in a liquid form, soaked into a matrix or immobilized within a matrix as a solid alkaline membrane in advanced designs (Ferriday & Middleton, 2021). One challenge with liquid KOH electrolytes is their sensitivity to carbon dioxide (CO_2), which reacts to form potassium carbonate (K_2CO_3), reducing electrolyte conductivity and causing performance degradation (Appleby & Foulkes, 1988). Managing CO_2 contamination is a significant factor in AFC system design, requiring air purification or CO_2 scrubbing in practical applications.

5.2.3.2 Electrodes

The anode and cathode in AFCs are porous structures coated with catalysts to facilitate electrochemical reactions. Nickel-based materials are widely used as anode catalysts due to their high activity for hydrogen oxidation and robustness in alkaline environments (Barbir, 2006). For the cathode, silver is commonly employed because of its superior catalytic activity for oxygen reduction in alkaline media (Ferriday & Middleton, 2021). The porous electrode design ensures effective gas diffusion to catalytic sites and efficient removal of water produced during operation. Electrode durability and stability in the alkaline environment are critical for AFC longevity.

5.2.3.3 Cell Structure and Bipolar Plates

AFCs typically use a stacked cell design to achieve desired power outputs, with bipolar plates separating individual cells and providing electrical connections and reactant gas flow channels (Barbir, 2006). Bipolar plates must exhibit high electrical conductivity, chemical stability in alkaline media, and mechanical robustness. Materials such as graphite composites and coated metals are often used.

Alkaline Fuel Cells demonstrate high electrical efficiencies due to rapid electrode kinetics and low ohmic losses in the alkaline electrolyte. Efficiencies of 60–70% in converting chemical energy to electrical energy are achievable under ideal conditions, with potential for even higher efficiency in combined heat and power (CHP) applications (Appleby & Foulkes, 1988). The low operating temperature, typically between 60°C and 250°C, enables rapid start-up and load following, making AFCs suitable for portable and auxiliary power systems (Barbir, 2006). The open-circuit voltage (OCV) of AFCs can reach approximately 1.2 V, close to the theoretical maximum, reflecting the high thermodynamic efficiency of the system (Ferriday & Middleton, 2021). Cell voltage decreases under load due to activation, ohmic, and concentration losses, which are managed through electrode and cell design improvements.

AFCs operate efficiently with pure hydrogen fuel but can also tolerate hydrogen-containing fuels with some impurities (Barbir, 2006). However, they are sensitive to CO_2 in the oxidant, as the formation of potassium carbonate reduces electrolyte conductivity and damages electrodes, necessitating air purification in most applications (Appleby & Foulkes, 1988). AFCs have a rich history of application in aerospace and space exploration due to their high reliability, energy density, and efficiency. NASA extensively used AFCs in the Apollo missions to provide electrical power and water generation aboard spacecraft (Appleby & Foulkes, 1988). Their ability to operate at low temperatures and generate pure water makes them ideal for closed-environment life support systems.

In terrestrial applications, AFCs are attractive for stationary power generation, portable power units, and backup power systems. Their rapid start-up and load response make them suitable for intermittent power demands, and non-precious metal catalysts reduce cost relative to PEMFCs (Barbir, 2006). Emerging applications include unmanned aerial vehicles (UAVs), remote sensing stations, and microgrid power supplies.

AFCs face key challenges limiting broader commercialization. The primary limitation is their high sensitivity to carbon dioxide contamination. CO_2 reacts with the alkaline electrolyte to form potassium carbonate, which precipitates and reduces ionic conductivity, lowering cell performance and durability (Appleby & Foulkes, 1988). This requires either pure oxygen or CO_2-free air supply, complicating system design and increasing costs (Ferriday & Middleton, 2021). Electrolyte management presents another challenge, particularly with liquid alkaline electrolytes prone to leakage and drying out, necessitating careful system engineering to maintain electrolyte volume and concentration (Barbir, 2006). Solid alkaline membrane electrolytes are under development to mitigate these issues but face durability challenges.

Material degradation due to prolonged exposure to alkaline media, water management within the porous electrodes, and gas crossover through the electrolyte are additional operational concerns (Ferriday & Middleton, 2021). Overcoming these issues is vital to improving AFC lifetime and reliability. Research in AFC technology has focused on addressing CO_2 sensitivity, improving electrolyte stability, and developing solid-state alkaline membranes. Advances in membrane electrode assemblies (MEAs) with anion exchange membranes (AEMs) aim to replace liquid KOH electrolytes with solid, polymer-based alkaline conductors to reduce electrolyte management issues and CO_2 sensitivity (Ferriday & Middleton, 2021).

Novel cathode catalysts and electrode designs are being investigated to enhance oxygen reduction kinetics and durability under operating conditions (Barbir, 2006). Additionally, hybrid systems coupling AFCs with renewable hydrogen production methods and air purification technologies show promise for sustainable energy applications. Cost reduction through the use of non-precious metal catalysts and scalable manufacturing methods remains a priority to make AFCs competitive with other fuel cell technologies (Appleby & Foulkes, 1988).

5.2.4 Phosphoric Acid Fuel Cells (PAFC)

Phosphoric Acid Fuel Cells (PAFCs) represent one of the earliest commercially viable fuel cell technologies and have played a significant role in demonstrating the practical use of fuel cells for stationary power generation (Kordesch & Simader, 1996). PAFCs use liquid phosphoric acid (H_3PO_4) as the electrolyte and operate at intermediate temperatures, typically around 150–220°C. This moderate operating temperature allows PAFCs to tolerate fuel impurities better than low-temperature proton exchange membrane fuel cells (PEMFCs) and reduces catalyst poisoning, thus extending their operational lifespan (Appleby & Foulkes, 1988). PAFCs are characterized by their robust performance in combined heat and power (CHP) applications and distributed generation systems.

PAFCs operate on the electrochemical conversion of hydrogen and oxygen into electricity and water through reactions facilitated by phosphoric acid as the electrolyte (Kordesch & Simader, 1996). The electrolyte conducts protons (II+) from the anode to the cathode while preventing electron flow internally, forcing electrons to flow through an external circuit, thereby producing an electric current.

Anode reaction: $H_2 \rightarrow 2H^+ + 2e^-$
Cathode reaction: $12O_2 + 2H^+ + 2e^- \rightarrow H_2O$
Overall reaction: $H_2 + 12O_2 \rightarrow H_2O + \text{electricity} + \text{heat}$

Phosphoric acid serves as the proton conductor, enabling ionic transfer while acting as a barrier for electrons. Its liquid form, absorbed into a silicon carbide or Teflon-based porous matrix, creates a stable electrolyte structure that facilitates proton conduction at intermediate temperatures (Carrette *et al.*, 2001). The intermediate temperature range (150–220°C) enables PAFCs to tolerate carbon monoxide (CO) concentrations up to about 1.5%, unlike PEMFCs, which require much purer hydrogen (Appleby & Foulkes, 1988). This tolerance reduces the need for expensive hydrogen purification, improving cost-efficiency for stationary applications (Carrette *et al.*, 2001).

The key components of PAFCs are as follows.

5.2.4.1 Electrolyte

The electrolyte is concentrated liquid phosphoric acid (H_3PO_4), typically maintained at about 100% concentration by weight within a matrix to avoid acid leakage and evaporation (Kordesch & Simader, 1996). Phosphoric acid offers good proton conductivity at elevated temperatures and stability in fuel cell operating conditions. The acid is absorbed in a silicon carbide or Teflon-based porous ceramic matrix that provides mechanical support and prevents acid migration within the cell stack. Maintaining electrolyte stability and preventing acid loss through evaporation or leakage is critical to PAFC durability (Carrette *et al.*, 2001).

5.2.4.2 Electrodes and Catalysts

PAFCs use porous carbon electrodes coated with platinum catalysts to facilitate the electrochemical reactions (Appleby & Foulkes, 1988). Platinum is preferred due to its high catalytic activity for both hydrogen oxidation and oxygen reduction reactions at the moderate operating temperatures of PAFCs. The anode catalyst promotes the splitting of hydrogen molecules into protons and electrons, while the cathode catalyst facilitates the reduction of oxygen molecules to water. The catalysts are deposited on carbon supports to maximize surface area and maintain electrode porosity for gas diffusion (Carrette *et al.*, 2001).

A PAFC cell consists of the porous carbon electrodes on either side of the electrolyte matrix, separated and compressed within a bipolar plate assembly. Bipolar plates provide structural support, electrical conduction between cells, and gas flow channels for reactant distribution and product removal (Kordesch & Simader, 1996). PAFC stacks are designed to operate efficiently at intermediate temperatures with materials resistant to phosphoric acid corrosion and thermal cycling. The stacks include cooling systems to manage heat generated during operation and to maintain optimal cell temperature.

PAFCs typically operate at temperatures between 150°C and 220°C, which offers a good balance between catalyst activity and electrolyte conductivity (Appleby & Foulkes, 1988). This temperature range reduces CO poisoning of platinum catalysts, allowing PAFCs to utilize less pure hydrogen fuels derived from reformed hydrocarbons. Electrical efficiencies of PAFC systems generally range from 40% to 50%, which can be increased to over 80% when the waste heat is recovered for heating or cogeneration applications (Kordesch & Simader, 1996). This makes PAFCs particularly suited for combined heat and power (CHP) systems in commercial and industrial settings.

The open-circuit voltage of a typical PAFC cell is about 0.9 V, with operational cell voltages under load ranging from 0.7 to 0.8 V depending on current density and system design (Appleby & Foulkes, 1988). PAFCs exhibit stable voltage output under varying loads and have relatively good tolerance to fuel impurities compared to other fuel cells. A significant advantage of PAFCs is their ability to tolerate higher levels of carbon monoxide in the fuel gas (up to ~1.5%) without significant catalyst poisoning (Kordesch & Simader, 1996). This tolerance reduces the need for extensive fuel purification, allowing PAFCs to use reformed natural gas, methanol, or other hydrocarbons, enhancing their practicality for stationary power plants.

PAFCs operate at intermediate temperatures that facilitate heat recovery, enabling cogeneration systems with high overall efficiencies. The waste heat from PAFC operation can be utilized for space heating, water heating, or industrial processes, significantly improving total energy utilization (Appleby & Foulkes, 1988). PAFCs were among the first fuel cell technologies to be commercialized, with systems deployed since the 1990s for stationary power generation (Kordesch & Simader, 1996). Several commercial units have demonstrated long operational lifetimes and reliability, especially in combined heat and power applications.

One of the major challenges with PAFCs is managing the phosphoric acid electrolyte to prevent acid loss, drying, or flooding during operation (Kordesch & Simader, 1996). Acid migration and evaporation can reduce electrolyte concentration, impairing proton conductivity and cell performance. Maintaining stable acid levels within the matrix over long-term operation is critical for system longevity. Additionally, the corrosive nature of phosphoric acid requires corrosion-resistant materials for bipolar plates and other components, increasing system costs.

While the intermediate operating temperature provides advantages, it also limits PAFC applicability compared to high-temperature fuel cells such as Solid Oxide Fuel Cells (SOFCs) (Appleby & Foulkes, 1988). The 150–220°C range requires robust thermal management and limits integration with certain industrial processes that operate at higher temperatures. PAFCs rely on platinum catalysts, which contribute significantly to system cost (Kordesch & Simader, 1996). Although PAFCs tolerate impurities better than PEMFCs, platinum loading remains a critical cost driver. Research into catalyst reduction and alternative materials continues to be important for commercial competitiveness.

5.2.5 Molten Carbonate Fuel Cells (MCFC)

Molten Carbonate Fuel Cells (MCFCs) are among the advanced high-temperature fuel cell technologies that offer promising solutions for efficient and clean power generation. Operating typically between 600°C and 700°C, MCFCs utilize a molten carbonate salt mixture as the electrolyte, enabling high ionic conductivity and fuel flexibility (Singhal & Kendall, 2003). Their ability to operate on a variety of fuels, including natural gas, biogas, and coal-derived gases, combined with their potential for high electrical efficiencies and cogeneration, makes MCFCs attractive for stationary power applications and distributed generation systems (McPhail *et al.*, 2015).

The MCFC operates by electrochemically converting chemical energy from fuel into electricity through redox reactions involving hydrogen and oxygen (McPhail, 2011). The key distinguishing feature of MCFCs is the use of molten carbonate salts (typically a mixture of lithium carbonate and potassium carbonate) dissolved in a porous ceramic matrix as the electrolyte (Singhal & Kendall, 2003).

In MCFCs, carbonate ions (CO_3^{2-}) act as the charge carriers, moving through the molten carbonate electrolyte from the cathode to the anode. At the cathode, oxygen from the air combines with carbon dioxide (CO_2) and electrons (from the external circuit) to form carbonate ions:

Cathode reaction: $O_2 + 2CO_2 + 4e^- \rightarrow 2CO_3^{2-}$
Anode reaction: $H_2 + CO_3^{2-} \rightarrow H_2O + CO_2 + 2e^-$
Overall reaction: $H_2 + 12O_2 \rightarrow H_2O + \text{electricity} + \text{heat}$

Carbon dioxide is recycled internally within the fuel cell system, which distinguishes MCFCs from other fuel cell types and plays a crucial role in sustaining the carbonate ion transport.

The molten carbonate electrolyte provides high ionic conductivity at elevated temperatures (typically around 650°C), allowing efficient transport of carbonate ions. The electrolyte is immobilized in a porous ceramic matrix (usually lithium aluminate) to maintain the liquid salt in place and provide mechanical stability (McPhail *et al.*, 2015). Operating at high temperatures enhances reaction kinetics and enables internal reforming of hydrocarbon fuels, improving system efficiency and fuel flexibility. Key Components of MCFCs are as follows.

5.2.5.1 Electrolyte

The molten carbonate electrolyte is a eutectic mixture of lithium carbonate (Li_2CO_3) and potassium carbonate (K_2CO_3) or sodium carbonate (Na_2CO_3) (Singhal & Kendall, 2003). This molten salt mixture provides excellent ionic conductivity for carbonate ions at high temperatures but requires specialized materials to contain and support it without corrosion or degradation. The electrolyte is retained within a porous ceramic matrix, typically lithium aluminate ($LiAlO_2$), which prevents leakage and maintains electrolyte integrity (McPhail, 2011).

5.2.5.2 Electrodes and Catalysts

The anode is commonly composed of a porous nickel-based material that provides catalytic activity for fuel oxidation and mechanical strength to withstand the harsh conditions of high-temperature operation (McPhail *et al.*, 2015). Nickel's catalytic properties facilitate both hydrogen oxidation and internal reforming of hydrocarbon fuels. The cathode is typically made from nickel oxide (NiO) mixed with ceramic materials to provide stability and catalytic activity for oxygen reduction and CO_2 absorption.

Each MCFC cell consists of an anode, electrolyte matrix, and cathode sandwiched between electrically conductive bipolar plates that serve as current collectors and gas flow distributors (Singhal & Kendall, 2003). The cell stacks are designed to operate at high temperatures with materials resistant to corrosion by molten carbonate and mechanical stresses from thermal cycling. MCFCs operate at high temperatures (600–700°C), enabling fast electrode kinetics and the use of non-precious metal catalysts such as nickel, significantly reducing catalyst costs compared to low-temperature fuel cells (McPhail *et al.*, 2015). The high operating temperature also allows MCFCs to internally reform hydrocarbon fuels like natural gas, avoiding the need for external reformers. Typical electrical efficiencies for MCFC systems range between 45% and 55% in standalone mode and can exceed 85% when configured for combined heat and power (CHP) applications (Singhal & Kendall, 2003). This high overall efficiency makes MCFCs attractive for distributed power generation and industrial uses.

MCFCs can tolerate fuel impurities such as carbon monoxide and carbon dioxide, which are often detrimental to low-temperature fuel cells, due to their unique carbonate ion transport and high-temperature operation (Kulkarni & Giddey, 2012). One of the major advantages of MCFCs is their ability to operate directly on a variety of fuels, including hydrogen, natural gas, biogas, and coal-derived syngas (McPhail *et al.*, 2015). The high operating temperature facilitates internal reforming reactions, where hydrocarbons and steam react within the cell to produce hydrogen and carbon monoxide for electrochemical oxidation, increasing system efficiency and reducing system complexity. MCFCs deliver high electrical efficiencies and excellent heat recovery potential due to their high operating temperature (Singhal & Kendall, 2003). The heat generated can be used for steam production, heating, or industrial processes, leading to overall system efficiencies exceeding 80% when integrated in CHP configurations.

Unlike low-temperature fuel cells that rely on expensive platinum catalysts, MCFCs use nickel-based electrodes, which are less costly and more durable under high-temperature conditions (McPhail *et al.*, 2015). This cost advantage makes MCFCs economically attractive for large-scale stationary power applications (Kulkarni & Giddey, 2012).

MCFCs are primarily suited for stationary power generation in commercial, industrial, and utility-scale applications due to their high efficiency, fuel flexibility, and cogeneration potential (McPhail *et al.*, 2015). Their ability to operate on natural gas or biogas and produce both electricity and heat makes them ideal for distributed generation near load centers (Kulkarni & Giddey, 2012). Large MCFC plants have been installed worldwide for combined heat and power applications, providing reliable power and thermal energy for hospitals, office complexes, and manufacturing facilities. The potential to operate on coal-derived syngas offers opportunities for cleaner coal power generation with reduced emissions (McPhail, 2011).

5.3 OVERVIEW OF RENEWABLE ENERGY SOURCES

The increasing global demand for energy, coupled with growing concerns about climate change and environmental degradation, has placed renewable energy sources (RES) at the forefront of sustainable development strategies worldwide. The world's energy consumption continues to rise, driven by population growth, industrialization, and improving living standards, particularly in developing countries (Efficiency & Stimulus, 2021). Simultaneously, the adverse effects of fossil fuel combustion, including greenhouse gas emissions, air pollution, and ecological harm, have catalyzed a shift toward cleaner, sustainable alternatives (Jacobson *et al.*, 2017). Thus, decarbonizing the energy sector through the adoption of renewable energy is a critical step to mitigate climate change.

Conventional energy systems heavily reliant on coal, oil, and natural gas are not only environmentally detrimental but also subject to geopolitical tensions, price volatility, and finite reserves. These vulnerabilities threaten energy security and economic stability, particularly in regions dependent on energy imports (Sovacool, 2016). Moreover, the extraction, transportation, and combustion of fossil fuels inflict substantial environmental damage, including habitat destruction, water contamination, and air pollution (Ritchie & Roser, 2017). Given these limitations, the transition to renewable energy has emerged as an imperative. Renewable energy sources, characterized by their ability to replenish naturally and their low or zero emissions during operation, offer a viable pathway to achieve sustainable, resilient, and equitable energy systems (Ellabban *et al.*, 2014).

Renewable energy sources harness naturally occurring phenomena such as solar radiation, wind, biomass, geothermal heat, and hydrological cycles to generate energy in an environmentally friendly manner (Hassan *et al.*, 2024). Unlike fossil fuels, renewables do not emit significant greenhouse gases during electricity generation or direct use, making them essential in combating climate change (Jacobson *et al.*, 2017). Furthermore, renewables contribute to reducing air pollutants responsible for respiratory and cardiovascular diseases, thus improving public health outcomes (Markandya *et al.*, 2018). Beyond environmental benefits, renewable energy supports energy security by diversifying supply and reducing dependence on imported fuels, which can be subject to price shocks and geopolitical disruptions (Sovacool, 2016). The decentralized nature of many renewable technologies also facilitates rural electrification and energy access, promoting socio-economic development in underserved communities (Bhattacharyya, 2006).

Over the past decade, rapid technological advancements and economies of scale have significantly lowered the costs of renewable energy technologies. Solar photovoltaic (PV) and wind power costs have dropped by 80% and 60%, respectively, enabling widespread adoption across both developed and developing nations (Masson-Delmotte *et al.*, 2021). Such cost reductions are attributed to improvements in materials, manufacturing processes, and system integration, alongside supportive policies and market mechanisms (Masson-Delmotte *et al.*, 2021). This cost competitiveness increasingly challenges traditional fossil fuel-based power generation economically, even without subsidies. Many regions now find renewables to be the least-cost option for new electricity generation capacity, driving accelerated deployment (Masson-Delmotte *et al.*, 2021). The resultant increase in installed renewable capacity contributes to grid decarbonization, reducing carbon footprints and helping meet international climate commitments such as the Paris Agreement targets (Change, 2015).

The role of renewable energy is increasingly recognized not only in electricity generation but also in heating, cooling, and transportation sectors. Electrification powered by renewables, combined with energy efficiency measures, is central to achieving deep decarbonization pathways. Additionally, sector coupling linking electricity with gas, heat, and mobility sectors can enhance flexibility and optimize resource use (Khan *et al.*, 2025). Innovations such as energy storage, demand response, hydrogen production from renewable electricity, and hybrid renewable systems are gaining traction to overcome intermittency challenges and expand applications (Ugah *et al.*, 2025).

5.4 SOLAR ENERGY

Solar energy, derived from the sun's radiation, is one of the most abundant and promising renewable energy sources available on Earth. With an estimated 173,000 terawatts of solar energy striking the Earth continuously—more than 10,000 times the world's total energy use—solar power holds immense potential to meet global energy demands sustainably (Lewis & Nocera, 2006). Unlike fossil fuels, solar energy is inexhaustible, clean, and widely distributed, which makes it a cornerstone in the global transition toward renewable energy and climate change mitigation (Hassan *et al.*, 2024). Solar energy originates from nuclear fusion reactions occurring in the sun's core, where hydrogen atoms combine to form helium, releasing massive amounts of energy in the form of electromagnetic radiation (Green *et al.*, 2021). This radiation spans a broad spectrum, including ultraviolet (UV), visible light, and infrared (IR) wavelengths. Only a fraction of this solar radiation reaches the Earth's surface approximately 1,000 watts per square meter under optimal conditions due to atmospheric absorption and scattering (Luque & Hegedus, 2011). Harnessing this energy effectively requires technologies capable of converting solar radiation into usable forms, primarily electricity or heat.

5.4.1 Photovoltaic Technology

Photovoltaic technology directly converts sunlight into electricity through the photovoltaic effect, discovered in the 19th century (Khan *et al.*, 2013). PV cells, typically made from semiconductor materials such as silicon, generate electron-hole pairs when exposed to photons, producing an electric current (Green *et al.*, 2021). These cells are interconnected to form modules and arrays that can power residential, commercial, or utility-scale applications. The dominant PV technology today is crystalline silicon (c-Si), known for its relatively high efficiency (15–22%) and longevity (Luque & Hegedus, 2011). Emerging technologies such as thin-film solar cells (e.g., cadmium telluride, copper indium gallium selenide) offer advantages in flexibility and manufacturing cost but generally have lower efficiencies (Green *et al.*, 2021). Perovskite solar cells, a recent breakthrough, have shown remarkable efficiency improvements and low production costs, holding potential for disruptive changes in the solar industry (Park, 2015). PV systems are often complemented by power electronics—such as inverters and maximum power point trackers (MPPT)—to optimize electricity output and interface with electrical grids (Luque & Hegedus, 2011). With technological advancements, PV systems have become highly modular, scalable, and suitable for diverse applications, ranging from rooftop installations to large solar farms.

5.4.2 Concentrated Solar Power (CSP)

Concentrated Solar Power (CSP) technology uses mirrors or lenses to concentrate sunlight onto a small area, generating high temperatures that produce steam to drive turbines for electricity generation (Kalogirou, 2004). Unlike PV systems, CSP is a thermal technology that often incorporates thermal energy storage, allowing power generation even when sunlight is not available. CSP plants can be configured in various designs such as parabolic troughs, solar towers, linear Fresnel reflectors, and dish Stirling systems (Kalogirou, 2004). These systems operate at temperatures typically between 400°C and 1,000°C, making them suitable for large-scale power plants. Thermal storage using molten salts or other materials enhances dispatchability, addressing the intermittency challenges associated with solar energy (Zhang *et al.*, 2013). CSP's ability to provide stable, dispatchable power positions it as a valuable complement to photovoltaic technologies, particularly in regions with high direct normal irradiance (DNI) such as deserts and arid zones (Zhang *et al.*, 2013).

Significant progress in solar energy technologies has been achieved in the past decade, driven by research, development, and supportive policies globally. The cost of solar photovoltaic modules has plummeted by over 85% since 2010, largely due to improvements in manufacturing, economies of scale, and competitive markets (IREA, 2020). This cost reduction has accelerated global PV deployment, making solar power increasingly competitive with traditional fossil fuel generation (Lai & McCulloch, 2016). Technological innovations such as bifacial solar panels, which capture sunlight from both sides, and tandem solar cells combining multiple semiconductors, have further improved PV efficiency and energy yield (Jäger-Waldau, 2019). The integration of solar with energy storage—particularly lithium-ion batteries—enables more flexible and reliable energy systems, facilitating higher penetration levels in power grids (IREA, 2020).

For CSP, advancements focus on increasing operating temperatures, improving heat transfer fluids, and enhancing thermal storage materials and designs to boost efficiency and cost-effectiveness (Zhang *et al.*, 2013). Hybridization with fossil fuels or biomass also extends operational hours and provides grid stability. Digital technologies, including smart inverters, Internet of Things (IoT), and artificial intelligence, are increasingly deployed to optimize solar energy systems' performance, maintenance, and grid integration (Mateo Romero *et al.*, 2022).

Solar energy offers a range of environmental, economic, and social benefits. Its operation emits no direct greenhouse gases or air pollutants, significantly reducing the carbon footprint compared to fossil fuels (Markandya *et al.*, 2018). Solar energy's decentralized nature promotes energy access in remote and off-grid areas, enhancing energy equity and economic development (Bhattacharyya, 2006). Economically, solar power creates jobs in manufacturing, installation, and maintenance sectors, contributing to local economies (IREA, 2020). Moreover, solar systems often have low operating and maintenance costs after installation, providing stable electricity prices over the long term (Lai & McCulloch, 2016). The scalability and modularity of solar energy systems allow flexible deployment from small rooftop panels to utility-scale solar farms catering to diverse energy needs and settings (Luque & Hegedus, 2011). Furthermore, solar energy systems can be integrated with building designs (building-integrated photovoltaics) and combined with other renewable sources for hybrid energy solutions.

5.4.3 Wind Energy

Wind energy is a rapidly growing renewable energy source that harnesses the kinetic energy of air movement to generate electricity. It is considered a clean, sustainable, and inexhaustible resource that can significantly reduce dependency on fossil fuels and mitigate climate change impacts (Saidur *et al.*, 2011). As the world seeks to transition to low-carbon energy systems, wind energy has emerged as one of the most cost-effective and widely deployed renewable technologies globally (IREA, 2020). Wind energy originates from the uneven heating of the Earth's surface by the sun, which creates pressure differences in the atmosphere. These pressure gradients generate air movement or wind, which possesses kinetic energy proportional to the air mass and the cube of wind speed (Manwell *et al.*, 2010).

5.4.4 Wind Turbines

The primary technology for converting wind energy into electrical energy is the wind turbine. Modern wind turbines consist of rotor blades connected to a hub, which spins a shaft connected to a generator through a gearbox or directly in direct-drive systems (Manwell *et al.*, 2010). Wind turbines are categorized mainly by their axis of rotation: horizontal-axis wind turbines (HAWT) and vertical-axis wind turbines (VAWT) (Burton *et al.*, 2011). HAWTs dominate the industry due to their higher efficiency and scalability. They typically have three blades and are designed to face into the wind using a yaw system. VAWTs, with designs such as the Darrieus or Savonius turbines, offer advantages like omnidirectional wind capture and simpler mechanics but generally suffer from lower efficiency and are mainly suited for small-scale applications (Noman *et al.*, 2020). Wind turbines range from small units for residential or off-grid use (<100 kW) to large utility-scale turbines exceeding 10 MW, especially offshore (IREA, 2020). The swept area and blade length have increased substantially over the years to capture more wind and enhance energy production.

5.4.5 Offshore and Onshore Wind Farms

Wind energy installations can be located on land (onshore) or in bodies of water (offshore). Onshore wind farms have been widely deployed for decades and account for the majority of installed capacity globally due to lower costs and easier access (Lund, 2024). However, onshore sites often face challenges such as land use conflicts, noise concerns, and variable wind conditions. Offshore wind farms are gaining significant attention because of their higher and more consistent wind speeds, larger available areas, and reduced visual and noise impacts (Musial & Ram, 2010). Offshore turbines tend to be larger and more powerful, often exceeding 10 MW capacity per turbine, and require specialized foundations and installation vessels. Floating offshore wind technology is also emerging, enabling deployment in deeper waters (IREA, 2020).

Global wind power capacity has experienced exponential growth in recent decades, reaching over 837 GW by the end of 2022 (Council, 2021). Countries such as China, the United States, Germany, and India lead in installed capacity, reflecting strong policy support and technological advancements. Technological innovations have driven reductions in costs, with the levelized cost of electricity (LCOE) from wind falling by approximately 70% since 2010, making it competitive with fossil fuel-based generation in many regions (IREA, 2020). This cost decline results from improvements in turbine design, manufacturing efficiencies, and economies of scale. Hybrid energy systems that integrate wind with solar PV and storage are becoming popular to address intermittency and enhance grid reliability (Lund, 2024). Digital tools such as advanced wind resource assessment, predictive maintenance using artificial intelligence, and grid integration technologies are further optimizing wind energy performance and operational lifespan (Saidur *et al.*, 2011).

Wind energy's environmental benefits are substantial, primarily due to its zero greenhouse gas emissions during operation and minimal water use compared to thermal power plants (Mey & Diesendorf, 2018). The deployment of wind power significantly reduces carbon dioxide and air pollutant emissions, contributing to improved air quality and public health (Blanco, 2009). Economically, wind energy projects create jobs in manufacturing, installation, operation, and maintenance sectors. According to IREA (2020), wind energy employment reached 1.25 million jobs globally, with opportunities concentrated in regions with strong industry presence. Wind energy also supports rural development by providing additional income streams to landowners and diversifying local economies (Lund, 2024).

The scalability of wind energy enables deployment across a wide range of geographic scales—from small off-grid turbines supporting remote communities to large grid-connected farms that supply millions of households (Musial & Ram, 2010). Additionally, wind energy offers energy security by diversifying the energy mix and reducing dependence on imported fuels.

5.5 BIOMASS

Biomass energy is one of the oldest and most versatile forms of renewable energy, derived from organic matter such as plants, agricultural residues, forestry waste, and animal byproducts. It plays a significant role in the global renewable energy portfolio, offering a sustainable alternative to fossil fuels and contributing to energy security, waste management, and rural development (Demirbas, 2008). Biomass can be converted into heat, electricity, or biofuels through various biochemical and thermochemical processes, making it highly adaptable across different energy sectors (McKendry, 2002; Ogundele *et al.*, 2024). Biomass energy is generated through the chemical energy stored in organic materials, primarily from carbon fixed via photosynthesis (Sims *et al.*, 2006). The carbon in biomass originated from atmospheric CO_2, thus creating a closed carbon cycle when biomass is sustainably sourced and used. This cyclical nature differentiates biomass from fossil fuels and underpins its classification as a renewable energy source (Zhou *et al.*, 2011).

Common biomass feedstocks include dedicated energy crops (e.g., switchgrass, miscanthus), agricultural residues (e.g., corn stover, rice husks), forestry residues (e.g., wood chips, sawdust), animal manure, and organic waste (Chen *et al.*, 2015). The availability and characteristics of these feedstocks depend on regional climate, agricultural practices, and waste management systems, influencing the feasibility and efficiency of biomass energy projects (Demirbas, 2008). Biomass can be converted into useful energy forms through several primary pathways: direct combustion, biochemical conversion, and thermochemical conversion (Sims *et al.*, 2006).

5.5.1 Direct Combustion

The simplest and most traditional method involves burning biomass directly to produce heat or steam, which can generate electricity via steam turbines (McKendry, 2002). This technology is widely used in power plants and heating applications, especially in regions with abundant biomass residues. Modern biomass boilers have improved combustion efficiency and emission controls compared to traditional open fires or rudimentary stoves (Demirbas, 2008).

5.5.2 Biochemical Conversion

Biochemical processes use microorganisms or enzymes to break down biomass into biofuels such as biogas and bioethanol. Anaerobic digestion is a key biochemical technology that decomposes organic matter in the absence of oxygen to produce methane-rich biogas, which can be used for heat, electricity, or as vehicle fuel (Mahla *et al.*, 2022). Similarly, fermentation converts sugars from biomass into ethanol, a liquid biofuel widely used in transportation (Balat & Balat, 2009). These processes are particularly suitable for wet biomass feedstocks, such as animal manure, food waste, and energy crops with high moisture content (Chen *et al.*, 2015). Biochemical conversion also supports waste management by valorizing organic waste streams.

5.5.3 Thermochemical Conversion

Thermochemical methods involve the application of heat to biomass in the presence or absence of oxygen, leading to gasification, pyrolysis, or liquefaction. Gasification converts biomass into a combustible gas mixture called syngas (mainly CO, H_2, and CH_4), which can be burned for heat and power or upgraded to liquid fuels (Bhaskar *et al.*, 2011). Pyrolysis thermally decomposes biomass in the absence of oxygen, producing bio-oil, char, and syngas, which have various energy and material uses (Bridgwater, 2012). These technologies are attractive for their ability to handle diverse feedstocks and produce multiple energy carriers, enhancing flexibility and efficiency (Sims *et al.*, 2006). Advances in catalytic upgrading and integrated systems are improving the viability of thermochemical conversion for large-scale biofuel production.

Biomass energy contributes to climate change mitigation by reducing net greenhouse gas (GHG) emissions compared to fossil fuels. When sourced sustainably, biomass combustion or conversion releases CO_2 previously absorbed during plant growth, resulting in a lower or neutral carbon footprint (Zhou *et al.*, 2011). Furthermore, biomass utilization can decrease methane emissions from decomposing organic waste and reduce reliance on landfills (Mahla *et al.*, 2022). In addition to GHG benefits, biomass energy can improve soil quality and biodiversity when residues are used as biochar or organic amendments rather than being burned or left to decay (Lehmann & Joseph, 2024). Properly managed biomass systems also prevent open-field burning of agricultural residues, which is a significant source of air pollution in many developing countries (Demirbas, 2008).

The economic advantages of biomass energy are particularly evident in rural and agricultural communities, where biomass resources are abundant. Biomass energy projects create employment opportunities in feedstock production, harvesting, processing, and plant operation (Sims *et al.*, 2006). This decentralized energy production supports local economies, enhances energy access, and reduces energy import dependency (Chen *et al.*, 2015). Moreover, biomass energy contributes to waste management by converting organic residues into valuable energy products, reducing environmental liabilities, and improving public health (Mahla *et al.*, 2022). Governments worldwide have introduced policies and incentives to promote biomass energy, reflecting its role in sustainable development and energy diversification (Wu & Pfenninger, 2023).

5.6 HYDROPOWER

Hydropower, also known as hydroelectric power, is a renewable energy source that harnesses the energy of flowing or falling water to generate electricity. It is one of the oldest and most established renewable energy technologies, contributing a significant share of global electricity generation (IREA, 2020). Hydropower's appeal lies in its ability to provide reliable, flexible, and clean energy without direct greenhouse gas emissions during operation (Gracey & Verones, 2016). Hydropower exploits the potential and kinetic energy of water moving from higher to lower elevations. The energy conversion process typically involves capturing water in reservoirs or run-of-river systems and directing it through turbines connected to generators to produce electricity (Twidell, 2021). The amount of electricity generated depends on the water flow rate and the hydraulic head—the height difference between the water source and the turbine (Gracey & Verones, 2016).

The renewable nature of hydropower stems from the natural water cycle, driven by solar energy, which replenishes water resources continuously (Kibler & Tullos, 2013). Unlike fossil fuels, hydropower does not consume water; it merely converts water's mechanical energy into electrical energy. This positions hydropower as a sustainable and low-carbon energy source critical to climate change mitigation strategies worldwide (IREA, 2020). Hydropower technologies vary according to the scale, water flow characteristics, and site conditions. The main classifications include large-scale hydropower, small hydropower, pumped storage, and emerging innovations.

5.6.1 Large-Scale Hydropower

Large hydropower plants typically involve the construction of dams and reservoirs on rivers to store vast amounts of water. This stored water is released in a controlled manner to drive turbines and generate electricity on demand (Twidell, 2021). These plants offer significant capacity and reliability, often providing base-load power to national grids. The Three Gorges Dam in China and the Itaipu Dam on the Brazil-Paraguay border are iconic examples of large hydropower installations (Gracey & Verones, 2016).

5.6.2 Small and Micro Hydropower

Small hydropower systems have capacities below 10 MW and usually operate on run-of-river principles without significant water storage. Micro-hydropower plants, with capacities below 100 kW, are often used for decentralized rural electrification (Kibler & Tullos, 2013). These systems are less intrusive, have lower environmental impacts, and provide flexible energy solutions for remote communities.

5.6.3 Pumped Storage Hydropower

Pumped storage hydropower (PSH) functions as a large-scale energy storage system. During periods of low electricity demand, excess electrical energy pumps water from a lower reservoir to an upper reservoir. When demand increases, stored water is released to generate electricity (IREA, 2020). PSH is pivotal for grid stability and integrating intermittent renewable sources like solar and wind (Gracey & Verones, 2016).

Hydropower is crucial for electricity generation, accounting for approximately 16% of global electricity supply and about 60% of renewable electricity. Its operational flexibility allows rapid ramping up and down, which supports grid stability and the integration of variable renewables like solar and wind (Gracey & Verones, 2016). Pumped storage plants provide essential energy storage, helping balance supply and demand fluctuations. This capability is increasingly important as renewable energy penetration grows worldwide (IREA, 2020). Hydropower's capacity to provide ancillary services, such as frequency regulation and black start capability, enhances grid resilience (Twidell, 2021).

5.6.4 Geothermal

Geothermal energy is a sustainable and renewable energy source derived from the natural heat stored within the Earth's crust. This energy originates from the planet's formation, radioactive decay of minerals, and ongoing geothermal gradient processes (Lund *et al.*, 2011). Unlike solar or wind energy, geothermal power offers a consistent and reliable energy supply, independent of weather conditions, making it a valuable component of the renewable energy mix (Bertani, 2009). The Earth's internal heat is continuously generated and transported through conduction and convection within the crust and mantle. This geothermal heat manifests as hot rocks, steam, and hot water reservoirs located at various depths below the surface (Dickson & Fanelli, 2013). The accessible heat resource for geothermal power varies from shallow ground heat used for direct applications to deep hydrothermal reservoirs capable of electricity generation (Lund *et al.*, 2011).

Geothermal energy extraction typically involves drilling wells into geothermal reservoirs to access hot water and steam, which can then be used to drive turbines for electricity generation or provide direct heating (Bertani, 2009). The global average geothermal gradient is approximately 25–30°C per kilometer depth, but this varies by location, with some regions, such as volcanic zones, exhibiting significantly higher gradients conducive to geothermal exploitation (Dickson & Fanelli, 2013). Several technologies harness geothermal energy, each tailored to specific reservoir characteristics.

5.6.5 Dry Steam Power Plants

Dry steam plants directly utilize steam from geothermal reservoirs to rotate turbines connected to generators. They are the oldest and simplest geothermal power technologies and require high-temperature reservoirs with abundant steam (Lund *et al.*, 2011). Examples include the Geysers in California, the world's largest dry steam geothermal field.

5.6.6 Flash Steam Power Plants

Flash steam plants use high-pressure hot water extracted from deep wells. When this water is depressurized or "flashed" into steam, it drives turbines for electricity generation (Bertani, 2009). Flash steam technology is more widespread than dry steam due to its ability to utilize a broader range of high-temperature geothermal resources.

5.6.7 Binary Cycle Power Plants

Binary cycle plants are used for moderate to low-temperature geothermal fluids. They employ a secondary working fluid with a low boiling point, such as isobutane or pentane, in a closed-loop system. Heat from the geothermal fluid vaporizes the secondary fluid, which then drives turbines (Dickson & Fanelli, 2013). Binary plants can exploit resources previously considered unsuitable, increasing the geographic scope of geothermal development.

Beyond electricity generation, geothermal energy is widely used for direct heating purposes. These include district heating systems, greenhouse agriculture, industrial processes, and balneology (Lund *et al.*, 2011). Geothermal heat pumps extract shallow ground heat for heating and cooling buildings, offering energy-efficient temperature regulation by leveraging stable subsurface temperatures (Bertani, 2009). Geothermal energy currently contributes about 0.3% of global electricity generation, with installed capacity exceeding 15 GW worldwide (IREA, 2020). Leading producers include the United States, Indonesia, the Philippines, Turkey, and Kenya. These countries exploit high-temperature hydrothermal systems for electricity and direct-use applications. The global geothermal potential is estimated at hundreds of gigawatts, far exceeding current utilization levels. Enhanced geothermal systems (EGS), which involve artificially creating permeability in hot dry rock formations, represent a promising technology to unlock vast untapped resources beyond conventional hydrothermal fields (Bertani, 2009).

5.7 THE NEED FOR INTEGRATION OF FUEL CELLS WITH RENEWABLE ENERGY SOURCES

The global energy sector is undergoing a transformative shift aimed at achieving sustainability, reducing carbon emissions, and meeting increasing energy demand. Renewable energy sources (RES) such as solar, wind, and biomass have become critical components of this transition. However, the inherent intermittency and variability of these sources pose challenges to reliable and continuous energy supply (Chu & Majumdar, 2012). Fuel cells, with their high efficiency, scalability, and ability to provide clean energy, offer a promising solution to complement renewable energy systems. Integrating fuel cells with renewable energy sources addresses limitations of standalone renewables and advances the goals of energy security, environmental sustainability, and grid stability (Sadeghian *et al.*, 2025).

Standalone renewable energy systems have inherent technical and operational limitations. Solar photovoltaic (PV) and wind turbines produce energy intermittently, often in times and locations that do not coincide with consumption peaks. This variability can lead to energy curtailment or reliance on backup fossil fuel generators, reducing the environmental benefits of renewables (Denholm & Hand, 2011). Energy storage technologies, such as batteries, have been employed to address intermittency but face challenges including high costs, limited lifespan, environmental concerns related to material extraction and disposal, and capacity constraints (Chen *et al.*, 2009). Moreover, battery storage is less suited for long-duration storage necessary to balance seasonal variations in renewable generation.

Grid infrastructure also faces stress due to the fluctuating inputs from renewables, requiring sophisticated management strategies and ancillary services to maintain frequency and voltage stability (Lund *et al.*, 2012). The variability in power generation can result in voltage fluctuations, frequency deviations, and potential blackouts if not properly managed (Kumar *et al.*, 2024). In this context, fuel cells offer advantages over traditional storage and backup solutions due to their ability to generate electricity on demand with high efficiency, scalability, and environmentally benign emissions (Sadeghian *et al.*, 2025).

Fuel cells convert chemical energy from fuels, such as hydrogen or natural gas, directly into electricity through electrochemical reactions. This process bypasses combustion, resulting in higher efficiency and near-zero harmful emissions when using clean fuels (Barbir, 2012).

Integrating fuel cells with renewable energy sources offers significant environmental advantages. It facilitates deeper decarbonization by enabling higher penetration of renewables without compromising reliability (Marouani *et al.*, 2023). The use of green hydrogen produced from renewable electricity in fuel cells eliminates fossil fuel combustion emissions, contributing to air quality improvement and climate change mitigation (Barbir, 2012). Moreover, hybrid fuel cell-renewable systems reduce the need for conventional peaker plants, which are often inefficient and polluting. This contributes to lowering overall lifecycle greenhouse gas emissions and reducing dependence on fossil fuel infrastructure (Dunn *et al.*, 2011).

5.7.1 Solar Energy + Fuel Cells

Solar energy is inherently intermittent due to diurnal and weather-related variations, resulting in variable power output that challenges grid stability (Denholm & Hand, 2011). To address this intermittency, integration with energy storage and conversion technologies like fuel cells enables a continuous and reliable power supply (Barbir, 2012). Fuel cells convert chemical energy from hydrogen into electricity electrochemically with high efficiency and low emissions (Wang *et al.*, 2011). When solar electricity powers water electrolysis to produce hydrogen, fuel cells can later convert stored hydrogen back into electricity, bridging gaps when solar energy is unavailable (Dunn *et al.*, 2011). Electrolysis, driven by excess solar power, splits water molecules into hydrogen and oxygen, effectively storing solar energy as hydrogen fuel (Lykas *et al.*, 2023). Proton Exchange Membrane (PEM) electrolyzers are particularly suited for solar integration due to their quick response to variable loads and ability to produce high-purity hydrogen, which is essential for fuel cell operation (Lykas *et al.*, 2023). Alkaline and solid oxide electrolyzers also contribute depending on operating temperature and system requirements (Dunn *et al.*, 2011). This solar-driven hydrogen production enables energy decoupling, allowing solar energy generation and consumption to be temporally separated (Denholm & Hand, 2011).

Hybrid systems combining solar PV arrays with fuel cells use the solar panels to meet immediate electricity demands and power electrolyzers, storing excess energy as hydrogen (Wang *et al.*, 2011). During periods without sunlight, stored hydrogen is fed into fuel cells to generate electricity on demand, ensuring uninterrupted power (Bigdeli, 2015). This hybrid approach supports peak shaving, load leveling, and ancillary grid services like frequency regulation (Bigdeli, 2015). The modular nature of fuel cells allows scalable system design from residential to utility scale (Barbir, 2012). In addition to electrical power, waste heat from fuel cell operation can be recovered for heating or industrial processes in combined heat and power (CHP) configurations, improving overall system efficiency (Waseem *et al.*, 2023). Such cogeneration applications enhance economic viability by utilizing both electricity and thermal outputs (Waseem *et al.*, 2023). The integration of solar energy with fuel cells thus extends beyond power generation to multifaceted energy solutions.

Off-grid and microgrid applications particularly benefit from solar-fuel cell integration by providing reliable, sustainable electricity in remote locations lacking grid access (Dagar *et al.*, 2021). These systems reduce dependence on fossil fuel generators, cut emissions, and enhance energy security (Dagar *et al.*, 2021). Solar-fuel cell microgrids can be designed for critical infrastructure, rural electrification, and disaster resilience (Dagar *et al.*, 2021). Fuel cells provide rapid response and ramping capabilities, allowing dynamic adaptation to fluctuations in solar power and load demands, thus improving power quality and grid stability (Lykas *et al.*, 2023). They can operate continuously with high efficiency over extended periods, contrasting with batteries that degrade with cycling (Barbir, 2012). Moreover, hydrogen produced from solar energy is a clean fuel, emitting only water upon use, thereby offering a zero-carbon energy cycle when combined with solar (Wang *et al.*, 2011).

5.7.2 Wind Energy + Fuel Cells

Wind energy is one of the fastest-growing renewable energy sources worldwide, driven by its abundant availability and zero carbon emissions during operation (Manwell *et al.*, 2010). However, similar to other renewable sources, wind power is inherently intermittent and variable due to fluctuating wind speeds, which creates challenges in maintaining a stable and reliable electricity supply (Krishna & Kumar, 2015). Integrating wind energy with fuel cell technologies offers a promising pathway to mitigate these limitations by providing efficient energy storage and continuous power generation (Krishna & Kumar, 2015). The variability of wind power can lead to mismatches between electricity supply and demand, causing grid instability, frequency fluctuations, and potential curtailment of generated power (Snyder & Kaiser, 2009). To address these issues, excess wind energy can be converted into hydrogen via water electrolysis, enabling energy storage in chemical form. This hydrogen can later be utilized by fuel cells to produce electricity when wind speeds are low, ensuring a steady power supply (Benghanem *et al.*, 2023).

Water electrolysis is a key process in wind-to-hydrogen systems. Proton Exchange Membrane (PEM) electrolyzers are favored for their rapid response and ability to handle dynamic loads from fluctuating wind output (Krishna & Kumar, 2015). Alkaline electrolyzers and Solid Oxide Electrolyzers (SOE) can also be employed depending on the operational requirements and temperature regimes (Dunn *et al.*, 2011). The integration of electrolyzers with wind turbines thus enables the decoupling of electricity generation from consumption, facilitating energy management and grid balancing (Manwell *et al.*, 2010). Fuel cells convert stored hydrogen back into electrical energy with high efficiency and low emissions, primarily producing water as a byproduct (Manwell *et al.*, 2010). When integrated with wind power, fuel cells serve as a dispatchable power source, providing electricity during periods of wind lull or peak demand (Benghanem *et al.*, 2023). This hybridization not only improves grid reliability but also enhances the overall utilization rate of wind energy assets (Krishna & Kumar, 2015).

One significant advantage of wind energy and fuel cell integration is the ability to form hybrid energy systems that can operate both on-grid and off-grid. In remote or islanded microgrids, where grid extension is costly or impossible, such systems offer sustainable, resilient power supply alternatives (Dagar *et al.*, 2021). These hybrid systems can also provide ancillary services, including frequency regulation, voltage support, and spinning reserve, thereby supporting grid stability (Lykas *et al.*, 2023). In addition to electricity generation, combined heat and power (CHP) configurations using fuel cells can recover thermal energy generated during electrochemical reactions, improving total system efficiency (Benghanem *et al.*, 2023). This heat can be used for residential heating or industrial processes, adding economic value and reducing reliance on fossil fuels for thermal energy (Benghanem *et al.*, 2023).

5.7.3 Multi-Renewable Energy Hybrids with Fuel Cells

The increasing global demand for sustainable, clean, and reliable energy has propelled the integration of multiple renewable energy sources with advanced energy conversion technologies like fuel cells. Multi-renewable energy hybrid systems, combining diverse sources such as solar, wind, biomass, and hydropower with fuel cells, present a comprehensive solution to address the intermittency, variability, and storage challenges associated with individual renewable technologies (Arul *et al.*, 2015). This integrated approach enhances system reliability, efficiency, and sustainability by synergistically exploiting the complementary characteristics of different energy sources and the high efficiency of fuel cells in converting stored fuels into electricity (Krishna & Kumar, 2015).

The rationale for integrating multiple renewables with fuel cells lies in overcoming the inherent limitations of single-source renewable energy systems. Solar and wind energies, for instance, exhibit significant fluctuations due to weather conditions and diurnal or seasonal cycles, resulting in unstable power output (Dagar *et al.*, 2021). Biomass and hydropower, on the other hand, offer relatively stable but geographically constrained energy availability (Lee *et al.*, 2023). The integration of these heterogeneous energy inputs with fuel cells, which can operate on hydrogen or other fuels derived from renewables, provides a buffering mechanism, ensuring a continuous and dispatchable power supply (Eriksson & Gray, 2017).

Fuel cells, particularly Proton Exchange Membrane Fuel Cells (PEMFCs), Solid Oxide Fuel Cells (SOFCs), and Molten Carbonate Fuel Cells (MCFCs), play a crucial role in multi-renewable energy hybrid systems by enabling efficient conversion of chemical energy into electrical energy with minimal environmental impact (Barbir, 2012). These fuel cells can utilize hydrogen produced via electrolysis powered by solar or wind energy, or directly reform biofuels generated from biomass, enabling a flexible and sustainable energy cycle (Salehmin *et al.*, 2022).

One of the most promising configurations in multi-renewable hybrid systems is the coupling of solar photovoltaic (PV) panels, wind turbines, and fuel cells, complemented by energy storage units such as batteries and hydrogen tanks (Krishna & Kumar, 2015). The excess electricity generated by PV and wind can be used to produce hydrogen through water electrolysis, storing energy for later conversion in fuel cells when renewable generation is insufficient (Herdem *et al.*, 2024). This hybridization not only smooths out supply variability but also optimizes overall system efficiency and lifespan by balancing loads and minimizing stress on individual components (Olatomiwa *et al.*, 2016).

Integrating biomass energy into multi-renewable hybrids with fuel cells adds an additional layer of reliability and carbon neutrality. Biomass can be converted into biogas or biohydrogen through anaerobic digestion or gasification processes, providing a continuous fuel source for fuel cells (Lee *et al.*, 2023). This is particularly valuable in regions with abundant agricultural residues or organic waste, enabling decentralized energy production and waste valorization (Eriksson & Gray, 2017). The use of bio-derived fuels in fuel cells contributes to a circular economy by reducing landfill usage and greenhouse gas emissions.

Hydropower, especially small-scale run-of-river and micro-hydropower, also complements multi-renewable hybrid systems by providing stable baseline power, reducing dependence on intermittent sources. Combined with fuel cells, hydropower can facilitate on-demand electricity generation, especially during prolonged low wind or solar periods (Wang *et al.*, 2017). The synergy of hydropower's steadiness and fuel cells' dispatchability enhances grid stability and energy security, making the system highly resilient (Sachdev *et al.*, 2015).

The architecture of multi-renewable energy hybrids with fuel cells necessitates advanced control strategies and energy management systems (EMS) to optimize operational efficiency, cost-effectiveness, and system longevity (Olatomiwa *et al.*, 2016). Real-time monitoring and forecasting algorithms for renewable generation and load demand enable predictive control, which dynamically adjusts hydrogen production, fuel cell operation, and battery usage (Wang *et al.*, 2017). Such EMS frameworks integrate power electronics, sensors, and communication networks to achieve optimal power flow and seamless switching among energy sources (Dagar *et al.*, 2021).

From a technological perspective, significant progress has been made in improving electrolyzer performance and reducing costs, which is vital for producing green hydrogen from multi-renewable sources (Salehmin *et al.*, 2022). PEM electrolyzers provide fast start-up times and flexible operation suitable for variable renewable inputs, whereas Solid Oxide Electrolyzers (SOE) offer higher efficiencies at elevated temperatures, beneficial in combined heat and power (CHP) applications (Dunn *et al.*, 2011). Similarly, advancements in fuel cell materials and manufacturing are enhancing durability, reducing precious metal catalyst loading, and enabling fuel flexibility (Barbir, 2012).

Economic feasibility remains a critical factor for the deployment of multi-renewable energy hybrids with fuel cells. Although capital costs are relatively high compared to conventional energy systems, declining costs of solar PV, wind turbines, and electrolyzers, coupled with policy incentives and carbon pricing, are improving competitiveness (Dunn *et al.*, 2011). Moreover, the ability of hybrid systems to provide grid services, reduce peak loads, and enhance energy security creates value propositions beyond mere electricity generation. Life cycle cost analyses indicate that integrated systems can achieve lower overall costs and emissions compared to standalone renewable or fossil fuel systems (Arul *et al.*, 2015).

The environmental benefits of these integrated systems are profound. By leveraging multiple renewables and fuel cells, greenhouse gas emissions are minimized through the displacement of fossil fuels and the efficient use of biomass wastes (Herdem *et al.*, 2024). Additionally, the reduction of air pollutants such as particulate matter, NOx, and SOx enhances public health and contributes to compliance with international climate goals (Arul *et al.*, 2015).

Pilot projects and demonstration plants worldwide illustrate the practical viability of multi-renewable hybrids with fuel cells. For instance, hybrid microgrids incorporating solar, wind, biomass, and PEM fuel cells have been successfully deployed in remote communities in Canada and Europe, providing reliable, clean energy and reducing reliance on diesel generators (Krishna & Kumar, 2015). These projects demonstrate the flexibility, scalability, and resilience of hybrid systems, paving the way for broader commercialization (Eriksson & Gray, 2017).

5.8 ENVIRONMENTAL AND ECONOMIC BENEFITS OF INTEGRATION OF FUEL CELLS WITH RENEWABLE ENERGY SOURCES

The integration of fuel cells with renewable energy sources is rapidly gaining traction due to their potential to offer significant environmental and economic benefits. This synergy leverages the advantages of both technologies, enabling cleaner, more efficient, and reliable energy systems.

Life Cycle Assessment (LCA) is a systematic methodology to evaluate the environmental impacts associated with all stages of a product or system's life, from raw material extraction through manufacturing, operation, and disposal (Arvanitoyannis, 2008). For hybrid systems integrating fuel cells with renewable energy, LCA is crucial in quantifying overall sustainability by capturing energy consumption, emissions, and waste generation across the system lifespan. Several studies have demonstrated that fuel cell-renewable hybrid systems exhibit lower environmental burdens compared to conventional fossil fuel-based power generation (Salehmin *et al.*, 2022). For instance, LCA of solar fuel cell hybrids indicates that although manufacturing fuel cell components involves energy-intensive processes and use of precious metals, the operational phase benefits from zero emissions and high efficiency, offsetting initial environmental costs (Mahesh & Sandhu, 2015). The operational efficiency of fuel cells often exceeding 50% compared to 30–40% for internal combustion engines contributes to reduced primary energy demand and lower pollutant releases (Barbir, 2012).

LCA results also vary by fuel cell type, renewable source, and system scale. Proton Exchange Membrane Fuel Cells (PEMFCs) paired with photovoltaic (PV) systems typically show better environmental performance than Solid Oxide Fuel Cells (SOFCs) integrated with biomass, due to lower material intensity and higher electrical efficiency in PEMFC-PV systems (Mosobi *et al.*, 2015). However, SOFCs provide benefits in combined heat and power (CHP) applications, improving overall system exergy efficiency (Kanani & Zahedi, 2024). End-of-life disposal and recycling of fuel cell components pose challenges due to scarce materials such as platinum catalysts, but advances in catalyst recovery and system modularity can enhance sustainability (Dunn *et al.*, 2011). LCA studies emphasize the importance of system design optimization, material selection, and renewable integration strategies to minimize life cycle environmental impacts (Dou *et al.*, 2022). Furthermore, LCA facilitates comparison between different energy storage options in hybrid configurations, including batteries, supercapacitors, and hydrogen storage, guiding choices that balance performance with environmental sustainability (Mahesh & Sandhu, 2015). In summary, LCA confirms that fuel cell-renewable hybrids have a favorable environmental profile, especially when powered by green hydrogen produced via renewable electricity.

A major driver for integrating fuel cells with renewable energy is the substantial potential to reduce greenhouse gas (GHG) emissions. Carbon footprint reduction refers to minimizing the total GHG emissions associated with energy production, use, and infrastructure. Fuel cells generate electricity through electrochemical reactions without direct combustion, producing only water as a byproduct when using hydrogen fuel (Barbir, 2012). When the hydrogen is sourced from renewable-powered electrolyzers, the entire energy cycle becomes nearly carbon-neutral (Salehmin *et al.*, 2022). This contrasts sharply with fossil-fuel-based systems that emit significant quantities of CO_2, methane, and other pollutants throughout fuel extraction, transportation, and combustion (Kanani & Zahedi, 2024).

Studies indicate that hybrid systems combining wind or solar power with fuel cells can reduce carbon emissions by 70–90% compared to conventional grid electricity (Mosobi *et al.*, 2015). For example, hydrogen generated via electrolysis powered by wind turbines, stored, and converted back to electricity in PEM fuel cells, achieves almost zero carbon footprint in operational phases (Mahesh & Sandhu, 2015). The carbon footprint reduction is enhanced in systems that utilize excess renewable generation, preventing curtailment and enabling energy storage for later use (Dou *et al.*, 2022). Furthermore, hybridization with batteries and supercapacitors optimizes system performance, reducing reliance on fossil backup and minimizing carbon-intensive ramping events (Dou *et al.*, 2022).

Beyond direct emissions, fuel cell-renewable hybrids contribute to lowering indirect emissions by displacing fossil-fuel infrastructures and reducing air pollutants such as NOx, SOx, and particulate matter, which have significant health and environmental impacts (Barbir, 2012). This reduction is critical in urban and industrial areas where air quality is a major concern. Policy incentives and carbon pricing mechanisms further incentivize adoption by translating carbon savings into economic benefits (Dou *et al.*, 2022). However, lifecycle carbon footprints must consider embedded emissions in manufacturing, transport, and disposal stages. Advances in low-carbon manufacturing techniques and recycling are critical to maximizing net carbon benefits (Dunn *et al.*, 2011).

Techno-economic analysis (TEA) evaluates the technical feasibility and economic viability of integrating fuel cells with renewable energy. This analysis combines capital expenditure (CAPEX), operational expenditure (OPEX), performance metrics, and revenue streams to assess cost-effectiveness over system lifetimes. Fuel cells have historically faced cost barriers due to expensive materials (platinum catalysts, membranes) and manufacturing complexity (Barbir, 2012). However, ongoing research and economies of scale have led to significant cost reductions, particularly for PEM fuel cells, which are widely used in renewable hybrids (Kanani & Zahedi, 2024).

Renewable energy sources like solar and wind have experienced dramatic cost declines in the past decade, with levelized cost of electricity (LCOE) dropping below fossil alternatives in many regions. Integrating these with fuel cells improves energy reliability and dispatchability, enhancing the overall value proposition (Mahesh & Sandhu, 2015). TEA studies highlight that the cost of hydrogen production via electrolysis remains a major factor affecting hybrid system economics. Advances in electrolyzer technology, including increased efficiency, reduced catalyst costs, and larger production scales, are pivotal for economic competitiveness (Salehmin *et al.*, 2022).

Hybrid system design must balance component costs with performance benefits. For example, incorporating batteries and supercapacitors increases upfront costs but reduces fuel consumption and extends fuel cell life, providing long-term savings (Mosobi *et al.*, 2015). Optimal system sizing and operational strategies minimize lifecycle costs while maximizing energy yield and reliability (Dou *et al.*, 2022). Financial incentives, carbon credits, and regulatory support improve economic feasibility. Additionally, techno-economic assessments consider externalities such as environmental benefits and social acceptance, which influence investment decisions (Dou *et al.*, 2022).

5.9 CHALLENGES AND LIMITATIONS OF INTEGRATION OF FUEL CELLS WITH RENEWABLE ENERGY SOURCES

The integration of fuel cells with renewable energy sources promises a clean and efficient energy future. However, several significant challenges and limitations impede the widespread adoption and optimization of these hybrid systems.

One of the foremost barriers to the integration of fuel cells with renewable energy systems is the substantial upfront capital cost. Fuel cells, especially Proton Exchange Membrane Fuel Cells (PEMFCs) and Solid Oxide Fuel Cells (SOFCs), rely heavily on expensive materials such as platinum-group metal catalysts, advanced membranes, and durable bipolar plates, all of which contribute to high manufacturing costs (Barbir, 2012). The catalyst cost alone can constitute up to 40% of the total fuel cell stack cost (Brandon *et al.*, 2003). Although research is ongoing to reduce precious metal loading and explore alternative catalyst materials, the balance between cost reduction and performance remains a significant hurdle (Ibrahim, 2024). Additionally, manufacturing processes for fuel cells are complex and require precision, which further elevates production costs (Dunn *et al.*, 2011).

Capital costs also escalate when integrating fuel cells with renewable energy sources because additional components such as electrolyzers for hydrogen production, power electronics, and energy storage units are required (Deshmukh & Deshmukh, 2008). These systems often require customized engineering and advanced control systems, leading to higher installation and commissioning expenses compared to conventional fossil fuel-based systems (Olatomiwa *et al.*, 2016).

Although the costs of fuel cells and renewable technologies like solar PV and wind have been decreasing, fuel cells still lag behind in terms of cost competitiveness (Pfriem, 1991). The overall system cost can limit deployment, particularly in cost-sensitive markets or regions lacking supportive policies or incentives. This economic barrier calls for continued innovation in materials, scalable manufacturing techniques, and supportive financing mechanisms to enhance adoption (Wang *et al.*, 2017).

Fuel cell durability and operational lifetime present another critical challenge for their integration with renewable energy. The dynamic nature of renewable energy sources, characterized by intermittent generation and fluctuating loads, imposes frequent start-stop cycles and variable operating conditions on fuel cells (Brandon *et al.*, 2003). These fluctuations accelerate degradation mechanisms such as catalyst poisoning, membrane thinning, and corrosion of cell components, reducing performance over time (Barbir, 2012). PEM fuel cells, which operate at relatively low temperatures, face membrane dehydration and catalyst sintering, especially under cyclic loading typical in renewable hybrids (Ibrahim, 2024). SOFCs, operating at high temperatures, suffer from thermal stresses and material compatibility issues, which can lead to cracks, delamination, and reduced structural integrity (Dunn *et al.*, 2011).

Moreover, impurities in hydrogen fuel—common when hydrogen is produced from renewable sources via electrolysis—can poison catalysts and reduce cell efficiency and durability (Deshmukh & Deshmukh, 2008). Contamination control and high-purity hydrogen supply are thus essential but add complexity and cost. The limited lifetime of fuel cells, typically ranging from 5,000 to 10,000 hours for PEMFCs and up to 40,000 hours for SOFCs, poses economic challenges due to frequent replacement or refurbishment requirements (Brandon *et al.*, 2003). This issue is compounded by the hybrid system's expected lifetime, which often exceeds the fuel cell's durability, complicating maintenance and operational planning.

Hydrogen storage and the associated infrastructure represent significant technical and logistical challenges in fuel cell-renewable integration. Hydrogen, being the primary fuel for many fuel cell systems, requires safe, efficient, and economical storage solutions that can meet variable demand profiles typical of renewable energy generation (Barbir, 2012). Current hydrogen storage methods include compressed gas tanks, cryogenic liquid storage, and materials-based storage such as metal hydrides and chemical carriers. Each method faces trade-offs in terms of energy density, weight, volume, cost, and safety. Compressed hydrogen tanks, the most mature technology, require high pressures (350–700 bar), posing safety concerns and infrastructure demands (Brandon *et al.*, 2003).

The integration of fuel cells with variable renewable energy sources leads to system complexity that poses challenges in design, operation, and control. Hybrid systems must coordinate multiple components, including fuel cells, electrolyzers, renewable generators, energy storage devices, and power converters, to ensure reliable and efficient operation (Deshmukh & Deshmukh, 2008). Renewable energy's intermittent and unpredictable nature results in fluctuating power input and load demand. Fuel cells, while efficient, have limitations in dynamic response and ramp rates compared to batteries or supercapacitors (Ibrahim, 2024). Balancing these characteristics requires sophisticated energy management systems that can optimize power flow, maximize efficiency, and protect system components.

Control strategies must address issues such as load following, peak shaving, fault detection, and degradation management (Olatomiwa *et al.*, 2016). Algorithms for energy management range from rule-based approaches to advanced model predictive controls and artificial intelligence techniques, each with varying degrees of complexity and computational demand (Wang *et al.*, 2017).

Integration complexity also increases due to the need to maintain system stability under variable operating conditions. Transients in renewable power generation can cause voltage and frequency fluctuations, necessitating fast-acting control systems and ancillary services (Brandon *et al.*, 2003). Maintenance and diagnostics are complicated by the multi-component nature of hybrid systems, requiring integrated monitoring and fault detection systems to minimize downtime and optimize lifecycle costs (Dunn *et al.*, 2011). Furthermore, interoperability issues among components from different manufacturers and standards add to system integration challenges.

5.10 WAY FORWARD ON THE CHALLENGES AND LIMITATIONS OF INTEGRATION OF FUEL CELLS WITH RENEWABLE ENERGY SOURCES

The integration of fuel cells with renewable energy sources holds great promise for a sustainable and decarbonized energy future. However, overcoming the existing challenges and limitations requires a multifaceted approach combining technological innovation, system optimization, and supportive policies.

A major way forward to reduce the high capital costs and improve the durability and performance of fuel cells lies in novel materials and innovative designs. Recent research emphasizes the development of low-cost and abundant catalyst materials to replace precious metals like platinum, which are primary contributors to fuel cell expenses. Transition metal-based catalysts, such as those incorporating iron, cobalt, and nickel, have shown promise in achieving comparable activity for oxygen reduction reactions while significantly reducing costs (Lang *et al.*, 2020).

Moreover, advancements in membrane technology, including the development of more chemically and mechanically robust proton exchange membranes, enhance the longevity and operational stability of fuel cells under variable load conditions typical of renewable integration (Zhang *et al.*, 2025). Novel polymeric materials and composite membranes are being explored to improve proton conductivity, reduce fuel crossover, and withstand dehydration or flooding issues (Zakaria *et al.*, 2020).

Fuel cell stack designs are also evolving toward modular and scalable architectures that facilitate easier manufacturing, maintenance, and integration with other system components. Lightweight and corrosion-resistant bipolar plates made from carbon-based composites or coated metals help reduce weight and enhance electrical efficiency (Pukha *et al.*, 2023). Additionally, micro-structured flow fields and optimized cell geometries improve reactant distribution and heat management, contributing to better performance and durability (Zhao *et al.*, 2024).

Solid oxide fuel cells (SOFCs), operating at high temperatures, are benefiting from materials with improved thermal stability and reduced degradation rates, which help address the durability issues associated with thermal cycling in hybrid systems (Fallah Vostakola & Amini Horri, 2021).

Efforts to lower SOFC operating temperatures also reduce material stress and expand their applicability in renewable-hybrid configurations.

The production of renewable hydrogen is fundamental to fuel cell integration with renewable energy sources. Green hydrogen produced via water electrolysis powered by renewable electricity has emerged as a key enabler for decarbonizing fuel cell supply chains and mitigating carbon footprint challenges (Deshmukh & Deshmukh, 2008).

Recent advances in electrolyzer technologies, including Proton Exchange Membrane (PEM), alkaline, and solid oxide electrolyzers, have improved efficiency, scalability, and operational flexibility (Sebbahi *et al.*, 2022). PEM electrolyzers are particularly suitable for dynamic operation with fluctuating renewable power, thanks to their fast response times and compact design (Koponen *et al.*, 2017). Innovations such as the use of non-precious metal catalysts and advanced membranes are reducing electrolyzer costs and enhancing durability (King *et al.*, 2019). Solid oxide electrolyzers (SOEs) offer high efficiency at elevated temperatures by utilizing waste heat from industrial processes or renewable thermal sources, which can further improve the overall system efficiency in integrated settings (Ferrete *et al.*, 2025). However, ongoing research focuses on extending SOE durability and lowering operating temperatures to enable broader adoption.

Beyond water electrolysis, photoelectrochemical (PEC) water splitting and biological hydrogen production methods are emerging as alternative green hydrogen production pathways, although these technologies are still largely in developmental stages (Zhao *et al.*, 2023). Hybrid renewable energy systems coupled with hydrogen production can utilize surplus renewable generation to produce and store hydrogen, which acts as an energy carrier and buffer for times when renewable generation is low (Deshmukh & Deshmukh, 2008). This sector stands to benefit significantly from scaling and cost reductions, facilitating the commercialization of fuel cell hybrids.

Artificial intelligence (AI) and smart grid technologies provide a vital pathway to address the complexity and control challenges of integrating fuel cells with renewable energy sources. The intermittent and variable nature of renewables requires advanced energy management systems that optimize generation, storage, and consumption in real-time to maximize efficiency and system stability (Khan *et al.*, 2022). AI algorithms, including machine learning and deep learning, are increasingly used for predictive maintenance, load forecasting, fault detection, and optimization of hybrid system operations (Okafor *et al.*, 2025). For instance, AI-driven prognostics can predict fuel cell degradation and schedule maintenance proactively, thus enhancing system reliability and reducing downtime (Ibrahim, 2024).

Smart grids enable two-way communication between energy producers, consumers, and storage systems, allowing dynamic adjustment of power flows and better integration of distributed energy resources, including fuel cells (Rehmani *et al.*, 2018). AI-powered controllers can respond to grid demands and renewable fluctuations by intelligently managing fuel cell operation alongside batteries and supercapacitors to provide a seamless power supply (Deshmukh & Deshmukh, 2008). Furthermore, integrating blockchain technology with AI and smart grids can enhance transparency, security, and efficiency in energy transactions within decentralized hybrid systems (Ghadi *et al.*, 2025).

Incorporating AI in energy management also supports demand response strategies, enabling flexible load adjustment to match renewable generation profiles, thus reducing curtailment and improving economic viability (Khan *et al.*, 2022). Together, AI and smart grids form the backbone for next-generation integrated fuel cell-renewable energy systems by enhancing control, flexibility, and resilience.

5.11 CONCLUSION

The importance of integrated energy systems cannot be overstated in the global quest for a sustainable energy future. As climate change concerns intensify and fossil fuel reserves dwindle, the need for clean, dependable, and adaptable energy solutions becomes increasingly urgent. Integrated systems utilizing fuel cells not only diversify the energy mix but also improve energy access, particularly in off-grid and remote areas where conventional energy infrastructure is limited or non-existent. Moreover, such systems contribute to decarbonizing critical sectors like transportation and heavy industry, which are traditionally hard to electrify with renewables alone. To realize the full potential of fuel cell-renewable integrations, several policy and research recommendations must be considered. Governments should prioritize the establishment of supportive regulatory frameworks, financial incentives, and infrastructure investments aimed at promoting fuel cell deployment and renewable energy development. Dedicated funding for demonstration projects and public-private partnerships can accelerate technology adoption and market penetration. On the research front, efforts should focus on improving fuel cell durability, reducing production costs, and developing non-precious metal catalysts to make the technology more economically competitive. Additionally, optimizing energy management systems, advancing hydrogen storage solutions, and enhancing system integration strategies remain essential areas for further study. Interdisciplinary collaborations among energy technologists, policy experts, and industry stakeholders will be pivotal in addressing the technical, economic, and policy challenges that currently limit the widespread adoption of integrated fuel cell systems. In conclusion, integrating fuel cells with renewable energy sources presents a viable and forward-looking strategy for achieving a clean, reliable, and sustainable global energy future. It is imperative that stakeholders collectively pursue innovation, supportive policies, and collaborative research to fully harness the environmental and economic benefits of these integrated energy systems.

REFERENCES

Abdelkareem, M. A., Elsaid, K., Wilberforce, T., Kamil, M., Sayed, E. T., & Olabi, A. (2021). Environmental aspects of fuel cells: A review. *Science of The Total Environment*, *752*, 141803.

Alamri, B. R., & Alamri, A. R. (2009). Technical review of energy storage technologies when integrated with intermittent renewable energy. In *2009 international conference on sustainable power generation and supply* (pp. 1–5). IEEE.

Appleby, A. J. (1988). *Fuel cell handbook*.

Appleby, A. J., & Foulkes, F. R. (1988). *Fuel cell handbook* (2nd ed.). Van Nostrand Reinhold.

Arul, P. G., Ramachandaramurthy, V. K., & Rajkumar, R. K. (2015). Control strategies for a hybrid renewable energy system: A review. *Renewable and Sustainable Energy Reviews*, *42*, 597–608.

Arvanitoyannis, I. S. (2008). ISO 14040: Life cycle assessment (LCA)–principles and guidelines. *Waste Management for the Food Industries*, 97–132.

Asif, M., & Muneer, T. (2007). Energy supply, its demand and security issues for developed and emerging economies. *Renewable and Sustainable Energy Reviews*, *11*(7), 1388–1413.

Balat, M., & Balat, H. (2009). Recent trends in global production and utilization of bio-ethanol fuel. *Applied Energy*, *86*(11), 2273–2282.

Barbir, F. (2006). PEM fuel cells. In *Fuel cell technology: Reaching towards commercialization* (pp. 27–51). Springer London.

Barbir, F. (2012). *PEM fuel cells: Theory and practice*. Academic Press.

Benghanem, M., Mellit, A., Almohamadi, H., Haddad, S., Chettibi, N., Alanazi, A. M., … & Alzahrani, A. (2023). Hydrogen production methods based on solar and wind energy: A review. *Energies*, *16*(2), 757.

Bertani, R. (2009). Geothermal energy: An overview on resources and potential. In *Proceedings of the international conference on national development of geothermal energy use, Slovakia*.

Bhaskar, T., Bhavya, B., Singh, R., Naik, D. V., Kumar, A., & Goyal, H. B. (2011). Thermochemical conversion of biomass to biofuels. In *Biofuels* (pp. 51–77). Academic Press.

Bhattacharyya, S. C. (2006). Energy access problem of the poor in India: Is rural electrification a remedy? *Energy Policy*, *34*(18), 3387–3397.

Bigdeli, N. (2015). Optimal management of hybrid PV/fuel cell/battery power system: A comparison of optimal hybrid approaches. *Renewable and Sustainable Energy Reviews*, *42*, 377–393.

Blanco, M. I. (2009). The economics of wind energy. *Renewable and Sustainable Energy Reviews*, *13*(6–7), 1372–1382.

Brandon, N. P., Skinner, S., & Steele, B. C. (2003). Recent advances in materials for fuel cells. *Annual Review of Materials Research*, *33*(1), 183–213.

Bridgwater, A. V. (2012). Review of fast pyrolysis of biomass and product upgrading. *Biomass and Bioenergy*, *38*, 68–94.
Brusso, B. C. (2021). Electrochemistry and the development of the hydrogen fuel cell [history]. *IEEE Industry Applications Magazine*, *27*(5), 8–13.
Burton, T., Jenkins, N., Sharpe, D., & Bossanyi, E. (2011). *Wind energy handbook*. John Wiley & Sons.
Carrette, L., Friedrich, K. A., & Stimming, U. (2000). Fuel cells: Principles, types, fuels, and applications. *ChemPhysChem*, *1*(4), 162–193.
Carrette, L., Friedrich, K. A., & Stimming, U. (2001). Fuel cells-fundamentals and applications. *Fuel Cells*, *1*.
Change, U. C. (2015). *The Paris agreement*. United Nations.
Chen, H., Cong, T. N., Yang, W., Tan, C., Li, Y., & Ding, Y. (2009). Progress in electrical energy storage system: A critical review. *Progress in Natural Science*, *19*(3), 291–312.
Chen, W. H., Peng, J., & Bi, X. T. (2015). A state-of-the-art review of biomass torrefaction, densification and applications. *Renewable and Sustainable Energy Reviews*, *44*, 847–866.
Chu, S., & Majumdar, A. (2012). Opportunities and challenges for a sustainable energy future. *Nature*, *488*(7411), 294–303.
Council, G. W. E. (2021). GWEC global wind report 2021. *Global Wind Energy Council: Brussels, Belgium*, 80.
Dagar, A., Gupta, P., & Niranjan, V. (2021). Microgrid protection: A comprehensive review. *Renewable and Sustainable Energy Reviews*, *149*, 111401.
Debe, M. K. (2012). Electrocatalyst approaches and challenges for automotive fuel cells. *Nature*, *486*(7401), 43–51.
Demirbas, A. (2008). Importance of biomass energy sources for Turkey. *Energy Policy*, *36*(2), 834–842.
Denholm, P., & Hand, M. (2011). Grid flexibility and storage required to achieve very high penetration of variable renewable electricity. *Energy Policy*, *39*(3), 1817–1830.
Deshmukh, M. K., & Deshmukh, S. S. (2008). Modeling of hybrid renewable energy systems. *Renewable and Sustainable Energy Reviews*, *12*(1), 235–249.
Dickson, M. H., & Fanelli, M. (2013). *Geothermal energy: Utilization and technology*. Routledge.
Dou, Z., Zhang, C., Wang, W., Wang, D., Zhang, Q., Cai, Y., & Fan, R. (2022). Review on key technologies and typical applications of multi-station integrated energy systems. *Global Energy Interconnection*, *5*(3), 309–327.
Dunn, B., Kamath, H., & Tarascon, J. M. (2011). Electrical energy storage for the grid: A battery of choices. *Science*, *334*(6058), 928–935.
Efficiency, I. E., & Stimulus, E. (2021). *International Energy Agency. 2021.*
Ellabban, O., Abu-Rub, H., & Blaabjerg, F. (2014). Renewable energy resources: Current status, future prospects and their enabling technology. *Renewable and Sustainable Energy Reviews*, *39*, 748–764.
Eriksson, E. L. V., & Gray, E. M. (2017). Optimization and integration of hybrid renewable energy hydrogen fuel cell energy systems–A critical review. *Applied Energy*, *202*, 348–364.
Fallah Vostakola, M., & Amini Horri, B. (2021). Progress in material development for low-temperature solid oxide fuel cells: A review. *Energies*, *14*(5), 1280.
Ferrete, F., Molina, A., Cabello González, G. M., Moreno-Racero, Á., Olmedo, H., & Iranzo, A. (2025). Solid Oxide Electrolyzers Process Integration: A Comprehensive Review. *Processes*, *13*(8), 2656.
Ferriday, T. B., & Middleton, P. H. (2021). Alkaline fuel cell technology-A review. *International Journal of Hydrogen Energy*, *46*(35), 18489–18510.
Ghadi, Y. Y., Mazhar, T., Shahzad, T., Jaghdam, I. H., Khan, S., Khan, M. A., & Hamam, H. (2025). A hybrid AI-Blockchain security framework for smart grids. *Scientific Reports*, *15*(1), 20882.
Goodenough, J. B., & Huang, Y. H. (2007). Alternative anode materials for solid oxide fuel cells. *Journal of Power Sources*, *173*(1), 1–10.
Graabak, I., & Korpås, M. (2016). Variability characteristics of European wind and solar power resources—A review. *Energies*, *9*(6), 449.
Gracey, E. O., & Verones, F. (2016). Impacts from hydropower production on biodiversity in an LCA framework—review and recommendations. *The International Journal of Life Cycle Assessment*, *21*, 412–428.
Green, M., Dunlop, E., Hohl-Ebinger, J., Yoshita, M., Kopidakis, N., & Hao, X. (2021). Solar cell efficiency tables (version 57). *Progress in Photovoltaics: Research and Applications*, *29*(1), 3–15.
Haider, R., Wen, Y., Ma, Z. F., Wilkinson, D. P., Zhang, L., Yuan, X., ... & Zhang, J. (2021). High temperature proton exchange membrane fuel cells: Progress in advanced materials and key technologies. *Chemical Society Reviews*, *50*(2), 113–187.
Hassan, Q., Viktor, P., Al-Musawi, T. J., Ali, B. M., Algburi, S., Alzoubi, H. M., ... & Jaszczur, M. (2024). The renewable energy role in the global energy Transformations. *Renewable Energy Focus*, *48*, 100545.

Herdem, M. S., Mazzeo, D., Matera, N., Baglivo, C., Khan, N., Congedo, P. M., & De Giorgi, M. G. (2024). A brief overview of solar and wind-based green hydrogen production systems: Trends and standardization. *International Journal of Hydrogen Energy*, *51*, 340–353.
Ibrahim, M. M. (2024). Energy management strategies of hybrid renewable energy systems: A review. *Wind Engineering*, *48*(1), 133–161.
IREA. (2020). Renewable power generation costs in 2020.
Jacobson, M. Z., Delucchi, M. A., Bauer, Z. A., Goodman, S. C., Chapman, W. E., Cameron, M. A., … & Yachanin, A. S. (2017). 100% clean and renewable wind, water, and sunlight all-sector energy roadmaps for 139 countries of the world. *Joule*, *1*(1), 108–121.
Jäger-Waldau, A. (2019). PV status report 2019. *Publications Office of the European Union: Luxembourg*, 7–94.
Jerez, S., Trigo, R. M., Sarsa, A., Lorente-Plazas, R., Pozo-Vázquez, D., & Montávez, J. P. (2013). Spatio-temporal complementarity between solar and wind power in the Iberian Peninsula. *Energy Procedia*, *40*, 48–57.
Kalogirou, S. A. (2004). Solar thermal collectors and applications. *Progress in Energy and Combustion Science*, *30*(3), 231–295.
Kanani, B., & Zahedi, A. (2024). Step toward sustainable development through the integration of renewable energy systems with fuel cells: A review. *Sustainable Energy Technologies and Assessments*, *70*, 103935.
Khan, A., Bressel, M., Davigny, A., Abbes, D., & Ould Bouamama, B. (2025). Comprehensive review of hybrid energy systems: Challenges, applications, and optimization strategies. *Energies*, *18*(10), 2612.
Khan, K. A., Paul, S., Zobayer, A., & Hossain, S. S. (2013). A study on solar photovoltaic conversion. *International Journal of Scientific and Engineering Research*, *4*.
Khan, M. A., Saleh, A. M., Waseem, M., & Sajjad, I. A. (2022). Artificial intelligence enabled demand response: Prospects and challenges in smart grid environment. *IEEE Access*, *11*, 1477–1505.
Kibler, K. M., & Tullos, D. D. (2013). Cumulative biophysical impact of small and large hydropower development in Nu River, *China. Water Resources Research*, *49*(6), 3104–3118.
King, L. A., Hubert, M. A., Capuano, C., Manco, J., Danilovic, N., Valle, E., … & Jaramillo, T. F. (2019). A non-precious metal hydrogen catalyst in a commercial polymer electrolyte membrane electrolyser. *Nature Nanotechnology*, *14*(11), 1071–1074.
Koponen, J., Kosonen, A., Ruuskanen, V., Huoman, K., Niemelä, M., & Ahola, J. (2017). Control and energy efficiency of PEM water electrolyzers in renewable energy systems. *International Journal of Hydrogen Energy*, *42*(50), 29648–29660.
Kordesch, K., & Simader, G. (1996). Fuel cells and their applications.
Kreuer, K. D. (2001). On the development of proton conducting polymer membranes for hydrogen and methanol fuel cells. *Journal of Membrane Science*, *185*(1), 29–39.
Krishna, K. S., & Kumar, K. S. (2015). A review on hybrid renewable energy systems. *Renewable and Sustainable Energy Reviews*, *52*, 907–916.
Kulkarni, A., & Giddey, S. (2012). Materials issues and recent developments in molten carbonate fuel cells. *Journal of Solid State Electrochemistry*, *16*, 3123–3146.
Kumar, K., Kumar, P., & Kar, S. (2024). A review of microgrid protection for addressing challenges and solutions. *Renewable Energy Focus*, 100572.
Lai, C. S., & McCulloch, M. D. (2016). Levelized cost of energy for PV and grid scale energy storage systems. *arXiv preprint arXiv:1609.06000*.
Lang, P., Yuan, N., Jiang, Q., Zhang, Y., & Tang, J. (2020). Recent advances and prospects of metal-based catalysts for oxygen reduction reaction. *Energy Technology*, *8*(3), 1900984.
Lee, J., Kim, S., You, S., & Park, Y. K. (2023). Bioenergy generation from thermochemical conversion of lignocellulosic biomass-based integrated renewable energy systems. *Renewable and Sustainable Energy Reviews*, *178*, 113240.
Lehmann, J., & Joseph, S. (Eds.). (2024). *Biochar for environmental management: Science, technology and implementation*. Taylor & Francis.
Lewis, N. S., & Nocera, D. G. (2006). Powering the planet: Chemical challenges in solar energy utilization. *Proceedings of the National Academy of Sciences*, *103*(43), 15729–15735.
Litster, S., & McLean, G. J. J. O. P. S. (2004). PEM fuel cell electrodes. *Journal of Power Sources*, *130*(1–2), 61–76.
Lund, H. (2024). *Renewable energy systems: A smart energy systems approach to the choice and modeling of fully decarbonized societies*. Elsevier.
Lund, H., Andersen, A. N., Østergaard, P. A., Mathiesen, B. V., & Connolly, D. (2012). From electricity smart grids to smart energy systems–a market operation based approach and understanding. *Energy*, *42*(1), 96–102.

Lund, J. W., Freeston, D. H., & Boyd, T. L. (2011). Direct utilization of geothermal energy 2010 worldwide review. *Geothermics*, *40*(3), 159–180.

Luque, A., & Hegedus, S. (Eds.). (2011). *Handbook of photovoltaic science and engineering*. John Wiley & Sons.

Lykas, P., Georgousis, N., Bellos, E., & Tzivanidis, C. (2023). A comprehensive review of solar-driven multigeneration systems with hydrogen production. *International Journal of Hydrogen Energy*, *48*(2), 437–477.

Mahesh, A., & Sandhu, K. S. (2015). Hybrid wind/photovoltaic energy system developments: Critical review and findings. *Renewable and Sustainable Energy Reviews*, *52*, 1135–1147.

Mahla, P. K., Vithalani, P. C., & Bhatt, N. S. (2022). Biomethanation: Advancements for upgrading biomethane using biogas technologies. *Industrial Microbiology and Biotechnology*, 487–504.

Manwell, J. F., McGowan, J. G., & Rogers, A. L. (2010). *Wind energy explained: Theory, design and application*. John Wiley & Sons.

Markandya, A., Sampedro, J., Smith, S. J., Van Dingenen, R., Pizarro-Irizar, C., Arto, I., & González-Eguino, M. (2018). Health co-benefits from air pollution and mitigation costs of the Paris Agreement: A modelling study. *The Lancet Planetary Health*, *2*(3), e126–e133.

Marouani, I., Guesmi, T., Alshammari, B. M., Alqunun, K., Alzamil, A., Alturki, M., & Hadj Abdallah, H. (2023). Integration of renewable-energy-based green hydrogen into the energy future. *Processes*, *11*(9), 2685.

Masson-Delmotte, V., Zhai, P., Pirani, A., Connors, S. L., Péan, C., Berger, S., … & Zhou, B. (2021). Climate change 2021: The physical science basis. *Contribution of Working Group I to the Sixth Assessment Report of the Intergovernmental Panel on Climate Change*, *2*(1), 2391.

Masson-Delmotte, V., Zhai, P., Pörtner, H. O., Roberts, D., Skea, J., Shukla, P. R., … & Pidcoc, R. (2018). Intergovernmental Panel on Climate Change (IPCC). Global Warming of 1.5o C: An IPCC Special Report on the Impacts of Global Warming of 1.5o C Above Pre-Industrial Levels and Related Global Greenhouse Gas Emission Pathways, in the Context of Strengthening the Global Response to the Threat of Climate Change. *Sustainable Development, and Efforts to Eradicate Poverty*. Available online: https://www.ipcc.ch/sr15/ (accessed on 18 October 2019).

Mateo Romero, H. F., González Rebollo, M. Á., Cardeñoso-Payo, V., Alonso Gómez, V., Redondo Plaza, A., Moyo, R. T., & Hernández-Callejo, L. (2022). Applications of artificial intelligence to photovoltaic systems: A review. *Applied Sciences*, *12*(19), 10056.

Mauritz, K. A., & Moore, R. B. (2004). State of understanding of Nafion. *Chemical Reviews*, *104*(10), 4535–4586.

McKendry, P. (2002). Energy production from biomass (part 1): Overview of biomass. *Bioresource Technology*, *83*(1), 37–46.

McPhail, S. J. (2011). Status and challenges of molten carbonate fuel cells. *Advances in Science and Technology*, *72*, 283–290.

McPhail, S. J., Leto, L., Della Pietra, M., Cigolotti, V., & Moreno, A. (2015). International status of molten carbonate fuel cells technology. *Advanced Fuel Cells Implementing Agreement, Annex*.

Mench, M. M. (2008). *Fuel cell engines*. John Wiley & Sons.

Mey, F., & Diesendorf, M. (2018). Who owns an energy transition? Strategic action fields and community wind energy in Denmark. *Energy Research & Social Science*, *35*, 108–117.

Mosobi, R. W., Chichi, T., & Gao, S. (2015). Power quality analysis of hybrid renewable energy system. *Cogent Engineering*, *2*(1), 1005000.

Musial, W., & Ram, B. (2010). *Large-scale offshore wind power in the United States: Assessment of opportunities and barriers* (No. NREL/TP-500-40745). National Renewable Energy Lab. (NREL), Golden, CO (United States).

Noman, F. M., Alkawsi, G. A., Abbas, D., Alkahtani, A. A., Tiong, S. K., & Ekanayake, J. (2020). Comprehensive review of wind energy in Malaysia: Past, present, and future research trends. *IEEE Access*, *8*, 124526–124543.

Ogundele, O. D., Amoo, I. A., Adesina, A. O., Abidemi, A., & Bisi-Omotosho, A. (2024). Biotechnology for renewable fuel and chemicals. In *Microbial biotechnology for bioenergy* (pp. 325–345). Elsevier.

Okafor, C. J., Adeniran, A. A., & Adeniran, A. O. (2025). Machine learning and AI for predictive maintenance and grid integration of wind farms. *Soft Computing Fusion with Applications*, *2*(2), 62–73.

Olatomiwa, L., Mekhilef, S., Ismail, M. S., & Moghavvemi, M. (2016). Energy management strategies in hybrid renewable energy systems: A review. *Renewable and Sustainable Energy Reviews*, *62*, 821–835.

Park, N. G. (2015). Perovskite solar cells: An emerging photovoltaic technology. *Materials Today*, *18*(2), 65–72.

Pfriem, H. J. (1991). Overview of the cooperative program on hydrogen storage, conversion and safety of the international energy agency. *International Journal of Hydrogen Energy*, *16*(5), 329–338.
Pukha, V. E., Glukhov, A. A., Belmesov, A. A., Kabachkov, E. N., Khodos, I. I., Khadem, M., ... & Karaseov, P. A. (2023). Corrosion-resistant nanostructured carbon-based coatings for applications in fuel cells based on bipolar plates. *Vacuum*, *218*(8), 112643.
Qasem, N. A., & Abdulrahman, G. A. (2024). A recent comprehensive review of fuel cells: History, types, and applications. *International Journal of Energy Research*, *2024*(1), 7271748.
Rehmani, M. H., Reisslein, M., Rachedi, A., Erol-Kantarci, M., & Radenkovic, M. (2018). Integrating renewable energy resources into the smart grid: Recent developments in information and communication technologies. *IEEE Transactions on Industrial Informatics*, *14*(7), 2814–2825.
Ritchie, H., & Roser, M. (2017). CO_2 and other greenhouse gas emissions. *Our World in Data.*
Sachdev, H. S., Akella, A. K., & Kumar, N. (2015). Analysis and evaluation of small hydropower plants: A bibliographical survey. *Renewable and Sustainable Energy Reviews*, *51*, 1013–1022.
Sadeghian, O., Shotorbani, A. M., Ghassemzadeh, S., & Mohammadi-Ivatloo, B. (2025). Energy management of hybrid fuel cell and renewable energy based systems-A review. *International Journal of Hydrogen Energy*, *107*, 135–163.
Saidur, R., Rahim, N. A., Islam, M. R., & Solangi, K. H. (2011). Environmental impact of wind energy. *Renewable and Sustainable Energy Reviews*, *15*(5), 2423–2430.
Salehmin, M. N. I., Husaini, T., Goh, J., & Sulong, A. B. (2022). High-pressure PEM water electrolyser: A review on challenges and mitigation strategies towards green and low-cost hydrogen production. *Energy Conversion and Management*, *268*, 115985.
Sebbahi, S., Nabil, N., Alaoui-Belghiti, A., Laasri, S., Rachidi, S., & Hajjaji, A. (2022). Assessment of the three most developed water electrolysis technologies: alkaline water electrolysis, proton exchange membrane and solid-oxide electrolysis. *Materials Today: Proceedings*, *66*(4), 140–145.
Sharaf, O. Z., & Orhan, M. F. (2014). An overview of fuel cell technology: Fundamentals and applications. *Renewable and Sustainable Energy Reviews*, *32*, 810–853.
Sims, R. E., Hastings, A., Schlamadinger, B., Taylor, G., & Smith, P. (2006). Energy crops: Current status and future prospects. *Global change biology*, *12*(11), 2054–2076.
Singhal, S. C., & Kendall, K. (Eds.). (2003). *High-temperature solid oxide fuel cells: Fundamentals, esign and applications*. Elsevier.
Snyder, B., & Kaiser, M. J. (2009). Ecological and economic cost-benefit analysis of offshore wind energy. *Renewable Energy*, *34*(6), 1567–1578.
Sovacool, B. K. (2016). How long will it take? Conceptualizing the temporal dynamics of energy transitions. *Energy Research & Social Science*, *13*, 202–215.
Srinivasan, S. (2006). *Fuel cells: From fundamentals to applications*. Springer Science & Business media.
Steele, B. C., & Heinzel, A. (2001). Materials for fuel-cell technologies. *Nature*, *414*(6861), 345–352.
Sun, C., Hui, R., & Roller, J. (2010). Cathode materials for solid oxide fuel cells: A review. *Journal of Solid State Electrochemistry*, *14*, 1125–1144.
Twidell, J. (2021). *Renewable energy resources*. Routledge.
Ugah, E. T. A., Ndubuisi, O. G., Ali, E. P. S. E., Obiorah, C. A. R., Nesiama, O., Agbakhamen, E. C. O., & Okoro, E. O. P. (2025). Renewable energy and sustainable development: Emerging trends and technologies. *IPHO-Journal of Advance Research in Science and Engineering (ISSN. 3050-8797)*, *3*(02), 16–30.
Wang, J., Zhong, H., Ma, Z., Xia, Q., & Kang, C. (2017). Review and prospect of integrated demand response in the multi-energy system. *Applied Energy*, *202*, 772–782.
Wang, Y., Chen, K. S., Mishler, J., Cho, S. C., & Adroher, X. C. (2011). A review of polymer electrolyte membrane fuel cells: Technology, applications, and needs on fundamental research. *Applied Energy*, *88*(4), 981–1007.
Waseem, M., Amir, M., Lakshmi, G. S., Harivardhagini, S., & Ahmad, M. (2023). Fuel cell-based hybrid electric vehicles: An integrated review of current status, key challenges, recommended policies, and future prospects. *Green Energy and Intelligent Transportation*, *2*(6), 100121.
Weschenfelder, F., Leite, G. D. N. P., da Costa, A. C. A., de Castro Vilela, O., Ribeiro, C. M., Ochoa, A. A. V., & Araújo, A. M. (2020). A review on the complementarity between grid-connected solar and wind power systems. *Journal of Cleaner Production*, *257*, 120617.
Widén, J., Carpman, N., Castellucci, V., Lingfors, D., Olauson, J., Remouit, F., ... & Waters, R. (2015). Variability assessment and forecasting of renewables: A review for solar, wind, wave and tidal resources. *Renewable and Sustainable Energy Reviews*, *44*, 356–375.
Wu, F., & Pfenninger, S. (2023). Challenges and opportunities for bioenergy in Europe: National deployment, policy support, and possible future roles. *Bioresource Technology Reports*, *22*, 101430.

Zakaria, Z., Shaari, N., Kamarudin, S. K., Bahru, R., & Musa, M. T. (2020). A review of progressive advanced polymer nanohybrid membrane in fuel cell application. *International Journal of Energy Research*, *44*(11), 8255–8295.

Zawodzinski, T. A., Derouin, C., Radzinski, S., Sherman, R. J., Smith, V. T., Springer, T. E., & Gottesfeld, S. (1993). Water uptake by and transport through Nafion® 117 membranes. *Journal of the Electrochemical Society*, *140*(4), 1041.

Zhang, F., Zu, B., Wang, B., Qin, Z., Yao, J., Wang, Z., … & Jiao, K. (2025). Developing long-durability proton-exchange membrane fuel cells. *Joule*, *9*(3)1132–1151.

Zhang, H. L., Baeyens, J., Degrève, J., & Cáceres, G. (2013). Concentrated solar power plants: Review and design methodology. *Renewable and Sustainable Energy Reviews*, *22*, 466–481.

Zhang, S., Yuan, X. Z., Hin, J. N. C., Wang, H., Friedrich, K. A., & Schulze, M. (2009). A review of platinum-based catalyst layer degradation in proton exchange membrane fuel cells. *Journal of Power Sources*, *194*(2), 588–600.

Zhao, G., Moulijn, J. A., Kapteijn, F., Dautzenberg, F. M., Xu, B., & Lu, Y. (2024). Monolithic fiber/foam-structured catalysts: beyond honeycombs and micro-channels. *Catalysis Reviews*, *66*(5), 1870–1950.

Zhao, Y., Niu, Z., Zhao, J., Xue, L., Fu, X., & Long, J. (2023). Recent advancements in photoelectrochemical water splitting for hydrogen production. *Electrochemical Energy Reviews*, *6*(1), 14.

Zhou, X., Wang, F., Hu, H., Yang, L., Guo, P., & Xiao, B. (2011). Assessment of sustainable biomass resource for energy use in China. *Biomass and Bioenergy*, *35*(1), 1–11.

6 Current Innovations in Microbial Fuel Cells (MFCs)

Joshua Ogboghena Akhigbe, Constance Omono Omokhodion, Aisha Eniola Adegbite, Amina Yahaya, Bolaji Oluwatosin Akanbi, Sesan Abiodun Aransiola, and Naga Raju Maddela

6.1 INTRODUCTION

The world is shifting towards natural and sustainable sources of energy due to the depletion of fossil fuels and the increasing urgency of reducing environmental damage (Maddela et al., 2024). The widespread use of coal and oil alongside natural gas in industries has resulted in severe side effects, such as environmental damage and resource scarcity, together with global climate change and air pollution (Wang *et al.*, 2020; Aransiola et al., 2024). As a result, scientists and engineers are exploring new sustainable and eco-friendly approaches to power generation.

As the world seeks suitable energy solutions to mitigate environmental degradation, Microbial fuel cells (MFCs) have been found as a groundbreaking technology that offers a suitable and eco-friendly means of energy generation (Jalili *et al.*, 2024). MFCs convert chemical energy into electrical energy by utilizing microorganisms' metabolic activities to create electricity from organic matter. This makes them a better alternative to conventional energy sources (Kwofie *et al.*, 2024). Over the past few decades, significant advancements in MFC technology have enhanced their efficiency, scalability, and potential applications. Researchers have focused on improving electrode materials, optimizing microbial communities, and developing innovative systems that increase energy (power) generation and expand their practical applications like wastewater treatment and bio-sensing applications (Sharif *et al.*, 2021; Auta et al., 2024).

The generation of electricity through the utilization of microorganisms was first conceived by Michael C. Potter in 1911 (Potter, 1911). And although he is credited for the idea, the notion of 'animal electricity' was first studied by Professor Luigi Galvani in his frog leg experiment. He observed the involuntary movements in a dissected frog leg when electrical current was passed through it for short periods (Galvani, 1791). This led to the conception of the term 'bioelectricity'. In later years, Michael Potter studied the movement of current through electrodes made of platinum when suspended in a solution of *Escherichia coli* and yeast. He observed that electricity could be generated by bacteria (Potter, 1911). Potter's incredible discovery was the catalyst for research of the MFC technology. Development was slow in the field, however, because of scepticism and pessimism. However, in 1931, Barnett Cohen resuscitated the idea in his experiment that showed that microbial cells could generate electricity. He was able to produce a voltage of over 35 volts in a biological fuel cell that was an amalgamation of a number of smaller cells with 10cc of culture. These small cells were connected in series with each on generating about 0.2 mA at a potential of 0.5 V (Cohen, 1931).

Microbial fuel cells are a great substitute to traditional energy sources, and although microbial fuel cells and traditional fuel cells follow similar principles, the difference lies in MFCs' inherent capability to utilize the energy stored in chemical bonds of organic fuel sources to generate electrical current instead of relying on something like a metal catalyst. They are capable of harnessing the

DOI: 10.1201/9781003645405-6

bio-catalytic abilities of viable microorganisms. Microorganisms can utilize substrates prone to biodegradation like wastewater, and other organic matter to create electricity while, incredibly, executing the biodegradation/processing of the biodegradable products like municipal wastewater, and it is this potential that MFCs explore.

A classic MFC, a variation of the traditional biological fuel cell, is a bio-electrochemical device capable of converting chemical energy, with microorganisms as biocatalysts, stored in substances like alcohols and sugars directly into bioelectric energy (Gupta et al., 2023). Typically, a microbial fuel cell setup is a conjunction of an aerobic chamber and an anaerobic one, both fitted with positive and negative electrodes and kept apart by a proton exchange membrane. Energy generated is harvested by an electric wire that connects the electrodes (anode and cathode). The role of the engine of the process is played the by biofilm on the anode, which breaks down the substrates, like carbohydrates (low redox potential, -Eh) under anaerobic conditions, while the electrons generated are taken up by an enzyme-active site, which plays the role of a reduced intermediate. The cathode acts as the electron acceptor, rerouting the electrons to an external circuit, and eventually, these electrons reduce molecular oxygen at the surface of the cathode, leading to the production of an electric current (Gupta et al., 2020; Li et al., 2018).

The production of bioelectricity via organic matter oxidation is the feature that stands out most about MFC technology. Pure compounds and complex mixtures, as well as a range of other sources of organic matter like glucose, acetate, alcohols, microalgae, protein-rich wastewater, sodium formate, starch, sucrose, brewery wastewater, etc., have been used for bioelectricity production (Mathuriya and Yakhmi, 2016).

The major differences between traditional fuel cells and MFCs are:

1. The electrocatalyst is biological (the biofilm on the anode or proteins) (Schroeder and Harnisch, 2014)
2. Optimum temperatures are close to ambient, and the temperature in the cell usually ranges from 15°C to 45°C (Tee et al., 2017)
3. pH in the cell is neutral (Santoro, et al., 2017a)
4. The MFC uses biomass as fuel for the anode (Pandey et al., 2016)
5. Minimal environmental impact (Pant et al., 2011)

Unlike conventional energy production methods, MFCs operate under ambient conditions, requiring minimal external energy inputs and producing little to no greenhouse gas emissions (Kwofie *et al.*, 2024). Microbial fuel cells can be said to function on the principle of bio-electrochemical reactions, where electroactive microorganisms break down organic compounds and facilitate electron transfer during the breakdown of organic compounds (Jalili *et al.*, 2024). After the breakdown of organic matter, the microorganism, e.g., bacteria such as *Geobacter* and *Shewanella* sp., transfers electrons to an anode. These electrons travel through an external circuit to a cathode, generating electricity (Elmazouzi *et al.*, 2022). The proton released in the process migrates through a proton exchange membrane to the cathode, where they, and oxygen, in combination, produce H_20. This mechanism allows MFCs to generate electricity while simultaneously treating wastewater or degrading organic pollutants (Eker, 2023).

The efficiency of an MFC largely depends on the following:

1. type of microorganism used
2. nature of the organic substrate
3. electrode materials and the system configurations (Prathiba *et al.*, 2021; Kwofie *et al.*, 2024).

Bacterial species like *Shewanella* and *Geobacter* have been studied extensively and have shown that they have the ability to transfer electrons extracellularly. These bacteria have a vital part to play in determining the amount of power an MFC can produce by forming biofilms on the anode and facilitating electron transfer. Also, environmental factors such as pH, temperature, and oxygen levels are said to influence MFC performance (Yuvraj and Aranganathan, 2017).

6.2 DIVERSITY OF MICROORGANISMS IN MICROBIAL FUEL CELLS PRODUCTION

To improve the performance of MFCs, the microorganisms utilized will have to undergo enrichments. Some enrichment techniques include:

- Chemical enrichment outside MFCs using a chemical electron acceptor
- Enrichment in the MFCs or in-cell methods
- Enrichment in microbial electrolysis cells, or potential poised MFCs (Roy *et al.*, 2023).

6.2.1 Classification of Microorganisms in MFCs

To fully understand the contributions of microorganism power production, substrate utilization, and overall MFCs performance, microorganisms can be classified (Table 6.1):

6.2.1.1 Based on Electron Transfer Mechanism (Exoelectrogens or Electroactive Bacteria)

They can be referred to as electricity-producing bacteria, as the name implies. These types of bacteria are more unique than others because, unlike others, they can transfer electrons outside their cells, to surfaces like electrodes after breaking down organic matter; this is why they are the most useful in MFCs. EAB not only produce electrical energy from organic matter, but they also contribute to the bioremediation process and pioneering biosensors. They are often found in the soil, sediment, wastewater, thermal vents, etc. (Garbini et al., 2023).

The major EAB includes:

- *Geobacter sulfurreducens* PCA and KN400 strains
- *Shewanella oneidensis* strain MR-1

Others are:

- *Pseudomonas mendocina* NR802
- *Pseudomonas pseudoalcaligenes*
- *Bacillus circulans*

(Kumari *et al.*, 2015; Gurav *et al.*, 2019).

Under extreme conditions, hyperthermophilic archaeon species like *Ferroglobus placidus* and *Geoglobus achangau* have been found to possess electroactive abilities (Yilmazel *et al.*, 2018).

6.2.1.2 Based on Metabolic Function

6.2.1.2.1 Methanogenic Archaea

This group of archaea can be referred to as 'stabilizers', and their role in MFCs can be double-edged. While they produce methane as a by-product under anaerobic conditions, they do not directly contribute to electricity production. Their activity prevents the toxic build-up of intermediates by stabilizing microbial consortia and, on the other hand, columbic efficiency by consuming electrons (Buan, 2018).

A recent study investigated the effect of these methanogenic species: *Methanosarcina barkeri, Methanobacterium formicicum, Methanothrix soehngenii*, and their interactions with *Geobacter* sp. in the anode. The result showed that *M. barkeri* initially reduced the electrochemical activity of

TABLE 6.1
Diversity of Microorganisms in Microbial Fuel Cells (MFCs)

Category	Microbial Role	Key Characteristics	Examples	Reference
Exoelectrogens (EAB)	Electricity producers via direct or indirect extracellular electron transfer.	Transfer electrons to external electrodes after breaking down organic matter. Can be used in biosensors and bioremediation.	*Geobacter sulfurreducens* (strains PCA and KN400) *Shewanella oneidensis* (strain MR-1) *Pseudomonas mendocina* *Pseudomonas pseudoalcaligenes* *Bacillus circulans*	Garbini et al., 2023; Kumari et al., 2015
Methanogenic Archaea	Stabilizers of microbial communities; methane producers.	Do not contribute directly to electricity production. Compete for electrons, possibly reducing MFC efficiency. Prevent toxic build-up of intermediates.	*Methanosarcina barkeri* *Methanobacterium formicicum* *Methanothrix soehngenii*	Ngoumelah et al., 2024
Fermentative Bacteria	Decomposers; support the activity of exoelectrogens.	Break down complex substrates (e.g., sugars, fatty acids) into simpler compounds like acetate and alcohol. Produce hydrogen and carbon dioxide, useful for biogas. Risk of outcompeting EAB for resources if not monitored.	*Clostridium acetobutylicum* *Clostridium tyrobutyricum* *Bacillus* spp. *Lactobacillus* spp. *Saccharomyces, Streptococcus, Micrococcus* spp.	Kumar et al., 2023; Melini et al., 2019; Feng et al., 2024
Natural Isolates	Serve as inoculum for diverse microbial consortia in MFCs.	Readily available and cost-effective. Contain native anaerobic and facultative anaerobic bacteria. Suitable for real-world applications such as wastewater treatment.	Wastewater; *Geobacter sulfurreducens* *Shewanella oneidensis* Anaerobic digester sludge; *Clostridium* spp. Contaminated sediments and topsoil (especially rhizosphere); *Rhodoferax ferrieducens*	Kifle et al., 2023; Passos et al., 2016; Kim et al., 2023
Enriched Media	Cultivate and select high-performing strains for specific functions in MFCs.	Grown in enriched media under anaerobic conditions. Contain specific nutrients (e.g., acetate, glucose, minerals, and buffers). Cultures undergo serial transfer and are introduced into MFCs for enhanced performance.	Enriched Geobacter strains and other electroactive microorganisms adapted to selective media.	Khoirunnisa et al., 2023

Geobacter sp. and measured up gradually, while *M. formicicum* decreased the current density by approximately 37%, and *M. soehngenii* reduced the total transfer charge by approximately 40% (Ngoumelah et al., 2024).

6.2.1.2.2 Fermentative Bacteria

These bacteria act as decomposers or auxiliary organisms. They break down complex organic substrates like sugar and fatty acids into simpler compounds like alcohol and acetate. These simpler compounds are primary electron donors for EAB, which would in turn transfer electrons to the anode, generating electricity. In simpler terms, the activity of fermentative bacteria supports the activity of EAB; there is a synergistic interaction that enables them to enhance substrate degradation. However, if they are not closely observed, they may outcompete EAB for resources and reduce MFCs' efficiency (Stratford *et al.*, 2014). Fermentative bacteria also produce hydrogen and carbon dioxide, which can be used as biogas, and their presence in the circuit can influence the overall power output of the MFCs (Feng *et al.*, 2024). By integrating fermentation and MFCs technology, a multifunctional system has been created where a wide range of bio-products can be formed, that is, biogas, biofuels, etc., to say the least (Christwardana *et al.*, 2021).

Some specific fermentative bacteria in MFCs include: *Clostridium acetobutylicum and Clostridium tyrobutyricum* (Kumar *et al.*, 2023). Others are species of *Bacillus, Lactobacillus, Saccharomyces, Streptococcus*, and *Micrococcus* (Melini *et al.*, 2019).

6.2.1.3 Based on Habitat/Source

6.2.1.3.1 Natural Isolate

They often consist of diverse microbial communities and are ready-to-use inocula. These sources include but are not limited to:

- **Wastewater**
  As a result of the high organic pollutant content in wastewater, it is the best source of microorganisms used in MFCs. Wastewater is readily available and cost-effective, and it contains a mixture of anaerobic and facultative anaerobic bacteria, making it an excellent spot for the Isolation of EAB (Kifle *et al.*, 2023).
- **Anaerobic Digester Sludge**
  Digester sludge is a by-product of wastewater treatment, and it involves a diverse microbial community that is enriched, which in turn enhances the performance of MFCs, generating higher power densities and facilitating effective wastewater treatment (Rabaey and Verstraete, 2005). The digester sludge is added to the anode chamber of the MFCs, acting as the substrate itself. The organisms from the digester sludge form a biofilm on the anode, and this encourages electron transfer and generation (Passos *et al.*, 2016).
- **Soil and Sediments**
  In contaminated underwater sediment, organic matter as well as anaerobic biogas like ammonia, hydrogen sulphide, etc., are present. These components are good substrates for electricity generation in MFCs. They are referred to as soil and sediment MFCs (SMFCs), and they are used for in-situ bioremediation, where pollutant degradation takes place (Kim *et al.*, 2023). Top soils and areas immediately next to plant roots (Rhizosphere) are the perfect spots to get the most amount of microorganisms in the soil (Magdoff and Van Es, 2021).

6.2.1.3.2 Enriched media

They contain specific nutrients to facilitate the growth of a specific microorganism. They are used to select, enrich, and cultivate the microbial communities that play a crucial role in electricity generation in MFCs, especially when fastidious organisms are in demand (Lagier *et al.*, *2015*).

This media is used specifically to isolate an enhanced strain of EAB. The media is prepared with the addition of a specific carbon source like acetate or glucose, trace minerals, and an anaerobic buffer like sodium bicarbonate. The sample is inoculated into the enriched media and incubated in an anaerobic chamber. A serial transfer of multiple generations is done, and finally, the enriched culture is introduced into the anode chamber of a working MFC while being closely monitored using a voltmeter (Khoirunnisa *et al.*, 2023).

6.3 MICROBIAL FUEL CELLS (MFCs) IN ENERGY PRODUCTION

The Microbial Fuel Cell is a promising technological advancement that generates electricity using microorganisms. Electricity production arises from microorganisms that decompose waste materials like waste, wastewater, and biomass (Bazina *et al.*, 2023). The distinctive feature of MFCs is their ability to simultaneously generate renewable energy while treating waste (Bazina *et al.* 2023). These devices find specific use in decentralized power generation, bioenergy production, and wastewater treatment operations because remote regions often lack primary energy infrastructure.

6.3.1 Mechanisms and Microorganisms in Microbial Fuel Cells

6.3.1.1 Electron Transfer Mechanism in MFCs

According to Bazina *et al.* (2023), Microbial Fuel Cells (MFCs) produce electricity by harnessing microbial metabolism to transfer electrons from organic matter to an electrode. The system requires three essential components in this process:

- **Anode**: This serves as the site where bacteria oxidize organic substrates which produce electrons.
- **Cathode**: This serves as the location where oxygen reduction takes place, completing the circuit.
- **Proton Exchange Membrane**: This performs proton transfer while keeping electrons on an external circuit (Vishwanathan, 2021).

Microorganisms facilitate electron transfer in three ways:

- **Direct Electron Transfer (DET)**: Some bacteria, like *Geobacter sulfurreducens*, form direct electrical connections with the anode using conductive pili (Santoro *et al.*, 2017b).
- **Mediated Electron Transfer (MET)**: Certain bacteria, such as *Shewanella oneidensis*, produce redox-active molecules that shuttle electrons to the electrode (Bazina *et al.*, 2023).
- **Biofilm Formation**: Bacteria form a dense biofilm on the electrode surface, improving electron transfer efficiency and conductivity (Roy *et al.*, 2022).

6.4 TYPES OF MICROBIAL FUEL CELLS

MFCs can be classified based on their design and function, based on a research done by Bhaduri and Behera (2024) into the following:

- **Single-Chamber Microbial Fuel Cells (SC MFCs)**: These systems have only an anode chamber, where atmospheric oxygen acts as the electron acceptor, making them easier to apply practically (Boas *et al.*, 2022).

- **Dual-Chamber Microbial Fuel Cells (DC MFCs)**: These contain separate anode and cathode chambers, allowing for improved electron transfer performance and higher energy output (Bhaduri & Behera, 2024).

6.4.1 Key Microorganisms in MFCs

According to Bazina *et al.* (2023), several bacteria play an important role in electricity generation within MFCs. Some of these include:

- ***Geobacter sulfurreducens*:** Uses conductive pili (nanowires) to directly transfer electrons to the anode.
- ***Shewanella oneidensis*:** Capable of both direct and mediated electron transfer, using secreted electron shuttles.
- ***Pseudomonas aeruginosa*:** Produces extracellular compounds that enhance electron movement.
- ***Escherichia coli*:** Can be genetically modified to improve electron transfer efficiency.
- ***Proteus vulgaris*:** Uses mediated electron transfer and biofilm formation to generate electricity.

6.4.1.1 Factors Affecting MFC Performance

The efficiency of an MFC depends on several factors (Aghababaie *et al.*, 2015):

- Microbial species
- Substrate availability
- Electrode materials
- System design
- Temperature
- pH

6.4.2 Applications of MFCs in Energy Production

1. **Wastewater Treatment and Bioelectricity**: MFCs break down pollutants while generating electricity, decreasing industrial costs and improving sustainability (Wang *et al.*, 2020).
2. **Remote Power Generation**: MFCs generate electricity in remote locations which supply power to biosensors, wearable medical devices like glucose monitors, and implantable medical devices such as pacemakers (Santoro *et al.*, 2017a).
3. **Renewable Energy Integration**: Renewable Energy Integration happens when MFCs team up with solar or wind power systems through hybrid setups that achieve higher performance by utilizing bioengineered bacteria (Boas *et al.*, 2022).
4. **Agricultural Waste Utilization**: MFCs convert organic waste from farms into electricity, reducing agricultural waste while producing renewable energy (Pandya *et al.*, 2024).
5. **Bioremediation of Contaminated Sites**: MFCs help clean up oil spills and heavy metal contamination while simultaneously generating electricity, making them useful for environmental restoration projects (Fang & Achal, 2019).

6.4.3 Challenges and Solutions in MFC-Based Energy Production

While MFCs offer great potential, several obstacles must be addressed before they can become widely used for large-scale energy production (Table 6.2).

TABLE 6.2
Challenges and Solutions in MFC-Based Energy Production

Challenge	Solution	Reference
Low Power Output		
MFCs generate little electricity.	Improve bacterial electron transfer, upgrade electrodes, and upgrade design.	Wang *et al.*, 2020
High Cost		
Expensive materials increase costs.	Use affordable, sustainable materials and explore membrane-less designs.	Boas *et al.*, 2022
Unstable Microbes		
Bacterial shifts reduce efficiency.	Use stable microbial communities and improve growth conditions.	Aghababaie *et al.*, 2015
Scaling Issues		
Hard to apply on a large scale.	Develop modular MFCs and integrate them into wastewater plants.	Wang *et al.*, 2020
Slow electron transfer		
Some bacteria transfer electrons inefficiently	Use genetically modified bacteria and efficient electron mediators.	Bazina *et al.*, 2023

6.5 NEW INNOVATIONS IN MFCs

Recent studies in the MFC differ significantly from traditional studies. These novel studies aim to proffer solutions to the limiting factors of the MFC technology and close the knowledge gaps in the field, to improve MFC efficiency (Al-sahari et al., 2022). A review of existing literature and published papers gives insights into some advancements in MFC technology and optimization of the general principle. Some of them are briefly discussed below.

Nguyen et al. (2021) designed a small unit of microbial fuel cells meant for use under the soil, not requiring any external microorganisms, although this was dependent on the soil microbes present. The system was constructed with acrylic boards, and the electrode chambers were separated by a Nafion proton exchange membrane. In order to increase the power generation of the MFC, the cathode was exposed partially to air while the anode was attached directly to the soil. Power was produced by the system within the first hour, and it ultimately produced 7.3 mW m^2 within 48 h. This small unit system is toward an MFC design that could have applications as a battery system for small operations such as soil sensing systems (Nguyen et al., 2021)

You et al. (2018) developed MFC systems that are easily movable in situ and allow for manipulating the size of the anode or the anode-to-membrane distance without using or disturbing other systems. A 50-ml polypropylene syringe was used as the MFC structure. The barrel of the syringe was cut and used as a chamber for the anode, and in order to separate it from the cathode chamber, a CEM membrane was used. The setup recorded a power density generation of 20.7 ± 1.9 μW. This is important because this type of system can be applied in trying to establish an adjustable, working prototype for various applications in order to maintain maximum efficiency before creating the final system. This would ultimately reduce the cost of manufacturing.

Zhou et al. (2018) designed a membrane-less microbial fuel cell system (ML-MFC) consisting of three chambers: the anode, cathode, and a nitrification chamber. The anode and cathode chambers were filled with carbon felt held by titanium mesh which was used as electrodes in those chambers and as a bacteria attachment surface in the nitrifying chamber. This ML-MFC design increased the system performance and brought down the usually high cost of ion exchange membrane.

Chen et al. (2018) employed the modification of carbon cloth (CC) anodes with the use of supercapacitor (SC) activated carbon (AC) carbon nanotubes (CNTs). This modification led to an increased anode surface area, and the MFC system with SC-AC-CNTs anodes recorded a power density of 899.54 mW m^{-2}, which is 2.9 times higher than the system with non-modified anodes.

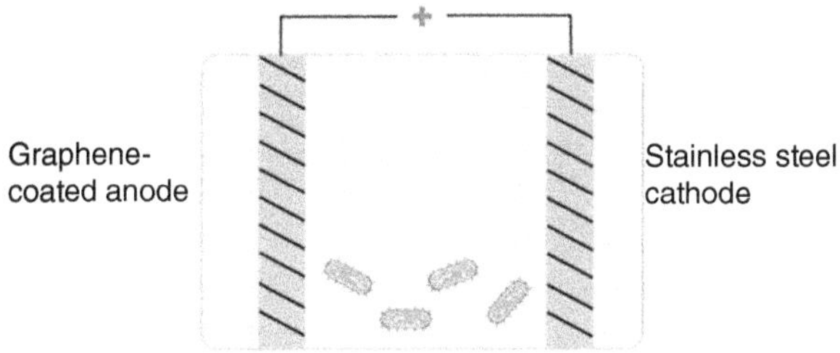

FIGURE 6.1 Graphite anode and stainless-steel cathode SC MFC.

One of the key areas of innovation in MFC technology has been the efficiency of electron transfer which can be done by the development of advanced electrode materials (Deninger *et al.*, 2022). Traditional carbon-based electrodes, such as graphite and carbon cloth, have been replaced with advanced materials such as graphene, carbon nanotubes, and metal-organic frameworks, due to their limitations related to surface area and microbial adhesions (Massaglia, 2017; Jafary *et al.*, 2018). Graphene-based electrodes have gained attention due to their high conductivity, large surface area, and excellent biocompatibility. Studies have shown that graphene-coated anodes significantly improve bacterial adhesion and electron transfer rates, leading to higher power densities (Olabi *et al.*, 2020). Similarly, carbon nanotubes (CNTs) have been incorporated into MFC electrodes to enhance conductivity and facilitate efficient microbial attachment. The nanostructured surfaces of CNTs create a favourable environment for biofilm formation, improving overall system performance (Wilberforce *et al.*, 2021; Aiswaria *et al.*, 2022). Gao et al. (2018) developed a low-cost cathode with porous polyethylene sheet (PE), blended activated carbon (AC), and highly conductive carbon back (CB). The electrode was able to generate a power density of 1400 mW m^{-2} despite its low manufacturing cost. Carbon fibre-coated zinc electrodes have been employed in DC MFCs to enhance the 3D electrode materials for the generation of power. An estimated 122 mW m^{-2} of power was generated by the electrodes (Lai et al., 2018). Erable et al. (2017) were able to successfully improve microbial activity and generation of power by the MFC with the use of a graphite anode and stainless-steel cathode in the SC-MFC system. The system generated a power density of 70 mW m^{-2} for over two weeks (Figure 6.1).

The development of bioengineered bacteria with enhanced electron transfer capabilities has further boosted MFC performance (Bashir *et al.*, 2021; Prathiba *et al.*, 2021). Genetic modifications have allowed the engineering of microbial strains with improved extracellular electron transport mechanisms, increasing the overall energy output (Kwofie *et al.*, 2024). The probability of a new modified electrophoresis strain of *Arcobacter butzleri* growing in acetate and lactate and its high adhesion to electrodes capability has been reported by Szydlowski et al. (2020). They showed that higher energy levels could be obtained from the catabolism of the organic compound through the editing of the genome (Figure 6.2).

In recent years, interdisciplinary cooperation has become more noticeable in innovative projects. A major area of innovation in MFCs is the integration of hybrid systems that combine microbial fuel cells with other renewable energy technologies (Jalili *et al.*, 2024). Hybrid MFCs, which integrate with solar cells, wind energy, or enzymatic fuel cells, have shown potential in enhancing energy output and providing a more stable power supply. These systems can be particularly useful in remote and off-grid locations where conventional energy sources are not available (Dwivedi *et al.*, 2021).

The integration of artificial intelligence and machine learning into MFC research is an emerging trend that holds significant potential for optimizing system performance. AI-driven models can analyse large datasets, predict system behaviour, and recommend strategies for improving efficiency in real time (Kwofie *et al.*, 2024). By leveraging these computational tools, researchers can accelerate the development of MFC technology, streamline system optimization, and enhance the reliability of MFC applications. This data-driven approach is expected to play a crucial role in the future

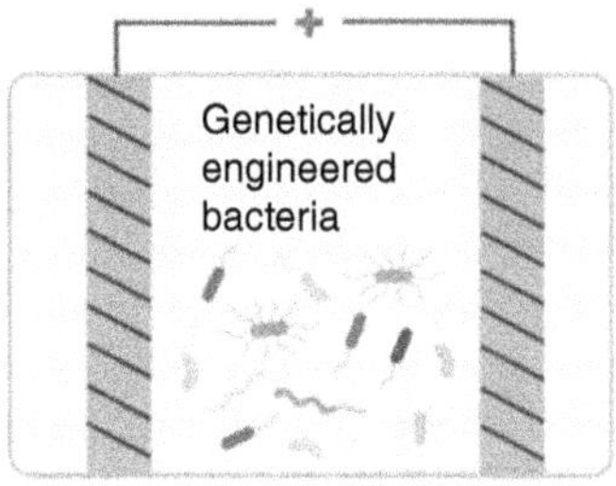

FIGURE 6.2 Genetically modified bacteria for MFC optimization.

commercialization of MFCs, making them more accessible and adaptable for various industries and environmental settings (Gyaneshwar *et al.*, 2022). These advancements have led to a significant increase in power density, making MFCs more competitive with other bio-electrochemical systems.

MFCs are being explored for their potential in bio-sensing and environmental monitoring. Paper-based MFCs and miniaturized devices have been designed for environmental monitoring and medical diagnostics (Kummari *et al.*, 2023). These low-cost, portable sensors can detect pathogens, toxins, and biochemical markers in water and biological samples. By integrating MFC technology with bio-sensing platforms, researchers are creating innovative tools for real-time environmental and healthcare monitoring (Jadhav *et al.*, 2020).

6.6 APPLICATIONS OF MFC TECHNOLOGY

6.6.1 Biosensors in BOD detection

Microbial fuel cells with a replaceable anaerobic consortium can be applied as a biosensor for online monitoring of organic matter (Ivars-Barceló et al. 2018). Traditional TOC and BOD calculation methods are mostly unsuitable for online screening and control of biological wastewater treatment processes. The linear correlation between the electrical current output of MFC and organic matter strength makes MFC a possible BOD sensor (Yao et al., 2023). BOD biosensors based on the MFC depend on this positive linear correlation. There have been several MFC-based biosensors developed for BOD determination. Commault et al. (2016) developed the MFC-based biosensor with *Geobacter*-dominated biofilms to determine the BOD of milk. The biosensor achieved high reproducibility of 94% with only a 7.4% error compared to the traditional BOD_5 method. Hsieh and Chung (2014) also developed an MFC-based biosensor, but they employed a mixed culture of six strains to determine the BOD concentrations in wastewater. They, however, observed high reproducibility for long operation periods. Pasternak et al. (2017) developed an autonomous floating MFC biosensor that could achieve independent operations of monitoring and timely detection of water pollutants for 150 days.

6.6.2 Robotics

In recent years, autonomous automated robots that can work independently and efficiently, especially in remote terrestrial or aquatic regions, are being used more often. Several robots currently use MFC technology as their major source of power. The Gastronome was the first recorded robot that made use of bacteria (Wilkinson, 2000). It used chemical fuels to charge its Ni-Cd batteries and operated on the electrical energy generated by *Escherichia Coli* consuming sugar through synthetic mediators. Several such robots have been engineered since, and they all belong to a class of intelligent bio-electrochemical machines that derive their operational power by utilizing the energy from the digestion of food by the catalytic activities of microorganisms. They are called gastrobots (Elhadad et al., 2024).

6.6.3 Wastewater treatment

Wastewater treatment is one of the applications of MFC technology where they serve the purpose of generating electricity, degrading organic pollutants, removing heavy metals, and mitigating the environmental impact of industrial and municipal wastewater (Dutta *et al.*, 2023). MFCs have the ability to generate electricity while simultaneously treating effluents and residues harbouring organic materials. Liu and Logan (2004) highlighted that since the introduction of for wastewater treatment in 1991 by Habermann and Pommer (1991), they have proven to generate up to 50% of the energy required for conventional wastewater treatment methods. They are also an environmentally sustainable approach as they contribute to a significant reduction of up to 90% in the production of solids for disposal.

Microbial fuel cells have great potential for treating sewage and generating electricity concurrently. They can potentially significantly reduce costs in comparison to traditional aerobic wastewater treatment processes, and this makes them a very attractive prospect for sustainable treatment of wastewater in the future (Jalili et al., 2024).

6.7 CONCLUSION

Although MFC technology is an interesting one, there are still several challenges to overcome. The energy output remains relatively low, and the materials used in MFC construction can be expensive. Additionally, maintaining a stable microbial community within the system is a complex process. However, ongoing research has shown promise to solve existing challenges since MFC technology continues to develop toward its future role in sustainable energy generation (Wang *et al.*, 2020). Microbial Fuel Cells (MFCs) offer a sustainable way to generate electricity while treating waste. Their applications in wastewater treatment, remote power, and environmental clean-up showcase their potential. However, challenges like low power output and high costs must be addressed. With ongoing advancements in microbial engineering and system design, MFCs could become a key player in renewable energy, supporting a cleaner and more sustainable future. As the world continues to prioritize sustainability and the transition to renewable energy sources, microbial fuel cells have emerged as a viable and innovative solution. Innovations in electrode materials, microbial engineering, and system configurations have significantly improved their efficiency and scalability. As research progresses, the commercialization and real-world implementation of MFCs will become more feasible, bringing us closer to a future where biological energy generation plays a crucial role in our energy infrastructure. With continuous advancements, microbial fuel cells have the potential to revolutionize energy production, wastewater treatment, and environmental monitoring, contributing to a more suitable and eco-friendlier environment.

REFERENCES

Aghababaie, M., Farhadian, M., Jeihanipour, A., & Biria, D. (2015). Effective factors on the performance of microbial fuel cells in wastewater treatment – a review. *Environmental Technology Reviews*, *4*(1):71–89.

Aiswaria, A. P., Mohamed, S. N., Singaravelu, D. L., Brindhadevi, K., and Pugazhendhi, A. (2022). A review on graphene / graphene oxide supported electrodes for microbial fuel cell applications: Challenges and prospects. *Chemosphere*, *296*, 133983. https://doi.org/10.1016/j.chemosphere.2022.133983

Al-Sahari, M., Al-Gheethi, A. A., Mohamed, R. M. S. R., Yashni, G., Vo, D. V. N., & Ismail, N. (2022). Microbial fuel cell systems; developments, designs, efficiencies, and trends: A comparative study between the conventional and innovative systems. *Chemosphere*, *298*, 134244.

Aransiola, S.A., Adeniyi, O.S., Omoregie, I.P., Akinhanmi, F.O., Oniha, M.I., Maddela, N.R. (2024). Microbial biotechnology for bioenergy: General overviews. In Naga Raju Maddela, Sesan Abiodun Aransiola, Chizoba I. Ezugwu, Lizziane Kretli Winkelstroter Eller, Laura Scalvenzi, Fangang Meng (eds.), *Microbial biotechnology for bioenergy*. Elsevier Science, ISBN: 9780443141133, 0443141134. https://www.sciencedirect.com/science/article/pii/B9780443141126000018

Auta, H.S., Aboyeji, D.O., Adelere, I.A., Babayil, H., Bala, J.D., Fauziah, S.H., Aransiola, S.A., (2024). Biotechnology for bioenergy production: Current status, challenges, and prospects. In Naga Raju Maddela, Sesan Abiodun Aransiola, Chizoba I. Ezugwu, Lizziane Kretli Winkelstroter Eller, Laura Scalvenzi, Fangang Meng (eds.), *Microbial biotechnology for bioenergy*. Elsevier Science, ISBN: 9780443141133, 0443141134. https://www.sciencedirect.com/science/article/pii/B9780443141126000146

Bashir, S., Mulvaney, S. P., Houf, W., Villanueva, L., Wang, Z., Buck, G., and Liu, J. L. (2021). Microbial Fuel cells: Design and evaluation of catalysts and device. In Gao, Y.J., Song, W., Liu, J.L., Bashir, S. (eds.), *Advances in sustainable energy*. Springer, Cham, pp. 681–764. https://doi.org/10.1007/978-3-030-74406-9_24

Bazina, N., Ahmed, T. G., Almdaaf, M., Jibia, S., & Sarker, M. (2023). Power generation from wastewater using microbial fuel cells: A review. *Journal of Biotechnology*, *374*(0618–1656):17–30.

Bhaduri, S., & Behera, M. (2024). From single-chamber to multi-anodic microbial fuel cells: A review. *Journal of Environmental Management*, *355*, 120465. https://doi.org/10.1016/j.jenvman.2024.120465

Boas, J. V., Oliveira, V. B., Simões, M., & Pinto, A. M. F. R. (2022). Review on microbial fuel cells applications, developments and costs. *Journal of Environmental Management*, *307*, 114525. https://doi.org/10.1016/j.jenvman.2022.114525

Buan N. R. (2018). Methanogens: Pushing the boundaries of biology. *Emerging Topics in Life Sciences*, *2*(4), 629–646. https://doi.org/10.1042/ETLS20180031

Chen, W., Liu, Z., Hou, J., Zhou, Y., Lou, X., & Li, Y. (2018). Enhancing performance of microbial fuel cells by using novel double-layer-capacitor-materials modified anodes. *International Journal of Hydrogen Energy*, *43*(3), 1816–1823. https://doi.org/10.1016/j.ijhydene.2017.11.034

Christwardana, M., Joelianingsih, J., & Yoshi, L. A. (2021). Performance of yeast microbial fuel cell integrated with sugarcane bagasse fermentation for cod reduction and electricity generation. *Bulletin of Chemical Reaction Engineering & Catalysis*, 16(3), 446–458.

Cohen, B. (1931). The bacterial culture as an electrical half-cell. *Journal of Bacteriology*, *21*(1), 18–19.

Commault, A. S., Lear, G., Bouvier, S., Feiler, L., Karacs, J., & Weld, R. J. (2016). Geobacter-dominated biofilms used as amperometric BOD sensors. *Biochemical Engineering Journal*, *109*, 88–95. https://doi.org/10.1016/j.bej.2016.01.011

Deninger, I. D., Sran, A. K., and Keleher, J. J. (2022). Enhancing electron transfer in a biomimetic nanocomposite electrode for microbial fuel cell applications. *Meeting Abstracts/Meeting Abstracts (Electrochemical Society. CD-ROM)*, *MA2022-02*(54), 2039. https://doi.org/10.1149/ma2022-02542039mtgabs

Dutta, A., Mukherjee, S. K., and Hossain, S. T. (2023). Microbial fuel cell assisted wastewater treatment: A review on current trends. In: Shah, M.P. (eds.), *Microbial technologies in industrial wastewater treatment*. Springer, Singapore, pp. 59–82. https://doi.org/10.1007/978-981-99-2435-6_5

Dwivedi, K. A., Huang, S., and Wang, C. (2021). Integration of various technology-based approaches for enhancing the performance of microbial fuel cell technology: A review. *Chemosphere*, *287*, 132248. https://doi.org/10.1016/j.chemosphere.2021.132248

Eker, S. (2023). Wastewater Treatment and Electricity Generation with Different Cathode Solutions in MFC. *Deu Muhendislik Fakultesi Fen Ve Muhendislik*, *25*(74), 513–523. https://doi.org/10.21205/deufmd.2023257420

Elhadad, A., Gao, Y., & Choi, S. (2024, January). A microfabricated gastrobot for sustainable on-water propulsion. In *2024 IEEE 37th international conference on micro electro mechanical systems (MEMS)*, (pp. 497–500). Austin, Texas, USA: IEEE.

Elmazouzi, S., Naimi, Y., and Zerdani, I. (2022). Microbial fuel cells for depollution of stagnant water and production of electrical energy. *2022 11th international conference on renewable energy research and application (ICRERA)*, 441–445. https://doi.org/10.1109/icrera55966.2022.9922813

Erable, B., Byrne, N., Etcheverry, L., Achouak, W., & Bergel, A. (2017). Single medium microbial fuel cell: Stainless steel and graphite electrode materials select bacterial communities resulting in opposite electrocatalytic activities. *International Journal of Hydrogen Energy*, *42*(41), 26059–26067. https://doi.org/10.1016/j.ijhydene.2017.08.178

Fang, C., & Achal, V. (2019). The potential of microbial fuel cells for remediation of heavy metals from soil and water—review of application. *Microorganisms*, *7*(12), 697. https://doi.org/10.3390/microorganisms7120697

Feng, F., Wu, C.-H., Li, F., Wang, X., Zhu, J., Zhang, R., & Chen, S.-C. (2024). Research on the integration of microbial fuel cells with conventional wastewater treatment technology: Advantages of anaerobic fermentation. *Energy Conversion and Management: X*, *23*, 100680. https://doi.org/10.1016/j.ecmx.2024.100680

Galvani, L. (1791). *De viribus electricitatis in motu musculari. Commentarius* [On the effects of electricity on muscular motion. A commentary]. *Bononiensi Scientiarum et Artium Instituto Academico Commentarii*, *7*, 363–418.

Gao, N., Qu, B., Xing, Z., Ji, X., Zhang, E., & Liu, H. (2018). Development of novel polyethylene air-cathode material for microbial fuel cells. *Energy*, *155*, 763–771. https://doi.org/10.1016/j.energy.2018.05.055

Garbini, G. L., Barra Caracciolo, A., & Grenni, P. (2023). Electroactive Bacteria in Natural Ecosystems and Their Applications in Microbial Fuel Cells for Bioremediation: A Review. *Microorganisms*, 11(5), 1255. https://doi.org/10.3390/microorganisms11051255

Gupta, S., Mittal, Y., Tamta, P., Srivastava, P., & Yadav, A. K. (2020). Textile wastewater treatment using microbial fuel cell and coupled technology: A green approach for detoxification and bioelectricity generation. In *Integrated microbial fuel cells for wastewater treatment* (pp. 73–92). Butterworth-Heinemann. https://doi.org/10.1016/B978-0-12-817493-7.00004-7

Gupta, S., Patro, A., Mittal, Y., Dwivedi, S., Saket, P., Panja, R., … & Yadav, A. K. (2023). The race between classical microbial fuel cells, sediment-microbial fuel cells, plant-microbial fuel cells, and constructed wetlands-microbial fuel cells: Applications and technology readiness level. *Science of the Total Environment*, *879*, 162757.

Gurav, R., Bhatia, S. K., Choi, T.-R., Jung, H.-R., Yang, S.-Y., Song, H.-S., Park, Y.-L., Han, Y.-H., Park, J.-Y., Kim, Y.-G., Choi, K.-Y., & Yang, Y.-H. (2019). Chitin biomass-powered microbial fuel cell for electricity production using halophilic Bacillus circulans BBL03 isolated from sea salt harvesting area. *Bioelectrochemistry*, 130, 107329. https://doi.org/10.1016/j.bioelechem.2019.107329

Gyaneshwar, A., Selvaraj, S. K., Ghimire, T., Mishra, S. J., Gupta, S., Chadha, U., Manoharan, M., and Paramasivam, V. (2022). A survey of applications of MFC and recent progress of artificial intelligence and machine learning techniques and applications, with competing fuel cells. *Engineering Research Express*, *4*(2), 022001. https://doi.org/10.1088/2631-8695/ac5fd9

Habermann, W., & Pommer, E. H. (1991). Biological fuel cells with sulphide storage capacity. *Applied Microbiology and Biotechnology*, *35*, 128–133.

Hsieh, M. C., & Chung, Y. C. (2014). Measurement of biochemical oxygen demand from different wastewater samples using a mediator-less microbial fuel cell biosensor. *Environmental Technology*, *35*(17–20), 2204–2211. https://doi.org/10.1080/09593330.2014.898700

Ivars-Barceló, F., Zuliani, A., Fallah, M., Mashkour, M., Rahimnejad, M., & Luque, R. (2018). Novel applications of microbial fuel cells in sensors and biosensors. *Applied Sciences*, *8*(7), 1184.

Jadhav, D. A., Chendake, A., Ghosal, D., Mathuriya, A. S., Kumar, S. S., and Pandit, S. (2020). Advanced microbial fuel cell for biosensor applications to detect quality parameters of pollutants. In Singh, L., Mahapatra, D.M. and Thakur, S. (eds.), *Bioremediation, nutrients, and other valuable product recovery, using bioelectrochemical systems*. Elsevier, pp. 125–139. https://doi.org/10.1016/b978-0-12-821729-0.00003-8

Jafary, T., Ghasemi, M., Alam, J., Aljlil, S. A., and Yusup, S. (2018). Carbon-Based polymer nanocomposites as electrodes for microbial fuel cells. In Ismail, A.F. and Goh, P.S. (eds.), *Carbon-based polymer nanocomposites for environmental and energy applications*, Elsevier, pp. 361–390. https://doi.org/10.1016/b978-0-12-813574-7.00015-0

Jalili, P., Ala, A., Nazari, P., Jalili, B., and Ganji, D. D. (2024). A comprehensive review of microbial fuel cells considering materials, methods, structures, and microorganisms. *Heliyon*, *10*(3), e25439. https://doi.org/10.1016/j.heliyon.2024.e25439

Khoirunnisa, N. S., Anwar, S., Sudadi, U., & Santosa, D. A. (2023). Enrichment media selection and co-culture potential among exoelectrogen bacteria vary with ecological factors. *Makara Journal of Science*, *27*(1), 5. https://doi.org/10.7454/mss.v27i1.1452

Kifle, T., Alemayehu, E., & Dereje Kitila, C. (2023). Development of microbial fuel cells for wastewater treatment and electricity generation using domestic wastes. *Environmental Health Engineering and Management Journal*, *10*(3), 273–279. http://ehemj.com/article-1-1140-en.html

Kim, K., Takahashi, T., Yumioka, R., & Hibino, T. (2023). Sediment microbial fuel cells in oxidative sedimentary environments using iron substrate as voltage booster. *Journal of Power Sources*, *557*, 232510. https://doi.org/10.1016/j.jpowsour.2022.232510

Kumar, K., Ding, L., Zhao, H., & Cheng, M.-H. (2023). Waste-to-energy pipeline through consolidated fermentation–microbial fuel cell (MFC) system. *Processes*, *11*(8), 2451. https://doi.org/10.3390/pr11082451

Kumari, S., Mangwani, N., & Das, S. (2015). Low-voltage-producing microbial fuel cell constructs using biofilm-forming marine bacteria. *Current Science*, *108*(5), 925–932.

Kummari, S., Panicker, L. R., Bommi, J. R., Karingula, S., Kumar, V. S., Mahato, K., and Goud, K. Y. (2023). Trends in paper-based sensing devices for clinical and environmental monitoring. *Biosensors, 13*(4), 420. https://doi.org/10.3390/bios13040420

Kwofie, M., Amanful, B., Gamor, S., and Kaku, F. (2024). Comprehensive analysis of clean energy generation mechanisms in microbial fuel cells. *International Journal of Energy Research, 2024*(1), 5866657. https://doi.org/10.1155/2024/5866657

Lagier, J. C., Edouard, S., Pagnier, I., Mediannikov, O., Drancourt, M., & Raoult, D. (2015). Current and past strategies for bacterial culture in clinical microbiology. *Clinical Microbiology Reviews, 28*(1), 208–236. https://doi.org/10.1128/CMR.00110-14

Lai, M. F., Lou, C. W., & Lin, J. H. (2018). Improve 3D electrode materials performance on electricity generation from livestock wastewater in microbial fuel cell. *International Journal of Hydrogen Energy, 43*(25), 11520–11529. https://doi.org/10.1016/j.ijhydene.2017.06.047

Li, M., Zhou, M., Tian, X., Tan, C., McDaniel, C. T., Hassett, D. J., & Gu, T. (2018). Microbial fuel cell (MFC) power performance improvement through enhanced microbial electrogenicity. *Biotechnology advances, 36*(4), 1316–1327. https://doi.org/10.1016/j.biotechadv.2018.04.010

Liu, H., & Logan, B. E. (2004). Electricity generation using an air-cathode single chamber microbial fuel cell in the presence and absence of a proton exchange membrane. *Environmental science & technology, 38*(14), 4040–4046.

Maddela, N. R., Aransiola, S. A., Ezugwu, C. I., Winkelstroter Eller, L. K., Scalvensi, L., Meng, F. (Eds.). (2024). *Microbial biotechnology for bioenergy*. (1st ed.). ISBN-9780443141126. https://doi.org/10.1016/C2022-0-02979-2

Magdoff, F., & Van Es, H. (2021). *Building soils for better crops: Sustainable soil management* (4th ed.). Sustainable Agriculture Research and Education (SARE) Program, National Institute of Food and Agriculture, USDA. https://www.sare.org/wp-content/uploads/Building-Soils-for-Better-Crops.pdf

Massaglia, G. (2017). *Development of new nanostructured electrodes in Microbial Fuel Cells (MFCs)*. Doctoral Program in Electronic Engineering, Scuola di dottorato - Doctoral School, pp. 1–218. https://doi.org/10.6092/polito/porto/2676549

Mathuriya, A. S., & Yakhmi, J. V. (2016). Microbial fuel cells–Applications for generation of electrical power and beyond. *Critical Reviews in Microbiology, 42*(1), 127–143. https://doi.org/10.3109/1040841X.2014.905513

Melini, F., Melini, V., Luziatelli, F., Ficca, A. G., & Ruzzi, M. (2019). Health-promoting components in fermented foods: An up-to-date systematic review. *Nutrients, 11*(5), 1189.

Ngoumelah, D., Heggeset, T. M. B., Haugen, T., Sulheim, S., Wentzel, A., Harnisch, F., & Kretzschmar, J. (2024). Effect of model methanogens on the electrochemical activity, stability, and microbial community structure of Geobacter spp. dominated biofilm anodes. *npj Biofilms and Microbiomes, 10*(1), 17. https://doi.org/10.1038/s41522-024-00490-z

Nguyen, H. U. D., Nguyen, D. T., & Taguchi, K. (2021). A novel design portable plugged-type soil microbial fuel cell for bioelectricity generation. *Energies, 14*(3), 553. https://doi.org/10.3390/en14030553

Olabi, A., Wilberforce, T., Sayed, E. T., Elsaid, K., Rezk, H., and Abdelkareem, M. A. (2020). Recent progress of graphene based nanomaterials in bioelectrochemical systems. *The Science of the Total Environment, 749*, 141225. https://doi.org/10.1016/j.scitotenv.2020.141225

Pandey, P., Shinde, V. N., Deopurkar, R. L., Kale, S. P., Patil, S. A., & Pant, D. (2016). Recent advances in the use of different substrates in microbial fuel cells toward wastewater treatment and simultaneous energy recovery. *Applied Energy, 168*, 706–723.

Pandya, R. S., Kaur, T., Bhattacharya, R., Bose, D., & Saraf, D. (2024). Harnessing microorganisms for bioenergy with Microbial Fuel Cells: Powering the future. *Water-Energy Nexus, 7*, 1–12.

Pant, D., Singh, A., Van Bogaert, G., Gallego, Y. A., Diels, L., & Vanbroekhoven, K. (2011). An introduction to the life cycle assessment (LCA) of bioelectrochemical systems (BES) for sustainable energy and product generation: Relevance and key aspects. *Renewable and Sustainable Energy Reviews, 15*(2), 1305–1313.

Passos, V. F., Aquino Neto, S., Andrade, A. R. de., & Reginatto, V.. (2016). Energy generation in a Microbial Fuel Cell using anaerobic sludge from a wastewater treatment plant. *Scientia Agricola, 73*(5), 424–428. https://doi.org/10.1590/0103-9016-2015-0194

Pasternak, G., Greenman, J., & Ieropoulos, I. (2017). Self-powered, autonomous Biological Oxygen Demand biosensor for online water quality monitoring. *Sensors and Actuators B: Chemical, 244*, 815–822.

Potter, M. C. (1911). Electrical effects accompanying the decomposition of organic compounds. *Proceedings of the Royal Society of London. Series B, Containing Papers of a Biological Character, 84*(571), 260–276. https://doi.org/10.1098/rspb.1911.0073

Prathiba, S., Kumar, P. S., and Vo, D. N. (2021). Recent advancements in microbial fuel cells: A review on its electron transfer mechanisms, microbial community, types of substrates and design for bio-electrochemical treatment. *Chemosphere*, *286*, 131856. https://doi.org/10.1016/j.chemosphere.2021.131856

Rabaey, K., & Verstraete, W. (2005). Microbial fuel cells: Novel biotechnology for energy generation. *TRENDS in Biotechnology*, *23*(6), 291–298.

Roy, A. S., Sharma, A., Thapa, B. S., Pandit, S., Lahiri, D., Nag, M., Sarkar, T., Pati, S., Ray, R. R., Shariati, M. A., Wilairatana, P., & Mubarak, M. S. (2022). Microbiomics for enhancing electron transfer in an electrochemical system. *Frontiers in Microbiology*, *13*, 868220.

Roy, H., Rahman, T. U., Tasnim, N., Arju, J., Rafid, M. M., Islam, M. R., Pervez, M. N., Cai, Y., Naddeo, V., & Islam, M. S. (2023). Microbial fuel cell construction features and application for sustainable wastewater treatment. *Membranes*, *13*(5), 490. https://doi.org/10.3390/membranes13050490

Santoro, C., Arbizzani, C., Erable, B., & Ieropoulos, I. (2017a). Microbial fuel cells: From fundamentals to applications. A review. *Journal of Power Sources*, *356*, 225–244. https://doi.org/10.1016/j.jpowsour.2017.03.109

Santoro, C., Kodali, M., Kabir, S., Soavi, F., Serov, A., & Atanassov, P. (2017b). Three-dimensional graphene nanosheets as cathode catalysts in standard and supercapacitive microbial fuel cell. *Journal of Power Sources*, *356*, 371–380.

Schroeder, U., & Harnisch, F. (2014). Biofilms, electroactive. *Encyclopedia of Applied Electrochemistry*, 120–126.

Sharif, H. M. A., Farooq, M., Hussain, I., Ali, M., Mujtaba, M., Sultan, M., and Yang, B. (2021). Recent innovations for scaling up microbial fuel cell systems: Significance of physicochemical factors for electrodes and membranes materials. *Journal of the Taiwan Institute of Chemical Engineers*, *129*, 207–226. https://doi.org/10.1016/j.jtice.2021.09.001

Stratford, J. P., Beecroft, N. J., Slade, R. C., Grüning, A., & Avignone-Rossa, C. (2014). Anodic microbial community diversity as a predictor of the power output of microbial fuel cells. *Bioresource Technology*, *156*, 84–91.

Szydlowski, L., Lan, T. C., Shibata, N., & Goryanin, I. (2020). Metabolic engineering of a novel strain of electrogenic bacterium Arcobacter butzleri to create a platform for single analyte detection using a microbial fuel cell. *Enzyme and Microbial Technology*, *139*, 109564. https://doi.org/10.1016/j.enzmictec.2020.109564

Tee, P. F., Abdullah, M. O., Tan, I. A., Amin, M. A., Nolasco-Hipolito, C., & Bujang, K. (2017). Effects of temperature on wastewater treatment in an affordable microbial fuel cell-adsorption hybrid system. *Journal of Environmental Chemical Engineering*, *5*(1), 178–188.

Vishwanathan, A. S. (2021). Microbial Fuel cells: A Comprehensive Review for Beginners. *3 Biotech*, *11*(5). https://doi.org/10.1007/s13205-021-02802-y

Wang, J., Zhang, X., Li, Z., Ma, Y., & Ma, L. (2020). Recent progress of biomass-derivedf carbon materials for supercapacitors. *Journal of Power Sources*, *451*, 227794.

Wilberforce, T., Abdelkareem, M. A., Elsaid, K., Olabi, A., and Sayed, E. T. (2021). Role of carbon-based nanomaterials in improving the performance of microbial fuel cells. *Energy*, *240*, 122478.

Wilkinson, S. (2000). "Gastrobots"—benefits and challenges of microbial fuel cells in foodpowered robot applications. *Autonomous Robots*, *9*, 99–111.

Yao, H., Xiao, J., & Tang, X. (2023). Microbial fuel cell-based organic matter sensors: Principles, structures and applications. *Bioengineering*, *10*(8), 886.

Yilmazel, Y. D., Zhu, X., Kim, K.-Y., Holmes, D. E., & Logan, B. E. (2018). Electrical current generation in microbial electrolysis cells by hyperthermophilic archaea Ferroglobus placidus and Geoglobus ahangari. *Bioelectrochemistry*, *119*, 142–149. https://doi.org/10.1016/j.bioelechem.2017.09.012

You, J., Greenman, J., & Ieropoulos, I. (2018). Novel analytical microbial fuel cell design for rapid in situ optimisation of dilution rate and substrate supply rate, by flow, volume control and anode placement. *Energies*, *11*(9), 2377. https://doi.org/10.3390/en11092377

Yuvraj, C., & Aranganathan, V. (2017). MFC—An approach in enhancing electricity generation using electroactive biofilm of dissimilatory Iron-Reducing (DIR) bacteria. *Arabian Journal for Science and Engineering*, *42*(6), 2341–2347. https://doi.org/10.1007/s13369-017-2529-8

Zhou, Y., Zhao, S., Yin, L., Zhang, J., Bao, Y., & Shi, H. (2018). Development of a novel membrane-less microbial fuel cell (ML-MFC) with a Sandwiched Nitrifying chamber for efficient wastewater treatment. *Electroanalysis*, *30*(9), 2145–2152. https://doi.org/10.1002/elan.201800232

7 Bioremediation Using MFCs for Cleaning Contaminated Environments

Mukul Machhindra Barwant, Balwant Singh, Mehtab Yasmeen, Odangowei Inetiminebi Ogidi, and Amir Vazirizadeh

7.1 INTRODUCTION

Over the past few decades, microbial fuel cells, or MFCs, have become incredibly important on a global scale. It is a device that produces bioelectricity and bioremediates pollution by using microorganisms to oxidize organic and inorganic materials in the anode. An external wire, an anode, a cathode, and a cation-sensitive membrane make up a microbial fuel cell (MFC) unit. The cathode and anode, respectively, are kept in anaerobic and aerobic conditions. With or without a mediator, MFC can function. In the case of mediated fuel cells, the oxidation of the substrate requires the addition of exogenous bacteria. Numerous man-made sources, such as the discharge of waste products from homes, businesses, farms, or municipalities, emit toxic metals into the environment. Because of metal buildup and toxicity in aquatic environments, heavy metal poisoning of the aquatic environment is a serious concern. Eventually, different levels of toxic metals that are discharged into the aquatic environment are absorbed by the aquatic sediments. River water quality and nearby populations are significantly impacted by the buildup of hazardous metals in sediments. For instance, sediment can be susceptible to the bioaccumulation of hazardous metals and is a food supply for a variety of freshwater invertebrates. Numerous species, particularly those at the top of the food chain like humans, fish, and birds, may be harmed by this bioaccumulation. Furthermore, the remobilization of hazardous metals from agricultural soil into crops due to the recovery of stream sediments and metal-contaminated rivers poses a significant risk to local consumers (Abbas et al., 2019).

Many anthropogenic reasons are primarily responsible for the pollution of soil and water environments, and the existence of organic contaminants—particularly those that are toxic, persistent, and bioaccumulative—raises concerns about their potential consequences on the environment and human health. Among water and soil biodegradation with pollution, microbial fuel cells (MFCs) are a natural approach that can be used. One of the significant features of an MFC, as far as energy harvesting is concerned, is that it does not emit any pollutants like CO, CO2, SOx, or NOx. The inherent ability of bioreactors to purify water and soils, along with this feature, is triggering research to build solutions that bank on MFC potential. Being bioelectrochemical systems (BESs), the various forms of MFCs include wastewater microbial fuel cells and terrestrial microbial fuel cells. These BESs utilize an organic substrate (biofuel) and a microorganism (biocatalyst) to transfer energy in basic environments like soil and water. Therefore, when combined with a high-end but more complicated electronic circuit, MFCs can be employed as power sources to power low-power and low-voltage sensors or an entire individual node in a distributed wireless sensor network (Ancona et al., 2020).

Microorganisms within the anodic chamber are supplied with a good substrate that is anaerobically degraded and releases electrons. Through an external circuit, the electrons are passed from the anode to the cathode by these electrons, and the protons thus formed are selectively allowed to pass through

 DOI: 10.1201/9781003645405-7

the membrane. Through the oxidation of numerous carbon sources or organic wastewaters, electrochemically active bacteria in MFCs produce electrical energy from chemical energy. Based on the chambers, MFCs can have different configurations, that is, single-, two-, or stacked chambers. Scientists have taken an interest in the invention of the microbial fuel cell (MFC) over the last 20 years. As a form of bio-electrochemical platform, MFCs are characterized by such properties as power generation, wastewater treatment, and use in biosensors. Trending dynamics of MFC studies on its synthetic, electrochemical, and microbiological perspectives have recently caused it to become particularly discernible in uses. They are in a way the state of affairs, and the MFCs originated as a rational curiosity.

The main causes of environmental pollution of surface and groundwater bodies are the use of MFCs to degrade various environmental pollutants, including antibiotics, artificial dyes, phenolic compounds, nitrogen-containing compounds, ethyl acetate, toluene, polycyclic aromatic hydrocarbons, perchlorate, pesticides, sulfur, and emerging contaminants. Researchers from all over the world have thought of using MFCs in the treatment of different wastewaters, including domestic or commercial wastewater and food processing wastewater.

This is specifically the result of this eco-friendly innovation's diversified uses. The creation is founded upon electroactive microorganisms, also called exoelectrogens. Most importantly, it is the primary invention which has the ability to produce energy from garbage without inputting additional or outside energy. The advancement of MFC performance for power generation and wastewater treatment has shown promising outcomes if electrodes are nanomaterials such as iron oxide and gold nanoparticles coated or pretreatment processes such as autoclave disinfection and sonication. The MFC technology has also been studied for the removal of various heavy metals and toxic compounds, as well as the detection of toxic compounds in wastewater. The MFCs can also be tuned to be employed as microbial electrolysis cells, which generate hydrogen energy from various natural sources. The cooperative associations of artificial and natural environmental microbial consortia have opened up research on microbial interactions and numerous engineering disciplines. Three various microbial fuel cells (MFCs) were employed here to treat various industrial wastewaters like dairy, pharmaceutical, and distillery spent wash (DSW) by using single and combined culture bacteria. Using single as well as mixed cultures, MFCs' performance was assessed based on its key characteristics such as biomass yield, voltage generation, and organic matter removal. After performing a first-order kinetic model, the model came out to be fitting and validated the experimental results. According to this study, it is possible to utilize modified microbial consortia in MFCs to effectively bioremediate organic and inorganic contaminants while concurrently producing bioelectricity from various industrial effluents. Microbial fuel cells, or MFCs, are a new biotechnology for environmental remediation that simultaneously can treat waste and produce energy. Taking advantage of the metabolic activity of electroactive bacteria (EAB) to break down organic substrates and transfer the electrons produced to an external circuit, MFCs are bio-electrochemical systems that produce electricity (Logan et al., 2006). Two electrodes, a cathode and an anode, separated by a proton exchange membrane or some other separator, constitute the overall working mechanism of an MFC. Anaerobic degradation of organic materials by microorganisms at the anode releases protons and electrons. Protons travel through the membrane, whereas electrons are passed over to the anode and then continue on to the cathode via an external circuit.

Water or other innocuous byproducts are produced when the electrons and protons at the cathode interact with oxygen (or other electron acceptors) to complete the circuit. It has proved that microbial fuel cells (MFCs) are a difficult but promising method for managing a wide range of environmental issues. Although the use of naturally occurring bacteria for bioremediation has been achieved worldwide as a potential alternative, its implementation is occasionally disrupted by the extremely low efficiency and poor controllability of bacteria. In this context, the MFC technology can be utilized as a tool for bioremediation to facilitate the efficient removal of a broad range of contaminants. In comparison with traditional technologies, MFCs are an incredible technology because of their outstanding characteristics, such as reduced sludge, energy production, and energy conservation. (Lu et al. 2016; Mohamed and Kim, 2015).

MFCs can also be used particularly to treat industrial wastewaters because of their microbe-enriched anode, which oxidizes the organic compounds, and their electron-carrying circuit to the porous carbon cathode structure with an oxygen reduction catalyst. MFCs are reactors that house the anode and cathode in either a single chamber or two chambers. Depending on the system architecture, they can function with or without a membrane (often Nafion-117). Microorganisms, mostly bacteria, are developing on the anode of the reactor when wastewater is added. Electrons are released to the anode by the bacteria as they break down the organic stuff in the wastewater stream (Bose et al., 2023).

Using the natural metabolic properties of microorganisms, the microbial fuel cell (MFC) is a bioelectrochemical method that generates power. The microorganisms in the MFC release some electrical energy when they take the nutrients in their medium. When biodegradable organic material corrodes due to the presence of electrogenic bacteria, MFCs work as a reassuring technology for the production of renewable power. Synchronous bioremediation and power generation are the precise uses of MFC. The process of bioremediation involves altering environmental conditions to encourage the growth of microorganisms and reduce pollutants in contaminated medium, such as water, mud, and subterranean debris. Bioremediation, which uses simply resided microorganisms, is widely acknowledged as a feasible option. Under these conditions, MFC may be used as a potential tool to stimulate bioremediation for the effective removal of a variety of pollutants. MFCs are an amazing technology as compared to older technologies because of their unique features, which include energy savings, less sludge, and electricity generation (Krithika et al., 2022).

This is a holistic review of MFC-based bioremediation, covering its principles, target pollutants, microbial processes, reactor configurations, recent case studies, and prospects. The worldwide tendency to mitigate pollution and encourage a circular bioeconomy concurs with the incorporation of bioelectrochemical systems, such as MFCs, into sustainable remediation approaches.

7.2 MICROBIAL FUEL CELLS: PRINCIPLES AND MECHANISMS

The bioelectrochemical devices known as MFCs convert the chemical energy in organic or inorganic compounds into electric current through the assistance of microorganisms as biocatalysts (Aelterman et al. 2006; Bermek et al. 2014). A double-chamber MFC normally consists of two compartments: the anode and the cathode. Most often, a proton exchange membrane (PEM) is used between these two chambers to enable protons produced by the anode to be transmitted through it to the cathode. A connection made via cathode and anode, usually via titanium or copper wires, completes the apparatus. The microorganisms within the chamber of the anode oxidize the organic substrates to yield electrons, protons, and carbon dioxide. Redox-active proteins or cytochromes first move the electrons produced by microbial metabolic activity to the anode surface; they then transfer them to the cathode via an electrical circuit (Borole et al., 2011).

Electrons are reduced in the cathode chamber. Usually, the cathode offers an electron acceptor, including oxygen or ferricyanide. Then, at the cathode, electrons join with protons and oxygen to create water. A catalyst like platinum could help this reaction to occur faster.

Anode reaction:

$$CH_3COO^- + 2H_2O \quad 2CO_2 + 7H^+ + 8e^-$$

Cathode reaction:

$$O_2 + 4e^- + 4H^+ \quad 2H_2O$$

MFC technology can be usefully combined with uses in wastewater treatment since it is a promising technology for electric current generation from various materials, including natural organic matter and complex organic waste (Chaudhuri and Lovley, 2003). Some particular qualities of MFCs that set this technology apart from others are:

1. The MFCs have a relatively higher ability to convert chemical energy into electric current.
2. The MFCs' ability to yield positive outcomes at different temperatures (between 20 and 40°C) sets them apart from other current bioenergy techniques.
3. Since the cathode can be passively aerated, MFCs can operate without the need for an external electric current to provide oxygen (as an electron acceptor). MFCs are built with a variety of materials and designs.
4. Although these systems can be operated under a variety of conditions, including low or high temperatures, acidic or basic pH, with different electron acceptors, etc., they are usually operated under optimal conditions for more efficiency.

(Amend and Shock 2001; Logan 2004; Oh et al., 2004)

7.3 MATERIALS FOR CONSTRUCTION OF MFC

7.3.1 Anode

It must be made of conductive materials. Additionally, the substance used in the anode chamber should be ecologically friendly and inert to the electrolyte. Carbon-based electrodes, such as graphite plates, rods, felt electrodes, granules, cloth, brushes, and stainless steel, are widely used in MFCs (Logan and Regan, 2006; Liu et al., 2014). In general, MFCs are very interested in electrode materials with a higher surface area and lower cost (Bergel et al., 2005). For better MFC performance, conductive nanomaterials with a larger surface area, like graphene, can be readily added to the low-surface-area bare electrodes.

The nanomaterial can be applied to the bare electrodes in a variety of ways. In the MFCs, the anodic modifications using the nanoparticles have so far produced positive outcomes. These nano-modifications typically promote the formation of biofilm on the anode, shorten the time it takes for the MFC to start up, facilitate the electron transfer mechanism, lower the system's internal resistance, and ultimately improve the MFC's overall performance (Bergel et al., 2005) (Figure 7.1).

7.3.2 Cathode

The cathode material, an electron acceptor, and a catalyst to enhance electron reduction are all found in the cathode compartment (Zhang et al. 2012). The above-mentioned anodc clectrode materials can also be used as the cathode. Furthermore, when oxygen is utilized as the electron acceptor, a catalyst (such as platinum) is applied onto the cathode electrode to speed up the rate of oxygen reduction. Because of its low price, universal availability, and high redox potential, oxygen is a very apt substitute for the electron acceptor (Bond and Lovley, 2003). Ferricyanide (K_3 [Fe $(CN)_6$]) is a common electron acceptor that has found extensive use in MFCs because of its good performance. When ferricyanide is employed as the electron acceptor in MFCs, no cathodic catalyst is needed. This is especially the case when a simple carbon electrode is used, which only shows a small overpotential. However, there are certain drawbacks to using ferricyanide in MFCs, such as its poor oxygen reoxidation, which necessitates frequent catholyte changes. Furthermore, ferricyanide can diffuse into the anode chamber via the ion exchange membrane; as a result, its use may impair MFC performance (Figure 7.1).

7.3.3 Membrane

An ion exchange membrane, such as a proton exchange membrane (PEM), which permits the passage of protons or particular cations from the anode to the cathode compartment, is typically utilized in an MFC between the anode and the cathode chamber (Heilmann and Logan, 2006). Nafion is the

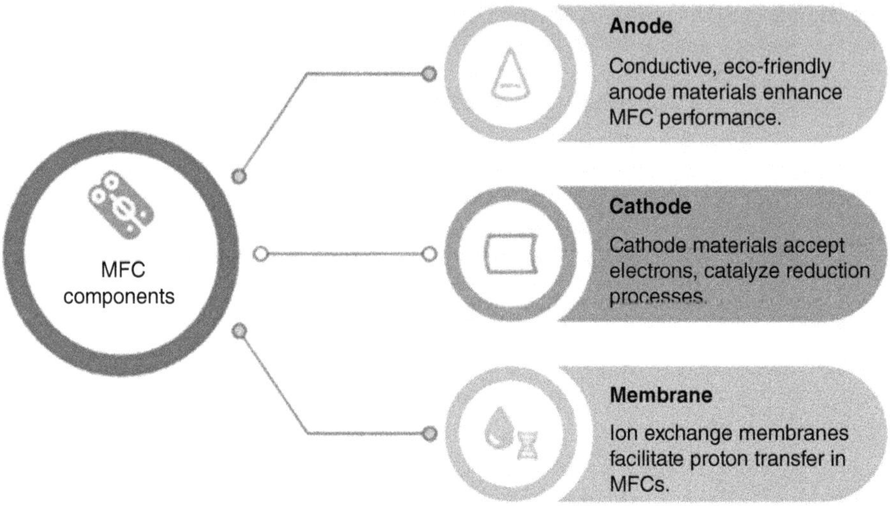

FIGURE 7.1 Materials for construction of MFC.

most popular and effective PEM. Ultrex CMI-7000, on the other hand, is ominously more cost-effective than Nafion and can also be used in MFC applications. Furthermore, PEM may leak oxygen, allowing bacteria or the anolyte to diffuse to the cathode and catholytes like ferricyanide to move to the anode. These effects can significantly reduce an MFC's performance. To evaluate the impact of the PEM on performance and long-term durability, however, more effective revisions are required (Min and Logan 2004) (Figure 7.1).

7.4 TYPES OF MFCs AND THEIR ROLE IN BIOREMEDIATION

7.4.1 Double-Chamber MFC

The simplest design of all MFCs is the double-chamber MFC (Niessen et al., 2004). Typically, a bottle which may have a different design is utilized as the anode and the other as the cathode, with PEM separating the two. Energy is typically produced in a two-chamber MFC using a defined medium (or substrate) in the anode and a defined catholyte solution. Stated differently, the double-chamber MFC is frequently run in batch mode. The double-chamber MFC can be shaped like a cube or a bottle. The nomenclature of the design can be defined by the choice of catholyte. For instance, if the air used in the cathode provides the electron acceptor, that is, oxygen, the MFC can be referred to as a two-chamber air-cathode MFC (Ringeisen et al., 2006). Such an MFC may be useful for producing electricity in remote sensing regions.

7.4.2 Single-Chamber MFC (SCMFC)

This type of MFC consists of a single chamber that houses both the anode and the cathode. PEM separates the anode from the cathode, which can be placed close to or far away. It has been said that internal ohmic resistance can be reduced by reducing the distance between electrodes. This can be accomplished by joining two chambers without using catholyte, which increases the power density. In comparison to double-chamber MFCs, such MFCs are straightforward, cost-effective, and generate a lot of power (Chaudhuri and Lovley, 2003; Ringeisen et al., 2006). However, the main issues in SCMFC, like reverse oxygen flow from cathode to anode and microbial adulteration, happen regularly. SCMFCs are the simplest and cost-effective designs. These MFCs typically only have an anodic chamber; a cathodic chamber not necessarily filled with air (Rabaey et al., 2004; Rabaey et al., 2005).

7.4.3 Upflow MFC

A new design called upflow MFC was created with the growing interest in the research of MFC. According to He et al. (2005), the upflow MFC has a cylinder-like shape. The anode at the bottom and the cathode chamber at the top make up the MFC. Layers of glass wool and glass beads divide the two chambers. The anode's bottom supplies the substrate, which travels upward to the cathode and exits at the top. Between the electrodes, a gradient is created that contributes to the fuel cell's advantageous operation (Cheng et al., 2006). In this design, distinct anolyte and catholyte and physical parting are not provided. Hence, proton transmission-related issues are easily handled (Venkata et al., 2014). Compared to other MFC designs, MFCs are easily scalable, which makes them interesting for wastewater treatment. The primary disadvantage of the fuel cell, however, is that the energy required to pump the substrate is significantly higher than the power produced. Thus, rather than producing electricity, the primary function of upflow MFC is to treat wastewater (Brutinel and Gralnick 2012). These MFCs are frequently used in basic research, and the studies suggest that their complex design, high internal resistance, and electrode-based losses are the reasons for their low power densities.

7.4.4 Stacked MFC

For increased power generation, stacked MFCs can consist of different MFCs in parallel or in series (Logan and Regan 2006; Aelterman et al., 2006). By using multiple MFCs and enhancing the power or current output of each single MFC, the MFC output is amplified. According to Logan et al. (2006), a single MFC unit with oxygen as an electron acceptor can typically produce a maximum of 0.8 V of OCV. The corresponding power output can then be obtained by multiplying the individual power outputs of different MFC unit cells that have been combined in a series or parallel combination. However, few other operating conditions also play a significant role in the present generation that can improve or worsen the performance of the MFC as a whole. Besides, as there would be a drop in voltage when each cell is stacked in series or parallel, the terminal voltage after stacking the single MFC in a single stacked MFC would not necessarily be precisely equal to the sum of the individual cell voltage. Parallel-connected MFCs and series-connected MFCs have been found to generate the same current.

Thus, we can conclude that a larger bioelectrochemical reaction rate is obtained in parallel-connected stack MFCs than in series-connected stack MFCs. Moreover, whenever the MFC units can't work separately, a parallel connection is more favorable for the maximum elimination of chemical oxygen demand (COD) to improve the efficiency of wastewater treatment. Copper wires were used to link six MFCs in parallel or series in an efficient experiment. Graphite rods served as the anode and cathode, and ferricyanide served as the catholyte, while acetate served as the substrate for the anode when powering the MFC. From the study findings, the power density (volumetric) of the stacked MFC was 59 W/m^3 when operated in series and 51 W/m^3 in parallel (Aelterman et al., 2006).

Electrons being transported from the anode, contributing to current generation, can be estimated through the coulombic efficiency (CE) of the MFC (based on the theoretically total electrons produced from substrate oxidation). Different CE rates can be obtained in stacked MFC by connecting the cells in different modes. Consistent with the above-mentioned study, when parallel mode was applied to activate the stacked MFC, a higher rate of CE (78%) was created than that activated by series mode, and only a CE of 12% was created. Voltage reversal is the main obstacle to higher voltage output in stack MFC. The depletion of the cell substrate can be the cause of the reversal of voltage, indicating that the capability of the bacteria to generate higher voltage has been lowered (Logan and Regan, 2006; Aelterman et al., 2006).

7.4.5 Other Designs

The majority of MFC studies use the basic designs, with little effort put into creating new designs that address the shortcomings of the current ones. Min and Logan (2004) designed a flat-plate MFC (FPMFC) to lessen one barrier by lowering the ohmic resistance brought on by greater interelectrode spacing. Chemical fuel cells typically employ this type of design, which also produces more power than earlier models. The anode and cathode of the FPMFC were composed of flat plates, each of which had a projected surface area of 225 cm^2. The two plates were separated by a Nafion membrane. Using household wastewater as a substrate, this reactor produced a power density of 56 mW/m^2, and the study found that 58% COD was attained (Min and Logan, 2004). Acetate, glucose, starch, and other substrates were also used to generate power in the reactor; however, the power output was lower than that of the other designs, such as the cube reactor. The growth of the microbial community may be impacted by oxygen passing through the membrane to the bacteria in the anode chamber, which could be caused by the electrodes being too closely joined

7.5 MICROBIAL COMMUNITIES IN MFC-BASED BIOREMEDIATION

Numerous researchers have conducted community analyses of anodic biofilms enriched with electrochemically active microorganisms. These investigations demonstrated majorly the prevalence of bacterial phyla. The dominant species and bacterial consortia change depending on the substrate type and inoculum (Logan and Regan, 2006; Kim et al., 2004, 2007a). Apart from the well-known iron-reducing bacteria, like *Shewanella* and *Geobacter*, there may be other unidentified species present in the anode biofilm due to the significant diversity of bacterial consortia.

Numerous reviews of MFC microorganisms have been published recently indicates the following: the parameters defining MFC performance and microbial metabolism (Rabaey and Verstraete, 2005), MFC characterizations and methodology (Logan et al., 2006) bacterial communities that generate electricity (Logan and Regan, 2006), the traits of bacteria that are electrochemically active in a system without a mediator (Chang et al., 2006), the main mechanism for electron transfer of electricigens to electrodes, microbial physiology and engineering approaches concerning Benthic Unattended Generators (BUGs) for powering remote sensors, the benefits and drawbacks of MFCs in contrast to traditional anaerobic digestion technology (Pham et al., 2006), the intrinsic limitations of MFCs and recommendations for enhancing their performance (Kim et al., 2007a), the latest developments in MFC configurations and capabilities, and the mechanisms of anodic electron transfer in bacteria that are electrochemically active.

Recently, Clauwaert et al. (2007), by using an open-air biocathode system with microorganisms that function as biocatalysts for oxygen reduction rather than chemical catalysts, demonstrated the successful generation of electricity. Since aerobic bacteria can be used in place of noble catalysts, this is a promising discovery. Anode in MFCs have the electrochemically active bacteria and are generally known as anodophilic bacteria (Logan et al., 2006), electricigens and exoelectrogens (Logan and Regan, 2006) based on their capacity to transfer electrons exocellularly to the electrodes without interference of any exogenous mediators. They can completely oxidize organic compounds to carbon dioxide with the electrode serving as the sole electron acceptor. Electron transfer can occur either through membrane-associated components, soluble electron shuttles generated by specific bacteria, or highly conductive nanowires. The mechanisms for electron transfer to the electrodes were reviewed by Lovley (2006) and Schröder (2007).

Generally, only electrochemically active bacteria and those that are symbiotically related to them should proliferate when a mixed culture is introduced into an MFC. However, bacterial community analysis of enriched-anode biofilms from various MFCs has revealed significant phylogenetic diversity in the anode community, with no single dominant bacterial species. This considerable diversity suggests the existence of other candidates beyond the commonly occurring *Geobacter* and Shewanella species that contribute to electricity generation through various ways. Based on 16S rDNA sequencing analysis, the bacterial consortia and predominant species vary with the operational conditions, such as the inoculum and electron donors, as shown in Table 7.1.

Kim et al. (2004) used a two-chamber MFC inoculated with anaerobic sludge and fed starch processing wastewater to report the predominance of β-Proteobacteria clones in the phylogenetically diverse anode community. It used a river water-fed MFC with a river sediment inoculum, and

TABLE 7.1
Bacterial Communities in Different MFCs Operating in Different Inocula and Substrates

Sr No	Inoculum Source	Substrate Type	Dominant Bacterial Genera	Role in MFC	Reference
1	Wastewater sludge	Acetate	*Geobacter*, *Shewanella*	Direct electron transfer; efficient oxidation	Logan et al. (2006); Lovley (2008)
2	River/lake sediment	Complex organics, acetate	*Geobacter*, *Desulfuromonas*	Anaerobic respiration; exoelectrogenic	Rabaey et al. (2004); Holmes et al. (2004)
3	Soil	Glucose, cellulose	*Pseudomonas*, *Bacillus*, *Acinetobacter*	Fermentation and biofilm formation	Reguera et al. (2005); Zhang et al. (2012)
4	Pure cultures (lab strain)	Acetate, lactate	*Geobacter sulfurreducens*, *Shewanella oneidensis*	Model exoelectrogens; controlled electron transfer	Bond and Lovley (2003); Marsili et al., (2008)
5	Industrial wastewater	Aromatics, phenol	*Pseudomonas*, *Comamonas*	Degradation of toxic compounds; electron transfer	Liu et al., (2014)

Kim et al. (2007b) used ethanol-enriched MFC inoculated with anaerobic sludge. Nevertheless, δ-Proteobacteria predominated in MFCs inoculated with aquatic sediments, and Logan et al. (2006) documented a significant abundance of γ-Proteobacteria in a cystein-enriched MFC inoculated with sediments. It's interesting to note that in a special seawater and marine plankton-filled MFC, ε-Proteobacteria—which were either absent or uncommon in other studies were predominant. Further research is required to fully understand the functional significance of these dominant community members. Because the bacterial communities are influenced by both the substrate and the inoculum, it was challenging to make a general comparison between the earlier studies on bacterial communities because they were typically conducted using a single substrate or a different inoculum.

In another study, in order to normalize the effect of the inoculum, MFCs inoculated with an identical anaerobic sludge were used to compare the effects of the substrate type (such as acetate, propionate, butyrate, and glucose) on the bacterial community. Except for the propionate-fed system, all MFCs showed a relative abundance of β-Proteobacteria but a lack of γ-Proteobacteria. On the other hand, Firmicutes (59.3% of clones) and γ-Proteobacteria (18.5%) dominated a propionate-fed MFC.

In a similar vein, Firmicutes have been described as essential components of the MFC bacterial community; they have the ability to move electrons recovered from organic matter metabolism to the anode. With a comparatively wide distribution of Proteobacteria and Firmicutes α-, β-, and δ-subclasses, the glucose-fed MFC displayed the most diverse community. The diversity of bacteria that can metabolize the soluble fermentation end products of glucose, such as acetate, propionate, and butyrate, is most likely the result of glucose's intrinsic fermentable properties. Because they have been widely observed in a variety of MFCs using aquatic sediments, the electrochemically active bacteria in MFCs are thought to be iron-reducing bacteria, including *Shewanella* and *Geobacter* species. In MFCs fed acetate (31.7%), butyrate (13.6%), and glucose (14%), *Geobacter*-like species—the most representative of which is *Geobacter sulfurreducens*—were found to be an essential component of the bacterial community, with the exception of propionate-enriched MFC. *Shewanella*-like sequences, another model strain, were not found, though. Even with lactate, one of the substrates that is favorable for Shewanella species, Jung and Regan (2007) reported that different strains of Geobacter-like bacteria predominated in their different substrate-fed MFCs, but no Shewanella sp. were reported even in lactate medium.

According to Lovle (2006), because of the substrates they use, *Geobacter* species, which can transfer electrons directly to electrodes, have an advantage over *Shewanella* species in the competition for space on the anode. This is most likely due to the fact that *Shewanella* species use lactate, which is not an essential extracellular intermediate in the anaerobic breakdown of organic compounds, and only partially oxidize it to acetate, which results in ineffective electricity production. The Geobacteraceae family of iron-reducing bacteria has the ability to transfer electrons straight onto an electrode (Bond and Lovley, 2003). However, simple organic acids like acetate are the only electron donors that these bacteria can use. Therefore, they depend on fermentative bacteria to provide electron donors from complex organic compounds, like glucose. This is the reason that there is more diversity of the bacterial community in glucose-fed anode biofilms than with other simple substrate-fed MFCs.

Numerous reports indicate that *Geobacter* species are essential to the exocellular electron transfer to the MFC electrodes. According to 16S rDNA sequencing, these bacteria's limited substrate versatility necessitates the presence of syntrophic or symbiotic bacteria, resulting in a range of bacterial community structures. The efficient conversion of organic matter to electricity seems to require pathways and cooperative consortia. Because of the synergistic interactions within the more complex cooperative anode communities, which include species or mechanisms that are currently unknown, mixed cultures seem to be more advantageous than pure cultures for the practical application of MFCs. This is based on both their much wider substrate specificity and power output (Kim et al., 2007a) seen in every MFC except the propionate-fed system. However, a propionate-fed MFC was dominated by Firmicutes (59.3% of clones), followed by γ-Proteobacteria (18.5%). Similarly, Firmicutes have been reported to be integral members of the MFC bacterial community; they can transfer the electrons retrieved from the metabolism of organic matter to the anode49,61-63). The glucose-fed MFC showed the most diverse community with a relatively broad distribution of α-, β-, and δ-subclasses of Proteobacteria and Firmicutes. This was most likely due to the inherent fermentable characteristics of glucose, which result in the diversity of bacteria that can metabolize the soluble fermentation end products of glucose (e.g., acetate, propionate, and butyrate).

One of the most important variables affecting an MFC's performance is the amount of biocatalyst present. Because an improvement in MFC performance can be achieved not only by increasing the anode's surface area, which merely increases the total number of bacteria, but also by increasing the number of metabolically viable cells that contribute to electron transfer on a given surface, bacterial viability data has significant implications for MFC design. Bacteria distant from the anode remained viable in an MFC with highly structured, multilayer biofilms by *Geobacter sulfurreducens* on the anode surface, and the efficiency of current production did not decrease with biofilm thickness.

An electronic network that penetrates the biofilm and can facilitate long-range electrical transfer in an energy-efficient way is most likely provided by highly conductive nanowires made by bacteria or conducting pili. Because they are still alive and the per-cell efficiency of current production does not drop as the biofilm forms, the cells that are not in direct contact with the anode thus seem to contribute to current production. According to Logan and Regan (2006), a pile of living bacteria may be essential components of the community because they can transfer electrons to the electrode even when they are far from the electrode surface. A study examining how electron donors affect bacterial viability (Chae et al., submitted) found that overall cell viability was comparable for all electron donors, including glucose, propionate, butyrate, and acetate. Because of their normal metabolic life cycle or because many of the bacteria in the inoculum may have adhered to the electrode surface but were unable to use it as an electron acceptor, the significant presence of dead bacteria was also noted.

Because bacteria like *Geobacter* species can both donate and accept electrons from the cathode if the cathode is positioned at a negative potential, an MFC can be modified in an appealing manner for bioremediation. The most popular technique for effectively bioremediating contaminated environments is the addition of an electron acceptor or donor to encourage the intended biodegradation. For the bioremediation of pollutants like metals or organics in contaminated groundwater, electrodes may act as electron donors or acceptors. For instance, adding organic electron donors to promote biological reductive dechlorination is a common method of eliminating chlorinated solvents from contaminated groundwater. It is challenging to deliver the appropriate quantity of electron donors to the target site. By poising the electrodes with a negative potential (−500 mV), Gregory et al. (2004) reported that microorganisms reduce the oxidized form of soluble uranium U6+ to insoluble U4+ and precipitate U4+ onto a cathode when the electrode acts as an electron donor. By merely inserting electrodes into wells drilled into contaminated subsurface environments, this poised-electrode system can function as a viable (or better) method for the bioremediation of uranium-contaminated groundwater as opposed to the addition of organic electron donors. Nitrate can also be converted to nitrite when electrodes are used as electron donors. On the other hand, electrodes can also act as a more convenient electron acceptor for the bioremediation of hydrocarbons compared with diverse electron acceptors such as oxygen, Fe3+, and sulfate.

The electrodes in the modified MFC can be used as an efficient substitute for an electron source or sink, since the primary challenge in the bioremediation of contaminated subsurface areas with organic pollutants or metals is the optimal delivery of electron acceptors or donors to promote the desired biodegradation.

7.6 APPLICATIONS OF MFCS IN ENVIRONMENTAL CLEANUP

Microbial Fuel Cells (MFCs) are becoming increasingly highly promising biotechnology devices for environmental remediation with their dual functionality of power production and concomitant environmental pollutant degradation. Their applications toward environmental issues cut across a variety of fields such as wastewater treatment, sediment and soil remediations, and industrial effluent detoxification.Microbial Fuel Cells (MFCs) also emerged as an excellent green technology for environmental remediation due to their distinct capacity to couple microbial metabolism with electricity production. Their uses in cleaning the environment cover various fields such as treatment of wastewater, soil and sediment remediation, and decontamination of industrial effluents. Through the use of electrochemically active microorganisms, MFCs not only contribute to the degradation of pollutants, but also assist in the recovery of valuables such as clean water and bioenergy.

Applications in wastewater treatment, MFCs have proven enormous potential as strong agents to eliminate organic substrates, heavy metals, and even nutrients from residential wastewater and industrial effluent.

Anodic chambers are bioreactors where microbial communities oxidize substrates, give away electrons and protons. The electrons move through an external circuit to the cathode, producing a useful electric current. It has been demonstrated through experiments that MFCs are capable of attaining high chemical oxygen demand (COD) efficiencies, typically more than 80%, as well as the by-product of bioelectricity (Logan 2019). Moreover, MFCs have been shown to be efficient in removing nitrogen and phosphorus from municipal wastewater, thereby avoiding eutrophication of receiving water bodies. Integrating MFCs with traditional wastewater treatment systems has been suggested to maximize energy recovery and minimize the carbon footprint of treatment plants. MFCs have been used to biodegrade petroleum hydrocarbons, polycyclic aromatic hydrocarbons (PAHs), and heavy metals in contaminated soil and sediment for remediation of contaminated soils and sediments. Sediment microbial fuel cells (SMFCs), a type of MFC that is specifically suited for in situ applications, has been utilized in contaminated riverbeds and marine coastal sediments. The bioelectrochemical gradients established by SMFCs have the ability to initiate microbial oxidation of recalcitrant organic pollutants and increase the bioavailability and conversion of heavy metals like cadmium, lead, and arsenic. The systems are ideally utilized at remote or ecologically sensitive locations where minimal site disturbance is necessary. The use of MFCs to treat industrial effluent

TABLE 7.2
Applications of MFCs in Environmental Cleanup

Application Area	Pollutants Treated	Mechanism	Advantages	References
Wastewater Treatment	Organic matter, nitrogen, phosphorus, heavy metals	Oxidation of organics by electroactive microbes; electron transfer generates electricity	Simultaneous pollutant removal and energy recovery	Logan 2019
Soil Remediation	Petroleum hydrocarbons, PAHs, heavy metals	In situ bioelectrochemical degradation via SMFCs	Minimal disturbance to ecosystem, low energy input	Deng et al., 2020
Industrial Effluent Treatment	Azo dyes, phenolics, chlorinated hydrocarbons	Microbial oxidation of xenobiotics in anodic chamber	Detoxification of toxic industrial discharges with concurrent power generation	Feng et al., 2022
Agricultural Runoff Treatment	Nitrate, phosphate, pesticides	Nutrient and chemical removal via microbial metabolism and EET	Prevents eutrophication, supports nutrient recovery	Li et al., 2023a

has also been in the limelight, particularly from textile, pharmaceutical, and agrochemical industries whose effluent is laden with toxic dyes, solvents, and complex organic matter. MFCs have been used to break down azo dyes, phenol compounds, and chlorinated hydrocarbons with significant efficiency while, at the same time, generating electricity. The electrogenic microbial populations in MFCs have the ability to adjust to high levels of contaminants and are hence well-suited for the treatment of shock loads and fluctuating compositions characteristic of industrial effluents. In general, MFC integration in environmental cleanup planning is consistent with international sustainability aims, as not only do such systems treat the pollution, but they also provide added advantages of recovering energy, mitigating carbon footprints, and closing resource loops. Further developments in electrode material, reactor architecture, and microbial communities are predicted to further enlarge the scope for MFC usage in various bioremediation applications (Table 7.2).

7.6.1 Wastewater Treatment

Among the best-researched and most sophisticated applications of MFCs is wastewater treatment of urban and industrial water. MFCs tap electrochemically active bacteria (exoelectrogens) to oxidize organic pollutants in wastewater, passing electrons to the anode while producing current in the process. The concurrent pollution degradation and power production render MFCs an energy-saving option for energy-intensive traditional wastewater treatment processes.

The latest research has confirmed the effectiveness of MFCs in decolonization of chemical oxygen demand (COD), total suspended solids (TSS), and pathogens from wastewaters. It has also been established that integrating MFCs with anaerobic or aerobic systems is synergistic to amplify treatment efficiency and energy recovery (Logan 2019). Hybrid systems like microbial electrolysis cells (MECs) and microbial desalination cells (MDCs) expand their horizons even further by enabling the production of hydrogen gas as well as water desalination in addition to contaminant removal.

7.6.2 Soil and Sediment Remediation

Soil and sediment in situ contaminant remediation by MFCs, or sediment MFCs (SMFCs), is now an inorganic and passive bioremediation method. These systems engage indigenous microbial communities present in sediments to oxidize contaminants like petroleum hydrocarbons, heavy metals, and organic contaminants, electrons passing through to implanted in situ electrodes. SMFCs have been reported to be highly effective in lowering the levels of polycyclic aromatic hydrocarbons (PAHs), chlorinated solvents, and even persistent chemicals like polychlorinated biphenyls (PCBs) in polluted sediments (Viggi et al., 2017). The electric field produced by MFCs can induce electrokinetic migration of contaminants, which further elevates rates of bioavailability and degradability (Li et al., 2021). Furthermore, combining phytoremediation with SMFCs has achieved improved root growth and contaminant uptake through greater rhizosphere microbial activity and reduced soil toxicity.

7.6.3 Industrial Effluent Detoxification

Industrial effluents typically consist of high concentrations of toxic substances like dyes, phenols, heavy metals, and solvents, and are hazardous to the environment as well as human health. MFCs have been considered a sustainable way of detoxifying such effluents with the added benefit of power generation.

Textile dyeing and leather processes, for instance, release effluents with high concentrations of azo dyes and chromium salts. Experiments have established that MFCs can decolorize and make such effluents less toxic by enabling reductive cleavage of dye bonds and conversion of toxic metal ions into less toxic forms. In addition, electroactive biofilms in MFCs can oxidize phenolic and pharmaceutical pollutants to harmless metabolites, ensuring environmental safety (Ghasemi et al., 2018). The application of innovative materials such as carbon nanofibers and functionalized electrodes likewise improved the performance and detoxification capacity of MFCs (Table 7.3).

TABLE 7.3
Different Types of Fuel Cells, Their Nature, and Applications in Environmental Cleanup

Type of Fuel Cell	Electrolyte Material	Operating Temperature	Key Features	Applications in Environmental Cleanup	References
Microbial Fuel Cell (MFC)	Microbial biofilm + proton exchange or separator	Ambient to 40°C	Utilizes electroactive bacteria to degrade organics and produce electricity	Wastewater treatment, soil/sediment remediation, industrial effluent detoxification	Logan 2019
Proton Exchange Membrane Fuel Cell (PEMFC)	Nafion or polymer membrane	~60–80°C	High power density, rapid start-up, sensitive to CO poisoning	CO_2 removal systems, powering environmental sensors in remote areas	Baroutaji et al., 2019
Molten Carbonate Fuel Cell (MCFC)	Molten lithium/ sodium/ potassium carbonate	600–700°C	Can capture and use CO_2, high-temperature operation	CO_2 separation from flue gases, integration with waste-to-energy plants	Li et al., 2019

7.7 TECHNOLOGICAL ADVANCES AND ENHANCEMENTS IN MFCs

The area of microbial fuel cells (MFCs) has witnessed great technological advancement toward increasing their power density, rate of pollutant degradation, operating stability, and scalability to environmental remediation on a real-world scale. These include developments in the areas of material science, microbiology, system design, and integration concepts, which collectively define the overall performance of MFCs. Ethereal technological progress has taken place in microbial fuel cells (MFCs) in recent years to make them efficient, scalable, and applicable for environmental remediation. The developments are being realized on different fronts of MFCs, from electrode materials and reactor designs to microbial community engineering and integration with other technologies. These developments are crucial for overcoming the shortcomings of conventional MFCs, including low power density, expensive materials, and inconsistent performance under real-world conditions. The most significant frontier of innovation is the invention of high-performance electrode materials. Conventional carbon-based electrodes are being nanomaterials like carbon nanotubes, graphene, and metal oxides modified in an attempt to improve conductivity, surface area, and biocompatibility. These materials allow for enhanced electron transfer between the anode and microbial biofilm, thus increasing power density and system efficiency as a whole (Santoro et al., 2017). The cathode has also witnessed the incorporation of non-platinum catalysts such as manganese oxide and iron phthalocyanine, which provide cheaper substitutes for oxygen reduction reactions in air-cathode MFCs.

Reactor design itself has been optimized through improvements that have resulted in better MFC performance. Emergent configurations such as tubular, stacked, and membraneless systems present greater flexibility to scale up MFCs for industrial as well as municipal purposes. Stacked MFCs, specifically, have shown the potential to maximize voltage output through series or parallel configuration, making them more appropriate for powering small electronic devices or remote sensors in contaminated settings. Mixing of flow-through architectures and continuous-fed systems has also resulted in higher substrate utilization rates and more stable operation. Microbial engineering is also an emerging area that is advancing MFC performance. Synthetic biology strategies have made possible the creation of genetically engineered microorganisms with enhanced extracellular electron transfer (EET) efficiency. Engineered Shewanella oneidensis and Geobacter sulfurreducens strains have also exhibited enhanced electron shuttling and contaminant degradation performance, thereby enhancing both bioremediation and bioelectricity harvesting efficiency. Artificial microbial consortia directed at certain pollutants and conditions are also being researched in order to maximize microbial cooperation and system robustness. The integration of MFCs with other technologies, including microbial electrolysis cells (MECs), constructed wetlands, and photocatalytic systems, has paved the way for their use in the environment.

Hybrid systems involving the integration of a mixture of MFCs and MECs have been employed to facilitate concurrent treatment of wastewater and generation of hydrogen gas. Similarly, integrating MFCs with constructed wetlands increases the removal of nutrients and offers low-maintenance passive treatment for decentralized wastewater treatment (Li et al., 2023a). Moreover, the utilization of machine learning and Internet of Things (IoT) for monitoring MFC performance has newly paved the way toward real-time control and optimization of bioremediation processes.

In addition, these technological advances not only enhance the energy and treatment efficiencies of MFCs but also lead the way for their sustainable integration into current environmental management practices. Sustained interdisciplinary studies and field-scale testing are required to realize the full potential of MFCs as a next-generation technology for bioelectrochemical remediation.

7.7.1 Advanced Electrode Materials

Electrodes are basic parts of MFCs since they are used as the interface for the transfer of electrons by microbes. Recent research has involved designing high-performance, conductive, and biocompatible electrode materials to optimize electron recovery. Materials based on carbon, such as carbon nanotubes (CNTs), graphene, activated carbon, and carbon cloth, have been demonstrated to have better performance in enabling microbial colonization and extracellular electron transfer (EET). Electrode functionalization with metal oxide (e.g., MnO_2, TiO_2) or conductive polymer (e.g., polyaniline, polypyrrole) also enhances electrocatalytic activity and corrosion protection.

7.7.2 Nanotechnology Integration

Nanomaterials provide new properties to facilitate microbial-electrode interaction. Incorporating nanoparticles like gold, silver, or iron oxide into electrodes enhances electron conductivity and biofilm development. Moreover, nano-engineered surfaces enhance bacterial adhesion and stress stability. For example, research has shown that MFCs with nanostructured anodes exhibited 2–3 times more power densities than standard carbon-based systems.

7.7.3 Genetic Engineering of Microorganisms

Synthetic biology and genetic modification have allowed specially engineered microbial strains to be produced with improved bioelectrochemical performances. Conductive pili or overexpressed cytochrome pathways in gene-modified strains of Shewanella oneidensis and Geobacter sulfurreducens were shown to attain higher electron transfer and contaminant degradation efficiencies. Apart from that, microbial consortia for the co-metabolism of recalcitrant pollutants can enhance MFC performance under various environmental conditions.

7.7.4 New MFC Architectures

New designs like stacked MFCs, membrane-less MFCs, and air-cathode MFCs have been proposed to attain maximum volumetric energy density, minimize internal resistance, and decrease maintenance costs. Multi-chamber and modulated stacks of MFCs are beneficial in improving scalability and integration with wastewater treatment plants. Micro-scale MFCs (μMFCs) are being researched to be utilized for biosensing and pollutant monitoring.

7.7.5 Hybrid Systems and System Integration

Integration of MFCs with other technologies has been found effective in enhancing system robustness in addition to multifunctionality. MFC-MEC hybrids enable concurrent treatment of wastewater and hydrogen production. Integration with constructed wetlands, anaerobic digesters, and phytoremediation systems improves treatment efficiency and eco-benefits. Additionally, real-time data collection and control through IoT-based sensors make automatic operation and system optimization possible.

7.7.6 3D Printing and Biofabrication

Current advances in additive manufacturing enable the production of personalized and optimized MFC components. 3D-printed casings, flow channels, and electrodes minimize fabrication time and cost while facilitating design freedom. Printing methods based on bioinks that incorporate microorganisms into structural matrices provide new ways of developing living biocatalytic systems.

7.8 CHALLENGES AND LIMITATIONS

Even though Microbial Fuel Cells (MFCs) have promising prospects in sustainable bioremediation and energy recuperation, there are various issues that hinder their large-scale deployment and sustained functionality. These constraints are both biological and engineering based and must be addressed in an organized manner to enable MFCs to scale up from laboratory prototypes to economically viable technologies.

7.8.1 Low Power Output

The primary challenge of MFC technology is the relatively low power density relative to other energy production systems. The limitation inhibits MFCs from being utilized as standalone power sources, especially in high-energy-demand environments. The internal resistance, microbial electron transfer rates, and cathodic oxygen reduction reaction kinetics are generally inhibiting factors for higher power densities (Logan 2019). In addition, scale-up normally leads to a loss of performance through higher ohmic and mass transfer losses.

7.8.2 Electrode and Material Limitations

Although developments in advanced electrode materials have advanced, cost, durability, and biofouling issues remain major challenges. Traditional carbon-based electrodes, as biocompatible and inexpensive as they are, tend to lack high conductivity and mechanical strength. Conversely, metal-based electrodes like platinum or stainless steel are more conductive but are costly and corrode easily (Santoro et al., 2017). Additionally, fouling of the electrode surface by microbial growth or sedimentation decreases electron transfer efficiency over time.

7.8.3 Limited Substrate Versatility

MFCs function best with easily biodegradable organic substrates like acetate, glucose, or lactate. Nevertheless, most environmental pollutants and industrial effluents are made up of complex or recalcitrant compounds that cannot be easily metabolized by electroactive bacteria. Under such conditions, the requirement for pretreatment or co-metabolism lowers the cost-effectiveness and simplicity of the system. The presence of toxic compounds can also inhibit microbial activity, causing instability in performance.

7.8.4 Biofilm Management and Microbial Stability

A stable and active biofilm on the anode surface is critical for effective MFC operation. Biofilm formation, however, is affected by many factors, including temperature, pH, substrate composition, and operating conditions. Abrupt changes in the environment or invasion of non-electrogenic microorganisms can compromise biofilm integrity and lower MFC efficiency (Rabaey & Rozendal, 2010). Long-term operation also poses challenges in maintaining microbial diversity and electrochemical performance.

7.8.5 Scale-Up and System Integration

Scaling up lab-scale MFCs to practical applications is hindered by reactor design problems, hydraulic retention time, and scalability. MFCs at large scales frequently face uneven substrate distribution, ineffective electron transfer, and process complexity. Compatibility with current wastewater or soil remediation systems also requires strong, modular, and low-maintenance designs, which are still in development.

7.8.6 Economic and Regulatory Barriers

From an economic standpoint, the cost of the capital of MFC materials—primarily high-performance electrodes, membranes, and control systems—is a major deterrent. Furthermore, the rate of return via electricity production or pollutant removal is not yet adequate to support large-scale rollout without subsidies or co-benefits. Lack of clarity about the regulation of the use of MFCs via genetically modified microorganisms or waste-energy systems in open environments is also a deterrent for using MFCs (Santoro et al., 2017).

7.9 FUTURE PERSPECTIVES AND SUSTAINABLE INTEGRATION

Microbial Fuel Cells (MFCs) possess great potential in addressing two global issues: environmental pollution and the production of clean energy. As a bioelectrochemical technology that harnesses microbial metabolism for the concurrent waste degradation and electricity generation, MFCs align with circular economy and low-carbon development policies. However, this potential hinges on continuous research, intersectoral cooperation, and system integration into current environmental and energy systems.

7.9.1 Scaling Up and System Optimization

One of the greatest future challenges for MFC technology is cost-effective scaling up without energy loss or treatment performance. Modular configurations, stackable arrangements, and flow-through designs are being developed to permit application in large-scale applications, including municipal wastewater treatment plants and contaminated land. Optimization of operating conditions—pH, hydraulic retention time, and substrate loading rates—by real-time monitoring and control systems will enhance robustness and reduce operation costs.

7.9.2 Integration with Renewable Energy and Resource Recovery Systems

MFCs may also be used as integral modules in decentralized, multi-functional resource recovery systems. Integrate MFCs with anaerobic digestion, solar power, constructed wetlands, or microbial electrolysis cells (MECs) and synergistic advantages like improved energy output, more efficient contaminant removal, and nutrient recovery may be realized (Li et al., 2023b). For rural or off-grid locations, MFC-powered sensors and treatment devices may provide self-sufficient solutions for environmental monitoring and sanitation.

7.9.3 Synthetic Biology and Microbial Engineering

The use of synthetic biology presents revolutionary possibilities to improve the electrogenic potential of microbial strains. Future research can include the development of synthetic consortia with optimized electron transfer routes and increased tolerance to toxic substances. Genetically modified microbes may be designed to particular environmental contaminants to enable directed bioremediation methods with MFCs. Additionally, the integration of quorum-sensing capability may enable adaptive microbial regulation and enhanced system robustness.

7.9.4 Internet of Things (IoT) and Smart Monitoring

With the advances in digital technologies, the next-generation MFCs can be deployed on IoT-based platforms for real-time monitoring, predictive repair, and process improvement. These "smart MFCs" can not only be used for bioremediation but can also be utilized as biosensors to sense heavy metals, pathogens, or organic pollutants in situ. The convergence of digitalization and biotechnology will play a key role in amplifying the social effect of MFC-based systems.

7.9.5 Policy Support and Economic Viability

Policy frameworks supporting green technology and circular waste treatment will play a key role in the business use of MFCs. Public–private partnerships, government incentives, and regulatory certification of MFCs as a renewable bioenergy technology can aid in market penetration. Life cycle assessment (LCA) and techno-economic analysis need further development to offer evidence of the environmental and economic advantages of MFC incorporation over traditional remediation technologies (Logan 2019).

7.9.6 Education and Public Awareness

Scholarly participation, public participation, and sharing of information also depend on MFC technology development. The integration of MFC modules into environmental science education, public displays, and community demonstration projects can help create acceptance and awareness of this new biotechnological alternative.

7.10 CONCLUSION

Microbial Fuel Cells (MFCs) are a revolutionary technology for the abatement of many of the world's most obstinate 21st-century energy and environmental challenges. By integrating bioremediation with clean electricity production, MFCs are a double solution for sustainable waste treatment and site restoration. By the metabolic process of electroactive microorganisms, MFCs are capable of degrading broad spectra of organic and inorganic contaminants present in industrial wastewater, effluent, and soil and generating bioelectricity at the same time—facilitating circular economy models and sustainable development goals (Logan 2019).

MFCs' ability to remove contaminants, for example, heavy metals, dyes, and hydrocarbons, has been extensively demonstrated using laboratory cultures and pilot-scale installations. Their uses in wastewater treatment, sediment cleanup, and effluent detoxification show broad prospects for large-scale implementation. Also, advancements in technology—from novel electrode materials and improved biocatalysts to intelligent sensors and IoT integration—are opening up opportunities for more efficient and more responsive MFC setups. Nevertheless, there are important challenges still to overcome, especially in achieving cost-effective scalability, long-term operational reliability, and regulatory acceptability. Low power density, restricted microbial diversity, material deterioration, and cost considerations still limit broad applications. Future research directions need to be toward interdisciplinary research, R&D investments in green materials, and integration of other renewable devices in a system toward further enhancement of MFC performance and commercialization (Santoro et al., 2017). In the future, MFCs are to be used in decentralized environmental monitoring and cleanup systems, particularly in off-grid or resource-constrained areas. With more innovation and coordinated regulation, MFCs could be a life-changing technology in environmental biotechnology—fueling ecological recovery and green energy production. Their inclusion in circular economy frameworks not only adds to pollution control but also reframes waste as a resource, providing a pragmatic and green road to environmental sustainability.

REFERENCES

Abbas, S. Z., Rafatullah, M., Khan, M. A., & Siddiqui, M. R. (2019). Bioremediation and electricity generation by using open and closed sediment microbial fuel cells. *Frontiers in Microbiology*, *9*, 3348. https://doi.org/10.3389/fmicb.2018.03348

Aelterman, P., Rabaey, K., Pham, T. H., Boon, N., & Verstraete, W. (2006). Continuous electricity generation at high voltages and currents using stacked microbial fuel cells. *Environmental Science & Technology*, *40*(10), 3388–3394. https://doi.org/10.1021/es0525511

Amend, J. P., & Shock, E. L. (2001). Energetics of overall metabolic reactions of thermophilic and hyperthermophilic Archaea and Bacteria. *FEMS Microbiology Reviews*, *25*(2), 175–243. https://doi.org/10.1111/j.1574-6976.2001.tb00576.x

Ancona, V., Caracciolo, A. B., Borello, D., Ferrara, V., Grenni, P., & Pietrelli, A. (2020). Microbial fuel cell: An energy harvesting technique for environmental remediation. *International Journal of Environmental Impacts*, *3*(2), 168–179. https://doi.org/10.2495/EI-V3-N2-168-179

Baroutaji, A., Wilberforce, T., Ramadan, M., & Olabi, A. G. (2019). Comprehensive investigation on hydrogen and fuel cell technology in the aviation and aerospace sectors. *Renewable and Sustainable Energy Reviews*, *106*, 31–40. https://doi.org/10.1016/j.rser.2019.02.027

Bergel, A., Feron, D., & Mollica, A. (2005). Catalysis of oxygen reduction in PEM fuel cell by seawater biofilm. *Electrochemistry Communications*, *7*(9), 900–904. https://doi.org/10.1016/j.elecom.2005.07.004

Bermek, H., Catal, T., Akan, S. S., Ulutas, M. S., Kumru, M., Ozguven, M., Liu, H., Ozcelik, H., & Akarsubasi, A. T. (2014). Olive mill wastewater treatment in single-chamber air-cathode microbial fuel cells. *World Journal of Microbiology and Biotechnology*, *30*(4), 1177–1185. https://doi.org/10.1007/s11274-013-1541-8

Bond, D. R., & Lovley, D. R. (2003). Electricity production by *Geobacter sulfurreducens* attached to electrodes. *Applied and Environmental Microbiology*, *69*(3), 1548–1555. https://doi.org/10.1128/AEM.69.3.1548-1555.2003

Borole, A. P., Reguera, G., Ringeisen, B., Wang, Z. W., Feng, Y., & Kim, B. H. (2011). Electroactive biofilms: Current status and future research needs. *Energy & Environmental Science*, *4*(11), 4813–4834. https://doi.org/10.1039/C1EE02511B

Bose, D., Bhattacharya, R., Gopinath, M., Vijay, P., & Krishnakumar, B. (2023). Bioelectricity production and bioremediation from sugarcane industry wastewater using microbial fuel cells with activated carbon cathodes. *Results in Engineering*, *18*, 101052. https://doi.org/10.1016/j.rineng.2023.101052

Brutinel, E. D., & Gralnick, J. A. (2012). Shuttling happens: Soluble flavin mediators of extracellular electron transfer in *Shewanella*. *Applied Microbiology and Biotechnology*, *93*(1), 41–48. https://doi.org/10.1007/s00253-011-3653-0

Chang, I. S., Moon, H., Bretschger, O., Jang, J. K., Park, H. I., Nealson, K. H., & Kim, B. H. (2006). Electrochemically active bacteria (EAB) and mediator-less microbial fuel cells. *Journal of Microbiology and Biotechnology*, *16*(2), 163–177.

Chaudhuri, S. K., & Lovley, D. R. (2003). Electricity generation by direct oxidation of glucose in mediator-less microbial fuel cells. *Nature Biotechnology*, *21*(10), 1229–1232. https://doi.org/10.1038/nbt858

Cheng, S., Liu, H., & Logan, B. E. (2006). Increased performance of single-chamber microbial fuel cells using an improved cathode structure. *Electrochemistry communications*, *8*(3), 489–494. https://doi.org/10.1016/j.elecom.2006.01.010

Clauwaert, P., Rabaey, K., Aelterman, P., De Schamphelaire, L., Ham, T. H., Boeckx, P., Boon, N., & Verstraete, W. (2007). Biological denitrification in microbial fuel cells. *Environmental Science & Technology*, *41*(9), 3354–3360. https://doi.org/10.1021/es062580r

Deng, L., Li, F., & Zhang, X. (2020). Soil bioremediation using microbial fuel cells: Recent advances and challenges. *Journal of Cleaner Production*, *276*, 123204. https://doi.org/10.1016/j.jclepro.2020.123204

Feng, Y., Yang, Q., Wang, X., & Logan, B. E. (2022). Electricity generation and wastewater treatment using microbial fuel cells: A review. *Environmental Science & Technology*, *56*(3), 1453–1467.

Ghasemi, M., Yousefi, A. A., & Mirsoleimani, A. (2018). A review on the performance of microbial fuel cells for wastewater treatment and energy production. *Journal of Environmental Chemical Engineering*, *6*(4), 5001–5013. https://doi.org/10.1016/j.jece.2018.07.027

Gregory, K. B., Bond, D. R., & Lovley, D. R. (2004). Graphite electrodes as electron donors for anaerobic respiration. *Environmental Microbiology*, *6*(6), 596–604. https://doi.org/10.1111/j.1462-2920.2004.00604.x

He, Z., Minteer, S. D., & Angenent, L. T. (2005). Electricity generation from artificial wastewater using an upflow microbial fuel cell. *Environmental Science & Technology*, *39*(15), 5267–5272. https://doi.org/10.1021/es0502876

Heilmann, J., & Logan, B. E. (2006). Production of electricity from proteins using a microbial fuel cell. *Water Science and Technology*, *54*(8), 301–305. https://doi.org/10.2166/wst.2006.769

Holmes, D. E., Nevin, K. P., O'Neil, R. A., & Lovley, D. R. (2004). Potential for direct interspecies electron transfer in methanogenic wastewater digester aggregates. *Applied and Environmental Microbiology*, 70(10), 5974–5983.

Jung, S., & Regan, J. M. (2007). Comparison of anode bacterial communities and performance in microbial fuel cells with different electron donors. *Applied Microbiology and Biotechnology*, *77*(2), 393–402. https://doi.org/10.1007/s00253-007-1162-y

Kim, B. H., Chang, I. S., & Gadd, G. M. (2007a). Challenges in microbial fuel cell development and operation. *Applied Microbiology and Biotechnology*, *76*(3), 485–494. https://doi.org/10.1007/s00253-007-1027-4

Kim, B. H., Park, H. S., Kim, H. J., Kim, G. T., Chang, I. S., Lee, J., & Phung, N. T. (2004). Enrichment of microbial community generating electricity using a fuel-cell-type electrochemical cell. *Applied Microbiology and Biotechnology*, *63*(6), 672–681. https://doi.org/10.1007/s00253-003-1412-6

Kim, J. R., Jung, S. H., Regan, J. M., & Logan, B. E. (2007b). Electricity generation and microbial community analysis of alcohol powered microbial fuel cells. *Bioresource Technology*, *98*(13), 2568–2577. https://doi.org/10.1007/s00253-007-1027-4

Krithika, T., Iswarya, T., & Sowndarya, T. (2022). Bioremediation of organic pollutants through microbial fuel cells. In *Microbial Fuel Cells for Environmental Remediation* (pp. 183–194). Springer Nature Singapore. https://doi.org/10.1007/978-981-19-2681-5_10

Li, W., Han, H., Zhang, Y., & Zhao, F. (2023a). Integrated systems for simultaneous energy production and environmental remediation: The future of bioelectrochemical systems. *Renewable and Sustainable Energy Reviews*, *170*, 113041.

Li, W., Zhang, X., & Zhao, F. (2023b). Microbial fuel cells for sustainable nutrient removal from agricultural runoff. *Science of the Total Environment*, *860*, 160351.

Li, X., Li, D., Zhang, Y., Liu, Z., & Li, F. (2021). Enhanced remediation of petroleum hydrocarbon-contaminated soil using sediment microbial fuel cells and electrokinetics. *Chemosphere*, *263*, 127902. https://doi.org/10.1016/j.chemosphere.2020.127902

Li, Y., Zhao, C., Liu, S., & Wang, J. (2019). Advances of molten carbonate fuel cells for CO_2 utilization. *Renewable Energy*, *139*, 19–31. https://doi.org/10.1016/j.renene.2019.02.024

Liu, H., Cheng, S., & Logan, B. E. (2014). Power generation in fed-batch microbial fuel cells as a function of ionic strength, temperature, and reactor configuration. *Environmental Science & Technology*, 38(18), 5202–5206. https://doi.org/10.1021/es050316c

Logan, B. E. (2004). Biologically extracting energy from wastewater: Biohydrogen production and microbial fuel cells. *Environmental Science & Technology*, *38*(9), 160–167. https://doi.org/10.1021/es040190o

Logan, B. E. (2019). Scaling up microbial fuel cells and other bioelectrochemical systems. *Applied Microbiology and Biotechnology*, *103*, 7141–7154. https://doi.org/10.1007/s00253-009-2378-9

Logan, B. E., Hamelers, B., Rozendal, R., Schröder, U., Keller, J., Freguia, S., Aelterman, P., Verstraete, W., & Rabaey, K. (2006). Microbial fuel cells: Methodology and technology. *Environmental Science & Technology*, 40(17), 5181–5192. https://doi.org/10.1021/es0605016

Logan, B. E., & Regan, J. M. (2006). Electricity-producing bacterial communities in microbial fuel cells. *Trends in Microbiology*, *14*(12), 512–518. https://doi.org/10.1021/es0605016

Lovley, D. R. (2006). Microbial fuel cells: Novel microbial physiologies and engineering approaches. *Current Opinion in Biotechnology*, *17*(3), 327–332. https://doi.org/10.1016/j.copbio.2006.05.007

Lovley, D. R. (2008). The microbe electric: Conversion of organic matter to electricity. *Current Opinion in Biotechnology*, 19(6), 564–571. https://doi.org/10.1016/j.copbio.2008.10.005

Lu, L., Zeng, R. J., & Angelidaki, I. (2016). Microbial electrolysis cells for sustainable energy and bioremediation: A review. *Renewable and Sustainable Energy Reviews*, *60*, 1183–1196. https://doi.org/10.1016/j.rser.2016.03.005

Marsili, E., Baron, D. B., Shikhare, I. D., Coursolle, D., Gralnick, J. A., & Bond, D. R. (2008). Shewanella secretes flavins that mediate extracellular electron transfer. *Proceedings of the National Academy of Sciences*, 105(10), 3968–3973. https://doi.org/10.1073/pnas.0710525105

Min, B., & Logan, B. E. (2004). Continuous electricity generation from domestic wastewater and organic substrates in a flat plate microbial fuel cell. *Environmental Science & Technology*, *38*(21), 5809–5814. https://doi.org/10.1021/es0491026

Mohamed, S. S., & Kim, B. H. (2015). Electricity generation from wastewater in microbial fuel cells: Recent progress and challenges. *Renewable and Sustainable Energy Reviews*, *52*, 1623–1635. https://doi.org/10.1016/j.rser.2015.07.191

Niessen, J., Schröder, U., & Scholz, F. (2004). Exploiting complex carbohydrates for microbial electricity generation–a bacterial fuel cell operating on starch. *Electrochemistry Communications*, *6*(9), 955–958. https://doi.org/10.1016/j.elecom.2004.07.010

Oh, S., Min, B., & Logan, B. E. (2004). Cathode performance as a factor in electricity generation in microbial fuel cells. *Environmental Science & Technology*, *38*(18), 4900–4904.

Pham, T. H., Boon, N., & Van, V. (2006). Biocatalyzed electrodes and microbe–electrode interactions in microbial fuel cells. *Applied Microbiology and Biotechnology*, *73*(5), 961–972. https://doi.org/10.1007/s00253-006-0506-3

Rabaey, K., Boon, N., Siciliano, S. D., Verhaege, M., & Verstraete, W. (2004). Biofuel cells select for microbial consortia that self-mediate electron transfer. *Applied and Environmental Microbiology*, *70*(9), 5373–5382. https://doi.org/10.1128/AEM.70.9.5373-5382.2004

Rabaey, K., Clauwaert, P., Aelterman, P., & Verstraete, W. (2005). Tubular microbial fuel cells for efficient electricity generation. *Environmental Science & Technology*, *39*(20), 8077–8082. https://doi.org/10.1021/es050986i

Rabaey, K., & Rozendal, R. (2010). Microbial electrosynthesis — revisiting the approach for producing fuels from electricity and CO_2. *Nature Reviews Microbiology*, *8*, 706–716. (Review context re: bioelectrochemical systems). https://doi.org/10.1038/nrmicro2422

Rabaey, K., & Verstraete, W. (2005). Microbial fuel cells: Novel biotechnology for energy generation. *Trends in Biotechnology*, *23*(6), 291–298. https://doi.org/10.1016/j.tibtech.2005.04.008

Reguera, G., McCarthy, K. D., Mehta, T., Nicoll, J. S., Tuominen, M. T., & Lovley, D. R. (2005). Extracellular electron transfer via microbial nanowires. *Nature*, 435(7045), 1098–1101. https://doi.org/10.1038/nature03661

Ringeisen, B. R., Ray, R., Little, B., Biffinger, J. C., & Jones-Meehan, J. (2006). Microbial fuel cells for metabolite monitoring and power generation. *Biosensors & Bioelectronics*, *21*(12), 2057–2063. https://doi.org/10.1016/j.bios.2005.09.030

Santoro, C., Arbizzani, C., Erable, B., & Ieropoulos, I. (2017). Microbial fuel cells: From fundamentals to applications. *A review. Journal of Power Sources*, *356*, 225–244. https://doi.org/10.1016/j.jpowsour.2017.03.109

Schröder, U. (2007). Anodic electron transfer mechanisms in microbial fuel cells and their energy efficiency. *Physical Chemistry Chemical Physics*, 9(21), 2619–2629.

Venkata Mohan, S., Velvizhi, G., Modestra, J. Annie, & Srikanth, S. (2014). Microbial fuel cell: Critical factors regulating bio-catalyzed electrochemical process and recent advancements. *Renewable and Sustainable Energy Reviews*, 40, 779–797.

Viggi, C., Puig, S., Vallée, T., & Freguia, S. (2017). Enhancing microbial fuel cell performance by controlling biofilm structure and function. *Electrochimica Acta*, *246*, 1–10. https://doi.org/10.1016/j.electacta.2017.05.116

Zhang, F., Cheng, S., Pant, D., Van Bogaert, G., Logan, B. E., & Nerenberg, R. (2012). Advances in bioelectrochemical systems for wastewater treatment. *Energy & Environmental Science*, 4(7), 2214–2240.

8 Fuel Cells in Remote Sensing and Monitoring

Gloria Ginikanwa Ezeagu, Agabi Enaji Kanu, Martina Onyinye Eze, Sharafadeen Adetunji Adeniji, Kafilat Olayemi Suleiman, Josephine Madukaihe Madu, and Namla Djadjitti

8.1 INTRODUCTION

The increase in pollution over the past decades has increased worries about climate change and global warming. This necessitates continuous monitoring of the environment, hence the importance of sensors. Sensors are deployed in remote locations for environmental monitoring, habitat monitoring, weather monitoring, agricultural applications, fisheries research, and military surveillance (Yick et al., 2008). For most of the applications stated prior, ideally, remote sensors or sensor nodes are operated in the long term. Sensor networks are comprised of tiny, low-cost wireless electronic devices capable of gathering environmental information and forwarding it to a base station (Patil and Chen, 2013). Sensors carry out remote sensing or monitoring, and one major challenge in the operation of these sensors in remote locations is the lifetime of the sensors, which is limited by the power source to which they are connected. The power source becomes more critical when the network gets dense with many sensor nodes, because the energy source can hinder the reliability and the frequency of sensor operation.

Conventionally, batteries are used as the power source. The cost of replacing these batteries becomes problematic, making the sensor network financially unsustainable. Consequently, there is a need for alternative renewable power sources to take over from batteries and support long-term remote monitoring (Dewan et al., 2013). Fuel cells, which are static energy conversion devices or direct energy conversion devices, meaning that they convert the chemical reaction of fuels directly into electrical energy without using a heat engine (Kirubakaran et al., 2009) (Wailes et al., 1999) are therefore used.

8.2 REMOTE SENSING/SENSORS

Remote sensing is the process of obtaining information about an object without being in physical contact with the object. This information is obtained by determining and measuring the changes the object makes to a surrounding field, which could be electromagnetic, acoustic (sound), or potential (gravitational or magnetic) field (Elachi and Zyl, 2021).

A sensor can therefore be defined as an electronic device capable of detecting properties and events occurring in its surrounding environment and quantifying their changes. This detected information is converted into an electrical voltage or current that is then sent to other electronic devices, mostly a computer processor (Shao et al., 2010). Remote sensing requires an energy source, or source of illumination. Modern acoustic remote sensing involves array-recorded sounds and array signal processing to recover multidimensional results. The application realm for acoustic remote sensing spans an impressive range of signal frequencies (10^{-2} to 10^{7} Hz) and distances of

 DOI: 10.1201/9781003645405-8

(10^{-2} to 10^{7} m), and involves biomedical ultrasound imaging, nondestructive evaluation, oil and gas exploration, military systems, and Nuclear Test Ban Treaty monitoring (Dowling & Sabra, 2015).

Remote sensors can be grouped into two types, based on the requirement, and the types include:

1. Passive remote sensors
2. Active remote sensors

8.2.1 Passive Remote Sensors

Passive remote sensing systems require external light or energy sources. As a result, passive sensors can detect solar energy that is reflected or reemitted by objects being monitored. Reflected sunlight is the most common energy source for passive remote sensors. At night, when there is no reflection of solar energy, passive remote sensors sense naturally emit solar energy, such as thermal infrared radiation (IR) or heat radiation, which is available during the day and at night. Sensors can also detect radiation emitted by stars and other celestial bodies. Examples of passive sensors are the Landsat series, IRS series, SPOT series, IKONOS, Quickbird, etc. (Gupta, 2018).

8.2.2 Active Remote Sensors

Active remote sensors produce their own source of light or energy to illuminate the objects under study and record the amount of energy reflected from their target. Because of this, active sensors are not hindered by factors such as time of day or seasons. Instances of active sensors are synthetic aperture radar (SAR) and light detection and ranging (LiDAR) (Gupta, 2018). Active remote sensing can be likened to flash photography, while passive remote sensing can be likened to available light photography.

8.2.3 Biosensors

A Biosensor is an analytical device that detects changes in Biological processes and converts them into an electrical signal. The term Biological process can be any biological element or material, like enzymes, tissues, microorganisms, cells, acids, etc. So, a Biosensor is a combination of a Biological sensing element and a transducer, which converts the data into electrical signals (Teja, 2024). Biosensors are classified as shown in Table 8.1.

8.2.4 Classes of Remote Sensing

Remote sensing can also be classified into microwave and optical remote sensing.

TABLE 8.1
Classification of Biosensors

Based on Receptor Type	Based on Transduction Process
Enzymatic biosensors	Electrochemical
Immunosensors	Optical
Genosensors/ DNA biosensors	Piezoelectric
Tissue-based biosensors	Thermal/calorimetric

Source: (Ivars-Barcelo et al. 2018).

8.2.4.1 Microwave Remote Sensing

Microwave remote sensing operates in the microwave portion found in the electromagnetic spectrum. It encompasses active and passive forms of remote sensing and generally uses wavelengths of 1 cm up to 1 m. Such wavelengths are considered long compared to the visible or infrared ones. Therefore, microwaves have unique properties that are crucial in remote sensing.

In microwave remote sensing, the active type receives the backscattering reflected from the transmitted microwave incident on the ground surface. Therefore, the characteristics can be derived from the radar cross-section and physical properties of an object. In comparison, the passive type receives the microwave radiation emitted from objects on the ground. Therefore, in passive microwave remote sensing, an object's characteristics are detected from the relationship between the object's received power and physical properties (Grind GIS, 2021).

8.2.4.2 Optical Remote Sensing

Optical remote sensing makes use of visible, near-infrared, and short-wave infrared sensors. They form images of the Earth's surface by detecting the solar radiation reflected from targets on the ground. Here, the Optical sensors gather light within the wavelength range a human eye can perceive (Grind GIS, 2021).

8.2.5 Remote Sensing Platforms

A platform is a vehicle from which a sensor can be operated (Lillesand et al., 2003). Sensors or cameras are usually installed on these platforms to obtain data about a specified target. It is imperative that sensors be set up on appropriate and sturdy platforms. These platforms can be ground-based, airborne, or spaceborne. As the platform moves farther away from ground level, the spatial resolution and observational area increase. Thus, the higher the sensor is mounted, the larger the spatial resolution and synoptic views that are obtained. The features of a platform are dictated by the type of sensor to be attached to it, and its use. Based on the task, the platform can vary from a ladder to a satellite.

Typically, airborne objects such as aircraft and satellites are used, as well as balloons and rockets. Sensors can also be placed on ground platforms such as handheld cameras, vehicles, and buildings (Nazim, accessed 21 March 2025). For remote sensing in aquatic environments, boats, buoys, submarines, and other submersibles are used in addition to balloons, aircraft, and satellites.

8.3 APPLICATION OF FUEL CELLS IN REMOTE SENSING AND MONITORING

The development of green energy has been continually emphasized recently due to increasing environmental protection consciousness. The fuel cell is one possible substitute since it produces no carbon dioxide but only water. Fuel cells can be characterized into several categories, such as alkaline fuel cells (AFC), phosphoric acid fuel cells (PAFC), solid oxide fuel cells (SOFC), direct methanol fuel cells (DMFC), and proton exchange membrane fuel cells (PEMFC) (Kuo et al., 2011). The key characteristic of fuel cells is their ability to directly transform the chemical potential energy of a fuel into electricity. This direct transformation of energy does not involve an intermediate step of heat generation (DeCicco, 2004).

8.3.1 Unmanned Aerial Vehicles (UAVs)/Drones

An Unmanned Aerial Vehicle (UAV) is a type of aircraft that operates without a pilot on board. UAVs can be controlled remotely by a human operator located at a ground control station or can navigate autonomously using an onboard computer. A UAV system comprises three main parts: the unmanned aircraft itself, sensor payloads (such as cameras), and the ground control system

(Rai et al., 2021). UAVs find applications across various sectors and are significant for both civilian and military operations. They are utilized for scientific research, monitoring wildfires, cartography, photography, search and rescue missions, and more.

Endurance in UAVs refers to the total duration a UAV can remain in the air, which is a key factor in improving its overall efficiency. Several elements influence endurance, including weight, size, payload capacity, and the power supply. Currently, most UAVs use combustion engines or batteries as their primary power sources. However, there is a growing trend in the use of fuel cell-powered UAVs (Cwojdziński & Adamski, 2014). The technical specifications of UAV batteries are generally adequate; however, these batteries often struggle to support stable, long-duration flights due to their limited energy density and the increased mass associated with additional payloads (Hwang et al., 2018). Recent research has suggested various strategies to overcome these challenges, emphasizing the urgency for innovative methods to improve UAV endurance. While these strategies show considerable potential, integrating fuel cells provides much higher specific energy and power, enabling UAVs to carry more substantial payloads or achieve longer distances.

8.3.1.1 Proton Exchange Membrane Fuel Cells

Proton exchange membrane fuel cells present several significant advantages as a power source for electric UAVs, such as high efficiency, zero emissions, elevated specific energy density, extended flight duration, adaptability to varying power needs, and no noise pollution.

8.3.2 Telecommunication/Transmission of Results

Telecommunication is essential in remote sensing as it enables the transmission of data acquired by sensors to ground stations for further analysis and interpretation. It allows for the effective and dependable transfer of this information, which is subsequently processed to yield valuable insights for various purposes such as weather prediction, disaster management, and environmental assessment (Shishkov & Garvanova, 2023).

8.3.2.1 Off-Grid Radio Base Stations

Off-grid radio base stations powered by fuel cells and locally sourced renewable energy are currently being developed and tested. A base station functions as a fixed communication point within a network, linking mobile devices to central data and voice networks, mainly utilized in telecommunications (Axell, 2024). In the area of GNSS (Global Navigation Satellite Systems) and GPS (Global Positioning System) antennas, base stations play a crucial role as reference points that enhance the precision and dependability of satellite data received by mobile devices (Novotech, n.d.).

Radio base stations are frequently positioned in remote locations where grid power may not be accessible, necessitating the local production of energy through low-efficiency and emission-heavy power supply options such as diesel generators. Alternative power supply methods that utilize renewable sources combined with electrical energy storage and programmable energy production are emerging as viable solutions to ensure reliable, efficient, and clean energy generation for remote radio base stations (Kusakana & Vermaak, 2012).

For this purpose, employing hydrogen and fuel cells for chemical energy storage and programmable energy generation has been recognized as a potentially competitive alternative to conventional diesel generators (Bruni et al., 2014). Hybrid Renewable Energy Systems (HRES), which incorporate hydrogen storage and fuel cells as programmable energy sources, are appealing because they can function unattended for extended periods due to their generally high conversion efficiency (i.e., low fuel use), low maintenance costs, minimal noise, quick and simple start-up capabilities, and excellent heat management in low-temperature environments. Additionally, they can meet the energy demands of radio base stations while producing very low CO_2 emissions, adhering to current and future greenhouse gas emission regulations (Cordiner et al., 2016).

Telecommunications standards regarding energy production are quite stringent (requiring continuous supply 24/7, high reliability, and long periods of operation without supervision), and under these circumstances, HRES that combine electrochemical energy storage with local hydrogen production and storage present an effective alternative compared to battery-only or hydrogen-only systems (Akarsu & Genç, 2022). In fact, the combined application of electrochemical batteries for short-term storage and locally generated hydrogen through electrolysis for medium- to potentially seasonal storage enables a more efficient use of local renewable energy sources without significantly compromising the lifespan of the batteries (Genç et al., 2012).

8.3.3 Microbial Fuel Cells for Sensing and Low-Power Devices

Recently, bioelectrochemical sensors based on Microbial Fuel Cells (MFCs) technology have been developed and implemented using both single-chamber and dual-chamber setups for real-time monitoring of various parameters, including dissolved oxygen, biological oxygen demand, volatile fatty acids (VFAs), toxins in wastewater, as well as serving as biosensors for heavy metals and the detection of hazardous chemicals, among other applications. Bioelectrochemical systems have proved to be relevant to the application of fuel cells in remote sensors. They are revolutionary new bioengineering technologies that combine microorganisms or enzymes with electrochemical methods to enhance the oxidation or reduction metabolism (Zheng et al., 2020). Effective operation of MFCs relies heavily on proper design, configuration, and fabrication. These fuel cells are tailored to meet various architectural specifications to attain optimal coulombic efficiency, durability, power output, and stability.

Assessing water quality indicators like pH, temperature, and heavy metal concentrations is a crucial use of wireless sensors today. In the case of biosensors, a biological or biomimetic (biomimicry) element serves as a sensitive detector for the target analyte. The interaction between the analyte and the sensor generates a signal that is converted into a physical, quantifiable form by a physicochemical component or transducer, which is then suitable for an electronic signal processor. The performance of a biosensor relies on the biochemical specificity of its active biological component or bioreceptor. Depending on the type of receptor, biosensors can be classified as enzymatic biosensors, immunosensors, genosensors or DNA biosensors, or tissue-based biosensors.

Biosensors can also be divided into several categories based on the transduction process, such as electrochemical, optical, piezoelectric, or thermal/calorimetric biosensors. Biosensors have important applications in several fields, such as the food industry, medicine, the marine sector, and others.

Microbial Fuel Cells can be employed as self-powered biosensors, thanks to their intrinsic sensitivity to biological and physicochemical parameters. Monitoring microbial activity, biocorrosion by microbial biofilms, biochemical oxygen demand (BOD), toxicants, and pH and temperature variations, are among the most interesting applications. Furthermore, MFC-based (bio)sensors provide higher stability and sensitivity compared with conventional (bio)sensors.

8.3.4 Sedimentary Microbial Fuel Cells in Environmental Sensing in High-Altitude Mountain Ecosystems

Sedimentary Microbial Fuel Cells (SMFC) produce little energy, and tests conducted in places close to sea level have shown the SMFC's potential as an energy source for monitoring systems in remote areas. When an electric load is applied to stack MFCs, the stack MFC's voltage does not remain constant over time, known as the reversal of voltage. Power Management Systems are designed to adapt for MFC optimal operation using ultralow power consumption components to raise the voltage of the MFC (Recalde et al., 2023). For example, SMFCs powered a sensor installed in the Palouse River. Also, SMFC with sediments extracted from Lake Michigan intermittently powered a temperature sensor. Many reports have shown the SMFC's potential to power environmental wireless sensor networks. The main point is to use the appropriate electronic circuit to transfer and store the electric energy produced by the SMFCs (Recalde et al., 2023).

8.3.5 Autonomous Underwater Vehicles (AUVs)

Autonomous underwater vehicles (AUVs) are programmable, robotic vehicles that, depending on their design, can drift, drive, or glide through the ocean without real-time control by human operators (Wynn et al., 2014).

AUVs have a wide range of applications in marine geoscience and are increasingly being used in the scientific, military, commercial, and policy sectors. Their ability to operate autonomously of a host vessel makes them well-suited to the exploration of extreme environments, from the world's deepest hydrothermal vents to beneath polar ice sheets (Wynn et al., 2014).

AUVs are used for geophysical surveys, subsea pipeline inspection, environmental monitoring, marine archaeology, and search for lost assets, to name a few examples. They are expected to play an even more important role in the future, for example, within deep-sea mining, subsea oil and gas installations, and large-scale subsea surveys (Weydahl et al., 2019).

Most AUVs are followed by a mother ship that aids with navigation and in-mission decision-making. After each mission, the AUV must be recovered by the mother ship or connect to a docking system for recharging and downloading of harvested data. Manned mother ships have high day rates and add significantly to the cost of an AUV mission. As an example, the day rate for a research vessel from the Marine Institute in Ireland is € 19 000 for non-commercial missions (Hagen et al., 2007).

Developing long-endurance AUVs is one possible pathway to reduce the need for an expensive mothership. A long-endurance AUV requires further improvements within autonomy, navigation, and energy storage. With these improvements in place, the AUV can be launched and operate independently for its full endurance, also in unknown waters. In the meantime, the mothership can be liberated to solve other tasks or launch AUVs in other areas. With today's level of AUV autonomy, this operation would introduce a risk of losing the AUV (Hagen et al., 2007). For many AUV users, this is an unviable option, and they therefore choose to follow the AUV with a mother ship.

Extending the AUV endurance to several weeks or even months could open up for more challenging mission types such as under ice seabed surveys, long-distance ocean current monitoring at constant depth, fish shoal tracking, and ultra-long pipeline inspection. The need for a mother ship could be eliminated altogether if the range of the AUV is sufficiently high to be launched from pier. Most commercial AUVs today are powered by batteries. Fuel cell systems have the potential to increase AUV endurance beyond what batteries can provide (Sawa et al., 2005)

Fuel cell systems for AUVs have been developed over the years by Japanese, German, French, and US research teams (Wynn et al., 2014). Japanese research teams developed URASHIMA, a deep-sea cruising AUV that makes use of a solid PEFC as its power source (Aoki, 2005). URASHIMA was an experimental AUV, with a hybrid power system. It is powered by PEFC and lithium-ion cells as an auxiliary power source. It started development in 1998 by the Japan Agency for Marine-Earth Science and Technology (JAMSTEC). Its fuel cell power system has two 2 kW fuel cell stacks at 120 V, and was developed by Mitsubishi Heavy Industries Ltd (Tokyo, Japan).

It is part of a hybrid system, which also has a 30-Ah lithium-ion battery, used for dealing with peak power demands that the fuel cell cannot supply.

JAMSTEC and Mitsubishi Heavy Industries Ltd. (Tokyo, Japan) also produced a second-generation long-cruising AUV. This fuel cell system is called HEML (High-Efficiency Multi-Less) fuel cell and is a novel concept that changes and is supposed to improve on standard fuel cell stack design and operation. The system eliminates recirculation blowers and replaces them with valves, eliminating also humidifiers by improving materials and MEA composition and reducing hydrogen leakage. After several revisions, developers reported that power was supplied stably and that the equipment being supplied functioned without interruption.

DeepC is an experimental German AUV development project. This AUV possessed a hybrid battery-PEM fuel cell system, with the batteries taking peak loads.

IDEF Ifremer is an experimental AUV powered by a PEM fuel cell adapted to an existing AUV. The fuel cell system was designed by the HELION firm, a subsidiary of the Areva group, a French multinational. The PEM fuel cell stack is composed of 63 cells and produces 1.5 kW. Its system efficiency, including auxiliaries' consumption, is 55%. Hydrogen and oxygen gases are recirculated, and full reactant conversion is achieved.

8.3.6 Robotic Rovers in Surface Exploration

Surface exploration is a broad term that refers to investigations carried out on the Earth's surface to collect data about the environment, geology, or subsurface conditions. It includes a range of approaches, from basic observations to sophisticated methods such as geological mapping, remote sensing, and geochemical surveys, along with subsurface exploration techniques like geophysical surveys and drilling (Bennet, 2024).

Fuel cells provide an effective power source for surface exploration, especially in space missions, by transforming the chemical energy from fuels like hydrogen or methanol into electricity. They are utilized in mobile rovers, fixed landers, and stationary power facilities, offering benefits such as prolonged operation and extended range (Camargo, 2025).

Robotic rovers are crucial components in the initial stages of surface exploration. Their capability to carry out investigations across multiple sample locations rather than just near the landing site gives them significant advantages over stationary landers and platforms.

On planetary surfaces like Mars, solar cells can effectively support a certain level of scientific exploration. However, solar technology will eventually reach a limit in the amount of power it can generate. Furthermore, solar panels are only functional during daylight hours, making them unsuitable as a long-term power solution for nighttime or low-light conditions.

Fuel cells present an alternative power generation technology. These devices convert energy stored in fuels (primarily hydrogen, oxygen, or methanol) into electricity and heat through an electrodynamic oxidation process. They are not dependent on light; they are fairly mass-efficient, have been validated in space missions (such as the NASA Space Shuttle), and possess high efficiency. They can be implemented in both mobile rovers and fixed landers on planetary surfaces as well as in outer space. Additionally, they can often be used in hybrid configurations with other power system components (e.g., solar panels, batteries, etc.) ("Fuel Cell Power System Options for Mars Rovers" accessed March 25, 2025).

The potential and constraints of fuel cells used in surface rovers on Mars have been explored. Factors such as the selection of fuel, type of fuel cell, and whether the system serves as primary (producing power without storing it) or regenerative (capable of both storing and generating electrical energy) need to be evaluated in developing fuel cells for surface exploration. The selection of fuel for a fuel cell application on Mars is mainly influenced by two considerations: the availability of fuels and oxidizers, and the specific type of fuel cell being employed.

Creating fuels and/or oxidizers directly on Mars's surface using In-Situ Resource Utilization (ISRU) technology is frequently considered for both robotic and crewed missions to the planet. Most current efforts related to ISRU focus on generating methane and oxygen by utilizing a supply of hydrogen transported from Earth. Therefore, using the same methane/oxygen propellants produced for the Mars ascent vehicle in the fuel cell system of a large surface rover is a logical option. Carbon monoxide is generated as a by-product in the Sabatier/Reverse Water Gas Shift (RWGS) process and is released into the atmosphere. Hence, using carbon monoxide for power generation in a fuel cell system presents a compelling opportunity. The application of direct-hydrogen fuel cell technology leads to a simple and lightweight overall system design, which has been validated in land-based applications. The challenge of selecting the optimal fuel alternative must be assessed in relation to the specific mission profile. Each fuel type provides unique benefits suitable for various applications. A strong connection among ISRU technology, ascent vehicle fuel production, and fuels for surface operations is both beneficial and achievable through the use of fuel cell technology. The two most promising fuel cell types are the Polymer Electrolyte Membrane Fuel Cell (PEMFC) and the Solid Oxide Fuel Cell (SOFC) (Fraser et al., 2004).

PEMFC technology is typically preferred for most applications where pure hydrogen serves as the fuel. Fuel cells have been employed in several space missions, producing electrical power and supplying liquid water for crew use. The first fuel cell to function in space was a solid polymer type developed by General Electric for power generation. Due to the focus on extending mission durations for the Apollo lunar missions, research quickly transitioned from solid polymer technology to alkaline fuel cell technology. Alkaline fuel cell systems successfully powered the Apollo missions and were later utilized in the Space Shuttle Orbiter. It is anticipated that PEMFC technology will replace alkaline fuel cells in the future. The current Space Shuttle Orbiter is still using alkaline fuel cells, but there are plans to upgrade these systems with new units utilizing Polymer Electrolyte Membrane Fuel Cell (PEMFC) technology (Fraser et al., 2004).

8.3.7 Underwater Sensor Systems

One type of sensor utilized for environmental monitoring is the submersible ultrasonic receiver (SUR), an underwater system designed for research in fisheries or wildlife. The SUR is capable of detecting ultrasonic signals and recording the movements of fish and other wildlife that are tagged with ultrasonic transmitters ("Submersible Ultrasonic Transducers: Applications, Design, and Advantages," accessed 20 May 2025).

The SUR operates as an independent device comprising a battery system, a microprocessor, flash memory, a real-time clock (RTC), a hydrophone, and an ultrasonic receiver. Among all the components of the SUR, the battery system is the primary limiting factor for how long the receiver can function. The batteries powering the SUR must be replaced every 5–10 months. Replacing batteries can be expensive and labor-intensive, particularly when managing a large network of sensors (Donovan et al., 2013). Consequently, alternative renewable power sources that enable consistent operation of the RTC are essential. Sediment microbial fuel cells (SMFCs) have the capability to provide uninterrupted low-level power and require no maintenance or replacement. Research has demonstrated that SMFCs can serve as a renewable power alternative for wireless sensors.

An SMFC consists of two electrodes made from inert conductive materials, such as graphite, stainless steel, and carbon cloth. One electrode, known as the anode, is submerged in the sediment, while the other electrode, referred to as the cathode, is situated in the water where oxygen is available. Electricity is produced by the oxidation of organic materials in the sediment by microorganisms that are naturally present there. This oxidation generates electrons at the anode, which then flow through an external circuit to the cathode, where oxygen is typically reduced to create water.

The primary benefit of SMFCs is that their power generation does not depend on a finite fuel supply, as the organic matter in the sediment is renewable. Furthermore, since the electrodes are made of inert materials, the longevity of power production is not constrained by the construction materials. Additionally, SMFCs can be deployed in many remote regions, including underwater locations, where replacing batteries would pose a challenge. However, SMFCs also have certain drawbacks, such as non-linear scalability, low and inconsistent power generation, and limited cell potential. Due to the challenges related to non-linear scaling of the electrodes, designing a large-scale microbial fuel cell for field use can be prohibitively expensive. As a result, an intermittent energy harvesting approach is more suitable, allowing for high power output in brief intervals. Previous reports indicate that intermittent energy harvesting employs the cyclic charging and discharging of a capacitor to gather energy. Capacitors are linked to SMFCs to collect and regulate microbial energy. When the stored energy in the capacitor reaches a specific threshold, it is used to engage the power management system (PMS) and operate the sensors. The PMS transforms low voltage into higher voltage to supply the energy needed by the sensors (Donovan et al., 2013).

8.4 CONCLUSION

There are countless ways fuel cells can be utilized. However, numerous technical challenges still exist between theoretical concepts and practical implementations. Continued research and advancement in fuel cell technology is certainly one avenue worth exploring to address our existing energy challenges.

REFERENCES

Akarsu, B., & Genç, M. S. (2022). Optimization of electricity and hydrogen production with hybrid renewable energy systems. *Fuel*, 324, 124465. https://doi.org/10.1016/j.fuel.2022.124465

Aoki, T. (2005). Cruising autonomous underwater vehicle "URASHIMA" powered with fuel cell. *Nihon Kikai Gakkaishi/Journal of the Japan Society of Mechanical Engineers*, 108(1039), 454–455. https://doi.org/10.1299/jsmemag.108.1039_454

Axell, P. (2024). What is a Base Station in Telecommunications? - Understanding Network Hubs. Retrieved from https://pbeaxell.com/about/glossary/what-is-a-base-station (Accessed May, 23, 2025).

Bennet, A. (2024). The role of robotics in space exploration: Current application and future prospects. *Journal of Space Exploration*, 13(12), 1–2.

Bruni, G., Cordiner, S., Mulone, V., Giordani, A., Savino, M., Tomarchio, G. and Picciotti, G. (2014). Fuel cell based power systems to supply power to Telecom Stations. *International Journal of Hydrogen Energy*, 39(36), 21767–21777. https://doi.org/10.1016/j.ijhydene.2014.07.078

Camargo, F. (2025). Hydrogen fuel cells: Powering the future of clean energy. Retrieved from https://stargatehydrogen.com/blog/hydrogen-fuel-cells/ (Accessed March 21, 2025).

Cordiner, S., Mulone, V., Giordani, A., Savino, M., Tomarchio, G., Malkow, T., … Jensen, J. (2016). Fuel cell based hybrid renewable energy systems for off-grid telecom stations: Data analysis from on field demonstration tests. *Applied Energy*, 192, 508–518. https://doi.org/10.1016/j.apenergy.2016.08.162

Cwojdziński, L., & Adamski, M. (2014). Power units and power supply systems in UAV. *Aviation*, 18(1), 1–8. https://doi.org/10.3846/16487788.2014.865938

DeCicco, J. M. (2004). Fuel cell vehicles. In *Elsevier eBooks*. Elsevier.

Dewan, A., Ay, S. U., Karim, M. N., & Beyenal, H. (2013). Alternative power sources for remote sensors: A review. *Journal of Power Sources*, *245*, 129–143. https://doi.org/10.1016/j.jpowsour.2013.06.081

Donovan, C., Dewan, A., Heo, D., Lewandowski, Z., & Beyenal, H. (2013). Sediment microbial fuel cell powering a submersible ultrasonic receiver: New approach to remote monitoring. *Journal of Power Sources*, 233, 79–85. https://doi.org/10.1016/j.jpowsour.2012.12.112

Dowling, D. R., & Sabra, K. G. (2015). Acoustic remote sensing. *Annual Review of Fluid Mechanics*, *47*(1), 221–243. https://doi.org/10.1146/annurev-fluid-010814-014747

Elachi, C., & Zyl, J. (2021). *Introduction to the physics and techniques of remote sensing*. Wiley.

Fraser, S., Monsberger, M., & Hacker, V. (Eds.). (2004). *Fuel cell power system options for mars rovers. 2nd international conference on green propellants for space propulsion* (Vol. 557). Retrieved from https://adsabs.harvard.edu/full/2004ESASP.557E..15F

Fuel cell power system options for Mars rovers. Retrieved from https://www.esa.int/Science_Exploration/Human_and_Robotic_Exploration/Exploration/Fuel_cell_power_system_options_for_Mars_rovers# (Accessed March 25 2025).

Genç, G., Çelik, M., & Genç, M. S. (2012). Cost analysis of wind-electrolyzer-fuel cell system for energy demand in Pınarbaşı-Kayseri. *International Journal of Hydrogen Energy*, 37(17), 12158–12166. https://doi.org/10.1016/j.ijhydene.2012.05.058

Grind GIS, (October, 2021). Optical Remote Sensing Vs. Microwave Remote Sensing. What is Remote Sensing? https://grindgis.com

Gupta, S. (2018). *Remote sensing & GIS applications in environmental sciences*. E-PG Pathshala.

Hagen, P. E., Midtgaard, O., & Hasvold, O. (2007). Making AUVs truly autonomous. *Oceans*, 1–4. https://doi.org/10.1109/oceans.2007.4449405

Hwang, M., Cha, H., & Jung, S. Y. (2018). Practical endurance estimation for minimizing energy consumption of multirotor unmanned aerial vehicles. *Energies* 11(9), 2221. https://doi.org/10.3390/en11092221

Ivars-Barceló, F., Alessio, Z., Marjan, F., Mehrdad, M., Mostafa, R., and Rafael, L. 2018. Novel applications of microbial fuel cells in sensors and biosensors. *Applied Sciences*, 8(7):1184. https://doi.org/10.3390/app8071184

Kirubakaran, A., Shailendra, J., & Nema, R. K. (2009). A review on fuel cell technologies and power electronic interface. *Renewable and Sustainable Energy Reviews*, 13(9), 2430–2440. https://doi.org/10.1016/j.rser.2009.04.004

Kuo, L., Huang, H., Yang, C. and Chen, P. (2011). Real-time remote monitoting of temperature and humidity within a proton exchange membrane fuel cell using flexible sensors. *Sensors* 11(9), 8674–8684. http://doi.org/10.3390/s110908674

Kusakana, K., & Vermaak, H. J. (2012). Hybrid renewable power systems for mobile telephony base stations in developing countries. *Renewable Energy*, 51, 419–425. https://doi.org/10.1016/j.renene.2012.09.045

Lillesand, T., Kiefer, R. W., & Chipman, J. (2003). *Remote sensing and image interpretation.* Wiley.

Nazim, M.D. (accessed 21 March, 2025). Remote sensing platforms. https://jagiroadcollegelive.co.in/attendence/classnotes/files/1624002934.pdf

Novotech. (n.d.). What is a base station and its role in enhancing GNSS/GPS antennas? Retrieved from https://novotech.com/pages/base-station (Accessed May 18, 2025).

Patil, H. K., & Chen, T. M. (2013). Wireless sensor network security. In *Elsevier eBooks*. Elsevier.

Rai, V. P., Ranjan, R., Gadhiya, A. R., & Mote, B. M. (2021). Use of modern physical tools for mitigating the effect of abiotic stresses. In *Elsevier eBooks* (pp. 387–397). https://doi.org/10.1016/b978-0-12-822849-4.00018-8

Recalde, C., Lopez, D., Aguat, D. and Garcia, V. J. (2023). Environmental sensing in high-altitude mountain ecosystems powered by sedimentary microbial fuel cells. *Sensors*, 23(40) 2101. https://doi.org/10.3390/s23042101

Sawa, T., Aoki, T., Yamamoto, I., Tsukioka, S., Yoshida, H., Hyakudome, T., … Nasuno, Y. (2005). Performance of the fuel cell underwater vehicle URASHIMA. *Nippon Onkyo Gakkaishi/Acoustical Science and Technology/Nihon Onkyo Gakkaishi*, 26(3), 249–257. https://doi.org/10.1250/ast.26.249

Shao, Y., Wang, J., Wu, H., Liu, J., Aksay, I. A., & Lin, Y. (2010). Graphene based electrochemical sensors and biosensors: A review. *Electroanalysis*, *22*(10), 1027–1036. https://doi.org/10.1002/elan.200900571

Shishkov, B., & Garvanova, M. (2023). Telecommunications and remote sensing: A public values perspective. In *Communications in computer and information science* (pp. 77–89). https://doi.org/10.1007/978-3-031-49263-1_6

Submersible Ultrasonic transducers: Applications, design, and advantages. Retrieved from https://www.omegasonics.com/knowledge-center/blog/submersible-ultrasonic-transducers-applications-design-and-advantages/ (Accessed 20 May, 2025)

Teja, R. (2024). What are biosensors? Principles, worjing, types and applications.

Wailes, R., Seale, R. L., Woodruff, E. B., Landis, F., Russell, & Charles, R. (1999). Energy conversion | Definition, examples, formula, principle, & facts. *Encyclopedia Britannica*. https://www.britannica.com/technology/energy-conversion

Weydahl, H., Gilljam, M., Lian, T., Johannessen, T. C., Holm, S. I., & Hasvold, J. Ø. (2019). Fuel cell systems for long-endurance autonomous underwater vehicles – challenges and benefits. *International Journal of Hydrogen Energy*, 45(8), 5543–5553. https://doi.org/10.1016/j.ijhydene.2019.05.035

Wynn, R. B., Huvenne, V. A., Bas, T. P. L., Murton, B. J., Connelly, D. P., Bett, B. J., … Hunt, J. E. (2014). Autonomous Underwater Vehicles (AUVs): Their past, present and future contributions to the advancement of marine geoscience. *Marine Geology*, 352, 451–468. https://doi.org/10.1016/j.margeo.2014.03.012

Yick, J., Mukherjee, B., & Ghosal, D. (2008). Wireless sensor network survey. *Computer Networks*, *52*(12), 2292–2330. https://doi.org/10.1016/j.comnet.2008.04.002

Zheng, T., Li, J., Ji, Y., Zhang, W., Xin, F., Dong, W., *et al.*, (2020). Progress and prospects of bioelectrochemical systems: Electron transfer and its applications in the microbial metabolism. *Frontiers in Bioengineering and Biotechnology*, *8*. https://doi.org/10.3389/fbioe.2020.00010

9 Fuel Cells in Wastewater Treatment

Emmanuel Ojo Ige and Emmanuel Adesuyi Omole

9.1 INTRODUCTION

Globally, wastewater is generated from vital sources, including households, industries, and agricultural activities, with approximate contributions of 55%, 25%, and 20%, respectively. Despite its ubiquity, only about 58% of wastewater is treated safely, while the remaining is discharged untreated, posing serious environmental and public health risks (*Jones et al., 2021*). Treatment capacity is unevenly distributed across regions; high-income countries treat approximately 74% of their wastewater (*UNESCO, 2024*). This disparity underscores the urgent need for improved infrastructure and policies in developing regions. Wastewater treatment not only protects water bodies and ecosystems from contamination but also supports water reuse, nutrient recovery, and climate resilience, making it a critical component of sustainable development.

Wastewater treatment is a critical process that removes contaminants from industrial, agricultural, and domestic wastewater before its release into the environment. The primary objective is to minimize the adverse effects of wastewater on human health, aquatic ecosystems, and water resources. Proper wastewater management is essential in addressing global challenges such as water scarcity, pollution, and climate change. Globally, untreated wastewater poses severe risks, including waterborne diseases, eutrophication, and chemical contamination. According to recent reports, nearly 80% of wastewater is discharged untreated into the environment, contributing to biodiversity loss and ecosystem degradation (*UNESCO, 2024*). Additionally, wastewater contains valuable resources, such as nutrients and organic matter, which can be harnessed for sustainable applications like bioenergy generation and nutrient recovery. Incorporating advanced wastewater treatment technologies, including fuel cell-based systems, can enhance sustainability by generating clean energy while treating wastewater. These innovative approaches align with Sustainable Development Goal (SDG) 6, which focuses on clean water and sanitation, and SDG 7, which promotes affordable and clean energy (*Shamshad & Rehman, 2025*).

9.2 CONVENTIONAL WASTEWATER TREATMENT CHALLENGES

Conventional wastewater treatment methods, including physical, chemical, and biological processes like sedimentation, coagulation-flocculation, and activated sludge systems, face significant sustainability and efficiency challenges. High energy consumption, particularly from aeration processes, contributes to substantial carbon emissions, undermining global climate mitigation efforts (*Grobelak et al.*, 2024). Additionally, biological treatments produce large volumes of sludge, which requires costly disposal and risks environmental contamination if mismanaged. Nutrient removal, especially for nitrogen and phosphorus, remains inefficient, while chemical disinfectants like chlorine generate harmful by-products such as trihalomethanes (THMs) and haloacetic acids (HAAs) (*Tafesse et al.*, 2023). Emerging contaminants, including pharmaceuticals and microplastics, further highlight the limitations of traditional methods, necessitating innovative and sustainable alternatives (*Emerging DBPs and Health Risks, 2025*).

DOI: 10.1201/9781003645405-9

Microbial Fuel Cells (MFCs) have emerged as a promising technology for sustainable wastewater treatment by harnessing microbial metabolism to convert organic pollutants into electricity (*Obileke et al.*, 2021). Integrated after secondary treatment, MFCs degrade remaining organic matter while generating energy, reducing both chemical oxygen demand (COD) and nutrient levels (*Bazina et al.*, 2023). This dual functionality not only enhances treatment efficiency but also addresses key challenges like sludge production and energy consumption (*Esfandyari et al.*, 2023). However, barriers such as high initial costs, low power density, and electrode material limitations hinder large-scale adoption (*Obileke et al.*, 2021). Ongoing research focuses on optimizing reactor designs, exploring advanced nanomaterials, and integrating MFCs with complementary technologies like constructed wetlands and anaerobic digestion to improve performance and scalability (*Arliyani et al.*, 2024).

The future of MFC technology holds immense potential for revolutionizing wastewater treatment by aligning with circular economy principles and global sustainability goals (*Arliyani et al.*, 2024). Innovations such as self-sustaining MFC systems, which recycle generated energy to power treatment processes, could significantly reduce reliance on external energy sources (*Esfandyari et al.*, 2023). Coupling MFCs with photoelectrochemical cells or anaerobic digestion may further enhance energy recovery and pollutant degradation (*Bazina et al.*, 2023). As advancements in materials and system design continue, MFCs are poised to become a cornerstone of next-generation wastewater management, offering a cleaner, more efficient alternative to conventional methods while contributing to environmental and economic benefits (*Obileke et al.*, 2021) (Figure 9.1).

9.3 ROLE OF FUEL CELLS IN SUSTAINABLE WASTEWATER TREATMENT

Fuel cells have gained significant attention in wastewater treatment due to their ability to degrade pollutants while simultaneously generating electricity. These bio-electrochemical systems, particularly microbial fuel cells (MFCs), utilize the metabolic activity of microorganisms to convert organic

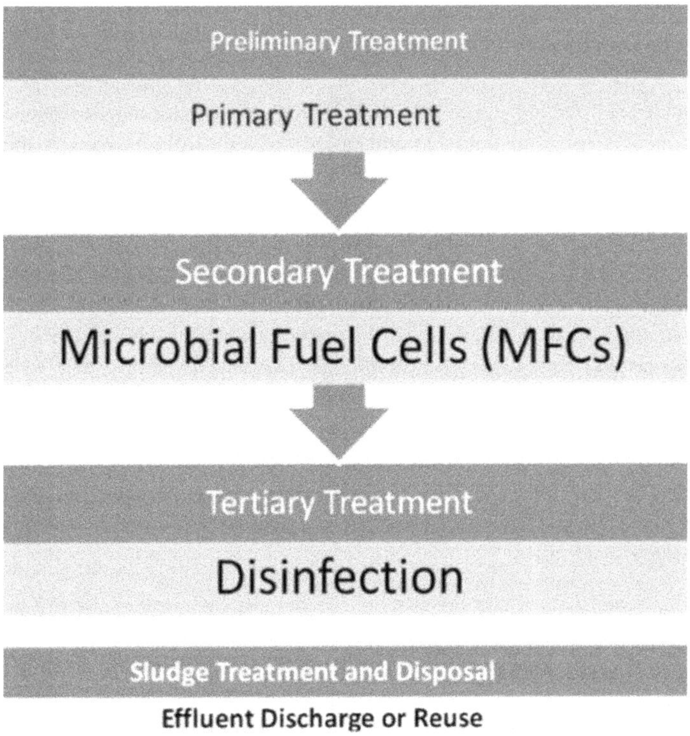

FIGURE 9.1 Integration of microbial fuel cells (MFCs) in wastewater treatment.

matter into electrical energy. Unlike conventional wastewater treatment methods, which are energy-intensive, fuel cells offer a sustainable and energy-efficient alternative (*Santoro et al., 2017*). One of the key advantages of fuel cells in wastewater treatment is their energy recovery potential, which reduces the operational costs of treatment facilities. In contrast to traditional approaches, MFCs can generate power from organic waste, making the treatment process more self-sufficient and cost-effective (*Grobelak et al., 2024*). Fuel cell technology offers the major benefit of reducing sludge production, a critical challenge in conventional biological treatment processes, as the anaerobic nature of microbial fuel cells minimizes the formation of excess sludge. Additionally, fuel cell systems have shown promise in removing emerging contaminants such as pharmaceuticals, pesticides, and heavy metals, which often persist in treated wastewater from traditional treatment plants. Their high adaptability allows for decentralized wastewater treatment applications, making them suitable for both small rural communities and large-scale municipal facilities. Furthermore, as a low-carbon technology, fuel cells contribute to greenhouse gas reduction by eliminating the need for fossil fuel-based energy inputs. Fuel cell-based wastewater treatment provides an environmentally sustainable solution to the increasing challenges of water pollution and energy demand while supporting circular economy principles by transforming waste into a valuable resource.

9.4 FUEL CELL TECHNOLOGY IN WASTEWATER TREATMENT

Fuel cell technology, particularly Microbial Fuel Cells (MFCs), represents a promising approach to sustainable wastewater treatment by simultaneously degrading pollutants and generating bioelectricity. MFCs operate as bio-electrochemical systems, utilizing the metabolic activities of microorganisms such as *Geobacter* and *Shewanella* species. These microbes metabolize organic pollutants under anaerobic conditions at the anode, releasing electrons and protons in the process (*Jalili et al.*, 2024; *Naaz et al., 2023*). The electrons are transferred through an external circuit to the cathode, generating an electric current, while protons migrate across a proton exchange membrane (PEM) to the cathode, where they react with oxygen to form water. This flow of electrons produces measurable electricity, contributing to energy recovery.

The efficiency of MFCs is largely dependent on electrode materials, with high-conductivity and biocompatible materials enhancing performance. Recent advances include the use of nanomaterial coatings to improve electron transfer rates and microbial adhesion, as well as the development of nickel-based electrodes, which have demonstrated significant bioenergy recovery during wastewater treatment (*Mahmoodzadeh et al., 2023*). Furthermore, the application of solar heterogeneous photo catalysis has shown potential in enhancing the degradation of persistent pollutants like trihalomethanes (THMs) and halo acetic acids (HAAs), integrating advanced oxidation processes into MFC systems (*Golfinopoulos et al.*, 2024). MFCs are highly effective in degrading organic pollutants, including chemical oxygen demand (COD), nitrogenous compounds, and heavy metals, making them versatile for various wastewater treatment applications. Despite these advancements, challenges such as electrode fouling, scalability limitations, and oxygen reduction efficiency at the cathode remain significant. Ongoing research is focusing on innovative cathode catalysts to boost power density and microbial engineering to optimize electron transfer pathways (*Ahmad et al.*, 2025). With continuous progress in material science and microbial optimization, MFCs hold immense potential for large-scale industrial wastewater treatment, offering a dual benefit of environmental remediation and sustainable power generation.

9.5 MECHANISMS IN MFCs FOR WASTEWATER TREATMENT AND ENERGY RECOVERY

Microbial fuel cells (MFCs) are an emerging bioelectrochemical technology that harnesses the metabolic activity of electrogenic bacteria to simultaneously treat wastewater and generate electricity. This dual-function system addresses two major global challenges: water pollution and the increasing need for renewable energy. The core mechanism involves exoelectrogenic microorganisms—such as *Geobacter* and *Shewanella* specie-oxidizing organic pollutants (e.g., carbohydrates, proteins, and lipids) in the anode compartment, thereby releasing electrons and protons through extracellular electron transfer (EET). These electrons are captured by the anode and travel through an external circuit, generating electricity, while the protons migrate across a proton exchange membrane (PEM) to the cathode. At the cathode, the electrons, protons, and terminal electron acceptors-typically oxygen- the oxygen reduction reaction (ORR) to form water. Depending on design goals, alternative acceptors like nitrate, sulfate, or metal oxides may be used, enhancing the removal of various contaminants, including nitrogenous compounds, heavy metals, and emerging micro pollutants (*Bird et al., 2022*). The fundamental structure of MFCs consists of an anode and a cathode chamber separated by the PEM. This configuration not only maintains redox balance but also enables effective pollutant degradation and energy recovery. Advancements in MFC technology have demonstrated promising results, with chemical oxygen demand (COD) removal efficiencies between 70% and 90% and power densities ranging from 0.5 to 2.5 W/m^2, influenced by factors like reactor design and substrate composition *(Jalili et al., 2024)*. Their low-energy requirement makes them ideal for decentralized wastewater treatment, especially in resource-constrained regions. However, key challenges remain, such as high costs of electrode materials, membrane fouling, and scaling limitations *(Chakma et al., 2025*). Ongoing research is exploring enhanced microbial consortia, cost-effective catalytic materials, and the integration of MFCs with technologies such as microbial electrolysis cells and constructed wetlands to improve efficiency and sustainability (*Arliyani et al., 2024*). As material science and process engineering evolve, MFCs could become integral to circular economy frameworks, transforming wastewater into a sustainable source of energy and nutrients. Future focus should be directed toward pilot-scale implementation and comprehensive life-cycle assessments to evaluate their viability in real-world municipal and industrial applications (*Roy et al., 2023*).

9.6 MICROBIAL ACTIVITY IN MFCS FOR ENERGY AND POLLUTANT CONTROL

Microorganisms serve as the fundamental drivers of bioelectrochemical processes in microbial fuel cells (MFCs), enabling simultaneous wastewater treatment and renewable energy generation. Electrogenic bacteria, particularly *Geobacter* and *Shewanella* species, play a pivotal role by catalyzing the oxidation of organic pollutants and facilitating extracellular electron transfer (EET) to the anode. These microorganisms transfer electrons derived from metabolic reactions to solid-state electrodes, converting chemical energy in wastewater into electrical energy (*Cao et al.*, 2019). The efficiency of electron transfer is governed by two mechanisms: direct electron transfer (DET), where bacteria physically contact the anode via conductive proteins (e.g., pili), and mediated electron transfer (MET), which relies on soluble redox mediators like flavins (*Naaz et al., 2023*). While DET minimizes energy losses, MET broadens microbial participation but introduces mediator instability (*Cao et al.*, 2019).

The microbial community composition in MFCs critically influences pollutant degradation and power output. Mixed consortia-common in wastewater-outperform pure cultures due to metabolic versatility, degrading complex hydrocarbons, aromatics, and xenobiotics through synergistic interactions among bacteria, archaea, and fungi. Syntrophic relationships enhance refractory organic matter breakdown, while electroactive species optimize electron harvesting (*Cao et al.*, 2019). This cooperation enables complete pollutant mineralization and efficient bioelectricity generation, key to real-world MFC applications.

Recent advances highlight microbial optimization strategies to boost MFC performance. Bioaugmentation with *Pseudomonas* improves phenolic compound degradation, while *Clostridium* enhances polysaccharide fermentation, supplying substrates for electrogenic bacteria. Integrating nitrogen- and sulfur-cycling microbes (e.g., *Nitrosomonas*, *Desulfovibrio*) extends MFC utility to nutrient-rich wastewaters, refining microbial synergy and electron flux (*Sriwichai et al.*, 2024; *Patel et al.*, 2021). These approaches strengthen MFCs as dual-function systems for energy recovery and advanced pollutant removal.

9.7 ENERGY RECOVERY AND ELECTRODE INNOVATION IN MFCs

The performance and efficiency of microbial fuel cells (MFCs) critically depend on advanced electrode materials and catalysts, which play a crucial role in enhancing pollutant degradation and energy recovery. Carbon-based anodes, such as modified carbon cloth and graphene with nanostructure surfaces, have demonstrated up to a 40% improvement in microbial adhesion and electron transfer efficiency compared to conventional materials. These innovations provide a larger surface area for biofilms formation, facilitating more effective current generation (*Olvera-Vargas et al., 2024*). Meanwhile, on the cathode side, researchers are shifting from expensive platinum-based catalysts to more cost-effective alternatives such as metal oxides (MnO_2, Fe_3O_4) and nitrogen-doped carbon materials. Enzymatic catalysts like lactase and peroxidase are also gaining attention for their ability to enhance oxygen reduction kinetics, further boosting MFC efficiency. MFCs exhibit significant potential for energy recovery, capable of extracting approximately 1.5 kWh/m^3 from domestic wastewater with conversion efficiencies of 5–10%. Integration with microbial electrolysis cells (MECs) further enhances performance, enabling simultaneous hydrogen production and nutrient recovery, thereby improving both energy output and wastewater treatment sustainability (*Lv et al., 2018*). Industrial-scale implementations have already shown promising results, with pilot systems achieving 85% chemical oxygen demand (COD) removal from brewery wastewater while generating 0.8 W/m^3 of power. Hybrid configurations, such as constructed wetland-MFC systems, have also proven effective, removing up to 90% of pharmaceuticals like sulfamethoxazole from wastewater. Despite these advancements, challenges remain in scaling MFC technology, particularly concerning electrode costs and optimal system design for large-scale deployment. Current research is addressing these limitations through the development of low-cost biochar electrodes, engineered microbial consortia for enhanced current output, and stackable MFC configurations suitable for municipal wastewater treatment. Emerging innovations, including AI-driven system optimization and hybrid MFC-reverse osmosis (RO) setups for zero-liquid-discharge applications, highlight the transformative potential of MFCs in sustainable wastewater management. By continuing to refine electrode materials, catalyst efficiency, and system integration, MFC technology holds immense promise for advancing both wastewater treatment and renewable energy generation in an environmentally sustainable and economically viable manner (*Ishaq et al., 2024*).

9.8 INNOVATIONS IN WASTEWATER MFCs

Microbial fuel cells (MFCs) have emerged as a promising bioelectrochemical technology that integrates wastewater treatment with renewable energy generation. The effectiveness of MFC systems is influenced by several critical operational parameters, such as pH (with an optimal range of 6.5–8.5), temperature (favoring mesophilic conditions between 20°C and 40°C), and the nature of the substrate-simple organic compounds like acetate often demonstrate superior degradation kinetics (*Singh et al., 2025*). Recent progress in material science has led to the development of cost-effective electrodes. Carbon-based anodes modified with nanostructure surfaces and alternative cathode materials, including metal oxides and conductive polymers, are showing particular promise for large-scale deployment (*Mandal et al., 2025*).

Field-scale implementations have validated the technical feasibility and adaptability of MFC technology across diverse wastewater treatment scenarios. For example, Denmark's Aarhus wastewater treatment plant has demonstrated energy-neutral operation by integrating MFC systems (*Roy et al., 2023*). In the United States, pilot-scale systems have achieved up to 85% chemical oxygen demand (COD) reduction, while specialized MFC configurations have proven effective in treating complex effluents in India's textile industry. Recent technological innovations include the adoption of graphene-based electrode materials and the application of artificial intelligence to optimize MFC system performance, marking significant steps forward in enhancing efficiency and scalability.

Despite these advancements, several challenges must be addressed for widespread adoption. Practical applications are often constrained by low power densities, typically below 3 W/m^2, ongoing issues with Biofouling, and high initial capital investment. Nonetheless, MFC technology offers substantial environmental benefits, such as reducing greenhouse gas emissions by approximately 50% and recovering energy from wastewater (*Youcef et al., 2025*). The modularity of MFC systems provides flexibility for both centralized upgrades and decentralized deployment in remote regions (*Roy et al., 2023*; *Ventura et al., 2024*). Current research is focused on improving electron transfer mechanisms and devising economically feasible scale-up strategies. To transition from laboratory prototypes to full-scale operations, interdisciplinary collaboration involving materials science, microbiology, and environmental engineering is essential. Future progress will hinge on overcoming technical barriers and establishing robust policy frameworks and economic models to support MFC adoption as a mainstream solution for sustainable wastewater treatment (*Roy et al., 2023*).

9.9 FUEL CELL WASTEWATER TREATMENT: TECHNOLOGY AND POLICY

Recent advancements in fuel cell technologies for wastewater treatment have been propelled by breakthroughs in material science and system integration. Graphene-modified electrodes and transition metal catalysts have enhanced charge transfer efficiency while reducing dependence on costly platinum-based components, with certain configurations maintaining stable performance for over 6,000 hours in municipal wastewater settings. Hybrid systems-such as microbial fuel cells (MFCs) combined with anaerobic digesters or constructed wetlands-have improved methane yields by 30-40% and achieved up to 90% removal of emerging contaminants like pharmaceuticals, underscoring their potential for sustainable wastewater remediation (*Olvera-Vargas et al.*, 2025).

The integration of advanced monitoring systems and machine learning tools has further optimized fuel cell performance. Neural networks have achieved prediction accuracies of up to 95%, allowing real-time adjustment of operational parameters to improve efficiency. Simultaneously, bioengineering advances have led to genetically modified electroactive bacteria capable of producing 50% higher current densities than their wild-type counterparts. These combined developments across materials science, artificial intelligence, and biotechnology are enabling more efficient, self-regulating treatment systems that balance pollutant removal with energy generation.

Despite these technological strides, implementation at scale continues to face regulatory and economic challenges. Delays in approval processes and inconsistent regulatory standards remain major barriers. Nevertheless, policy measures such as the European Union's fast-track regulatory pathways and financial mechanisms-like green energy credits-have helped reduce the payback period for MFC systems to under five years, making them more economically attractive (*Abu Sofian et al.*, 2024). Achieving widespread adoption will depend on coordinated collaboration between scientists, engineers, and policymakers. These partnerships are essential for refining scalable system designs, harmonizing regulations, and aligning efforts with global sustainability objectives such as SDG 6, ultimately supporting the transformation of wastewater treatment into a net energy-generating process (*Shamshad & Rehman*, 2025).

9.10 CHALLENGES IN IMPLEMENTING FUEL CELLS FOR WASTEWATER TREATMENT

The transition to sustainable energy solutions has become increasingly critical due to growing concerns over climate change and the environmental impact of fossil fuel consumption. The rapid depletion of natural gas and petroleum reserves, combined with their adverse environmental effects, has intensified the global energy crisis. Within this context, Microbial Fuel Cell (MFC) technology emerges as a promising solution that addresses both waste management and sustainable energy generation. By converting organic waste into renewable energy, MFCs bridge the gap between effective wastewater treatment and clean energy production. Despite its potential, the widespread implementation of MFCs faces several technical and economic challenges that must be overcome for large-scale adoption (*Olabi et al., 2023*).

9.10.1 Scaling Up from Laboratory to Industrial Applications

Microbial Fuel Cells (MFCs) have demonstrated efficiency and energy recovery potential at the laboratory scale, scaling up to industrial levels remains a significant challenge. Laboratory models operate under controlled conditions, which are difficult to replicate on a larger scale. Key issues such as hydraulic retention time (HRT), mass transfer limitations, and reactor design constraints become more prominent as system size increases. In industrial applications, maintaining biofilm stability and consistent electron transfer rates is substantially more complex, often resulting in efficiency losses. Achieving stable performance in larger reactors necessitates advanced electrode materials and optimized reactor configurations that can sustain high power densities and pollutant removal rates over extended periods (*Singh et al., 2025*)

9.10.2 High Initial Costs

One of the primary barriers to the large-scale implementation of MFCs in wastewater treatment is the high initial cost of materials and system setup. Traditional platinum-based electrodes, known for their high conductivity and catalytic efficiency, are prohibitively expensive. To address this, researchers have explored carbon-based materials, such as graphene and carbon nanotubes, which offer good conductivity and durability at a lower cost. Despite these advances, the cost of bioreactor construction, integration with existing wastewater infrastructure, and ongoing maintenance requirements still pose financial challenges. For MFC technology to become economically feasible, innovations in cost-effective electrode materials and scalable reactor designs are essential (*Ojha and Pradhan, 2025*).

9.10.3 Biofouling and Electrode Degradation

A major challenge in the practical application of MFCs is Biofouling and electrode degradation. Over time, microbial growth on electrode surfaces hinders electron transfer efficiency and reduces power output. This issue is particularly prominent in wastewater environments, where diverse microbial communities interact with the electrodes, leading to clogging and reduced conductivity. Additionally, materials used for electrodes often suffer from corrosion and loss of catalytic activity, especially in harsh wastewater conditions (*Chakma et al., 2025*). Strategies to mitigate these issues include the development of anti-fouling materials and the use of self-regenerating electrode surfaces that can maintain performance over long durations (*Santoro et al., 2017*).

9.10.4 Adapting to Variable Wastewater Characteristics

A significant challenge for MFC technology in wastewater treatment is its adaptability to variable wastewater compositions. Industrial and municipal wastewater streams can differ greatly in terms of organic load, pH, temperature, and contaminant types. These fluctuations affect the metabolic activity of electro active microbes and the overall bioelectrochemical reactions within MFCs. Advanced monitoring and control systems have been developed to adjust hydraulic retention times and organic loading rates in real time, optimizing both pollutant removal and energy recovery (*Ojha and Pradhan, 2025*).

9.10.5 Regulatory and Policy Barriers

Microbial Fuel Cell technology faces significant regulatory and policy barriers that impede widespread adoption, despite its potential for sustainable wastewater treatment and energy recovery. Existing regulatory frameworks are often not equipped to accommodate bio-electrochemical systems, leading to lengthy verification processes and inconsistent performance standards across different regions. While international initiatives, like the Sustainable Development Goals (SDG) 6, emphasize clean water and sanitation, national regulations lag behind in integrating MFC technology into wastewater treatment standards. To address this, some regions have begun exploring fast-track approval pathways and green energy incentives to promote innovative wastewater solutions (*Ali et al., 2025*).

9.11 CASE STUDIES AND REAL-WORLD APPLICATIONS

Microbial Fuel Cell (MFC) technology has progressed from laboratory research to practical implementation across various wastewater treatment applications. This section examines three primary implementation domains through documented examples, showcasing MFCs' dual potential for wastewater treatment and energy recovery.

9.12 MUNICIPAL WASTEWATER TREATMENT

Municipal wastewater treatment represents one of the most promising applications for MFC technology due to consistent flow characteristics and organic loading. Research has demonstrated that municipal systems can achieve chemical oxygen demand (COD) removal rates between 75 and 90% while simultaneously generating recoverable energy. The technology has shown particular effectiveness when integrated with existing treatment processes, with some implementations reporting energy savings in aeration requirements. Compatibility with conventional activated sludge systems has been established, though scale-up challenges remain (*Ge et al., 2014*).

9.13 INDUSTRIAL WASTEWATER TREATMENT

Industrial applications offer both potential and limitations for microbial fuel cell (MFC) deployment. MFCs have been successfully applied to treat a variety of industrial effluents, but their performance is largely influenced by the type and concentration of the organic matter present in the wastewater. Substrates with high biodegradability, such as those derived from the food processing and beverage industries, have demonstrated particularly promising outcomes in terms of both chemical oxygen demand (COD) reduction and energy recovery. Reported COD removal efficiencies typically range from 65% to 80%, though the actual energy yield depends on the substrate's composition and availability. Furthermore, MFCs are capable of handling variable organic loads, making them attractive for industrial sectors with fluctuating discharge profiles (*Pant et al., 2010*).

9.14 DECENTRALIZED AND RESOURCE-LIMITED APPLICATIONS

Microbial fuel cell (MFC) technology offers significant potential for decentralized wastewater treatment, particularly in remote or under-resourced regions where centralized infrastructure is unavailable. Small-scale MFC systems have demonstrated the capacity to simultaneously treat wastewater and generate low levels of electrical power suitable for basic needs, such as powering LEDs or sensors. Although these systems typically achieve lower treatment efficiencies compared to centralized plants—often reporting chemical oxygen demand (COD) removal rates in the range of 55–70%—they provide a unique dual function: environmental sanitation and renewable energy production. In resource-limited contexts, system designs often emphasize affordability, operational simplicity, and minimal maintenance over maximum energy output. Such features make MFCs especially attractive for applications in developing countries and off-grid settings (*Santoro et al., 2017*). Overall, while the technology shows promise across various settings, optimal deployment depends on site-specific conditions and clearly defined treatment and energy goals.

9.15 ECONOMIC AND ENVIRONMENTAL IMPLICATIONS

The integration of MFC technology presents transformative potential for sustainable water management. Economically, while initial costs exceed conventional systems, MFCs offer long-term savings through energy generation (1.8 kWh/m^3 potential) and reduced sludge handling (*Muñoz et al.*, 2021). Break-even periods of 5–7 years are achievable for municipal systems, with graphene electrodes and other advances improving affordability (*Fathima et al.*, 2024).

Environmentally, MFC systems demonstrate 40% lower carbon emissions compared to activated sludge processes (*Liu et al.*, 2021), approaching carbon-neutral operation when optimized. Plant-MFC hybrids show additional promise for CO_2 sequestration (*Chong et al.*, 2025). Policy support through initiatives like the EU Green Deal and US CWSRF program is accelerating adoption (*Kumar et al.*, 2024), particularly in developing nations where decentralized systems align with SDG 6 targets.

9.16 COST-BENEFIT ANALYSIS: ASSESSING LONG-TERM SAVINGS THROUGH ELECTRICITY GENERATION VS. INITIAL INVESTMENT

The economic viability of microbial fuel cell (MFC)-based wastewater treatment systems is influenced by their higher initial installation costs compared to conventional methods, but their long-term benefits in energy generation and reduced operational expenses present a compelling case. Traditional wastewater treatment consumes 2–3% of global electricity, while MFCs can convert organic waste into electricity, yielding up to 1.8 kWh/m^3 of treated wastewater and offsetting energy costs (*Muñoz et al., 2021*). Although capital costs remain high due to electrode materials and biofilms optimization, advancements like graphene-based electrodes are lowering expenses while improving efficiency (*Fathima et al., 2024*). Additionally, MFCs reduce sludge production by up to 30%, cutting disposal costs and environmental impact. Technological progress in electrode design and microbial bioengineering is further enhancing MFC affordability, making large-scale implementations increasingly feasible. Studies show that municipal MFC systems can achieve economic break-even within 5–7 years, particularly when combined with anaerobic digestion for biogas production. As these innovations drive down costs, MFCs are becoming a more attractive option for both municipal and industrial wastewater treatment, balancing higher upfront investments with long-term energy savings and reduced sludge management expenses (*Kumar et al., 2024*).

9.17 ENVIRONMENTAL IMPACT OF FUEL CELL WASTEWATER TREATMENT

Conventional wastewater treatment methods, particularly activated sludge processes, are associated with high carbon emissions due to energy consumption and methane leakage during anaerobic digestion, whereas fuel cell-based technologies offer a sustainable alternative by not only reducing energy demand but also minimizing greenhouse gas emissions, as MFCs achieve this through direct electricity generation from microbial metabolism, bypassing energy-intensive aeration processes typical of conventional methods (*Twi-Yeboah et al., 2024*). A comparative study by *Liu et al. (2021)* demonstrated that MFC-based treatment reduces carbon emissions by approximately 40% compared to activated sludge systems, primarily due to decreased reliance on external power sources, with the process being nearly carbon-neutral when powered entirely by the generated bioelectricity, thereby contributing to sustainable wastewater treatment. Additionally, the use of plant microbial fuel cells (PMFCs) has shown potential for both wastewater treatment and carbon sequestration, as studies indicate that PMFCs can absorb CO_2 during the microbial metabolism of root exudates, further reducing the carbon footprint of wastewater treatment plants positioning fuel cell-based wastewater treatment as a critical technology for achieving net-zero emissions in the sanitation sector (*Chong et al., 2025*).

9.18 GOVERNMENT SUPPORT FOR WASTEWATER TREATMENT INNOVATIONS

The successful deployment of microbial fuel cells (MFCs) and other fuel cell technologies in wastewater treatment relies not only on technological advancements but also on supportive government policies and financial incentives, with governments worldwide increasingly recognizing the role of bioelectrochemical systems (BES) in achieving sustainable energy goals and environmental protection. Policies such as the European Green Deal and the U.S. Infrastructure Investment and Jobs Act (2021) have allocated substantial funding toward research and the development of sustainable energy technologies, including MFCs, while the European Union's Horizon Europe Programme has prioritized sustainable wastewater treatment under its Green Technologies initiative, providing €95.5 billion for innovations that reduce carbon emissions and improve energy efficiency (*Kumar et al., 2024*). Similarly, China's 13th Five-Year Plan included targeted subsidies for wastewater treatment facilities adopting MFC technology, leading to a 15% increase in pilot projects, and financial incentives like feed-in tariffs for bioelectricity generated from wastewater treatment have been introduced in various jurisdictions, allowing plants to sell excess electricity back to the grid. In the U.S., the Clean Water State Revolving Fund (CWSRF) supports innovative wastewater treatment projects with low-interest loans, lowering financial barriers for municipalities to adopt MFCs. The convergence of technological advancements, supportive policies, and economic incentives is paving the way for widespread MFC adoption in wastewater treatment, addressing waste management and energy recovery while aligning with global sustainability targets like UN SDG 6, which emphasizes clean water and sanitation for all (*Twi-Yeboah et al., 2024*).

9.19 FUTURE PERSPECTIVES AND RESEARCH DIRECTIONS

The future of fuel cell-based wastewater treatment holds immense potential, driven by interdisciplinary advancements aimed at enhancing efficiency, scalability, and sustainability. Key research directions include the integration of these systems with renewable energy sources like solar and wind to ensure carbon-neutral operations, as well as the development of hybrid bioelectrochemical systems that maximize energy recovery while treating complex pollutants. Genetic engineering of electro active bacteria could further optimize microbial fuel cell performance by enhancing electron transfer rates and substrate utilization, while synthetic biology approaches may enable tailored microbial consortia for targeted pollutant degradation. Additionally, the application of artificial intelligence and machine learning can revolutionize real-time monitoring and adaptive control of operational parameters, improving process stability and energy efficiency (*Chakma et al., 2025*). Advances in nonmaterial and electrode design will also play a critical role in reducing costs and improving durability. Policy frameworks and economic incentives must evolve to support large-scale implementation, ensuring that these innovative systems transition from lab-scale prototypes to commercially viable solutions, ultimately contributing to a circular economy by transforming wastewater from a waste stream into a resource for energy, nutrients, and clean water. (*Chakma et al., 2025*).

9.20 FUEL CELLS AND RENEWABLE ENERGY FOR WASTEWATER TREATMENT

One of the most promising directions for the enhancement of fuel cell-based wastewater treatment lies in its integration with renewable energy sources such as solar and wind power, as the intermittent nature of renewable energy generation can be balanced with microbial fuel cells (MFCs) that continuously produce bioelectricity during wastewater treatment. Hybrid systems have demonstrated increased energy efficiency and stability, particularly in off-grid and decentralized treatment applications (*Heidrich et al., 2011*), with solar-powered MFCs enhancing electrochemical activity during daylight hours to boost pollutant degradation rates—evidenced by a pilot study where solar-assisted MFCs achieved a 30% increase in power generation and a 20% higher COD removal efficiency compared to traditional MFCs (*Wang et al., 2014*). Similarly, wind-supplemented MFC systems stabilize energy outputs during variable wind conditions, optimizing energy recovery while maintaining high pollutant removal rates, contributing to energy sustainability and minimizing the carbon footprint of wastewater treatment. Additionally, the emerging field of photo catalytic microbial fuel cells (PMFCs) harnesses both solar energy and microbial electrochemical activity, utilizing photoactive materials like TiO_2 nanoparticles to enhance electron flow during sunlight exposure, thereby improving power output and contaminant degradation, representing a next-generation solution for sustainable water treatment through the synergy of light-driven reactions and microbial metabolism (*Unuofin et al., 2023*).

9.21 GENETIC ENGINEERING OF MICROORGANISMS: ENHANCING THE EFFICIENCY OF BIOELECTROCHEMICAL REACTIONS

The performance of microbial fuel cells (MFCs) largely depends on the electro active properties of microbes in the anode chamber. Genetic engineering has become a key approach to improve these properties by modifying extracellular electron transfer (EET) pathways in bacteria like *Geobacter sulfurreducens* and *Shewanella oneidensis*. Such modifications have led to increased current densities and better electron recovery. For example, engineered *Geobacter* strains show up to 40% higher electron transfer rates and enhanced electricity generation during wastewater treatment (*Mkilima et al.*, 2025). Furthermore, synthetic biology has allowed the creation of microbial communities with complementary metabolic functions. These communities improve the breakdown of complex pollutants while boosting bioelectricity output through cooperative microbial interactions, enabling the design of tailored microbial systems for specific wastewater types (*Che and Men 2019*).

9.22 AUTOMATION AND AI IN WASTEWATER MANAGEMENT

The incorporation of artificial intelligence (AI) into MFC-based wastewater treatment represents a transformative shift toward real-time optimization and predictive maintenance. AI-driven control systems leverage machine learning algorithms to monitor operational parameters such as pH, temperature, COD levels, and power output, adjusting conditions dynamically to maintain optimal treatment efficiency. Machine learning models, such as artificial neural networks (ANNs), have been employed to predict MFC performance based on real-time data, achieving accuracy rates of over 95% in pilot-scale applications (*Baarrimah et al., 2024*). This predictive capability not only enhances system reliability but also minimizes human intervention, reducing costs associated with manual monitoring and maintenance. Moreover, Internet integration has facilitated the development of smart microbial fuel cells that communicate with cloud-based monitoring platforms. These platforms enable remote diagnostics, fault detection, and performance analytics, making MFC-based wastewater treatment more resilient and efficient. Such advancements are critical for scaling up MFC applications in urban and industrial wastewater management. The synergy of AI with advanced sensor networks allows for the continuous optimization of MFC conditions, dynamically responding to fluctuations in wastewater composition and external environmental factors. This real-time adaptability not only improves pollutant degradation efficiency but also maximizes energy recovery, making MFC systems a robust solution for sustainable wastewater management in smart cities (*Akintola, 2024*).

9.23 CONCLUSION

Fuel cell-based wastewater treatment represents a pivotal advancement in sustainable water and energy management, offering a dual solution to pollution and energy scarcity. By converting organic waste into electricity or hydrogen, these systems not only reduce the environmental footprint of wastewater treatment but also contribute to renewable energy generation. The integration of microbial fuel cells (MFCs) and microbial electrolysis cells (MECs) with renewable energy sources, smart technologies, and advanced materials underscores their potential to revolutionize both municipal and decentralized wastewater treatment. However, overcoming scalability challenges, optimizing cost efficiency, and ensuring long-term durability remain critical for widespread adoption. Continued innovation in electrode materials, genetic engineering of microbes, and AI-driven process optimization will be key to unlocking their full potential. The future of fuel cell-based wastewater treatment hinges on collaborative efforts among researchers, policymakers, and industry stakeholders. Supportive policies, financial incentives, and international partnerships are essential to accelerate deployment and make these systems economically viable. As global water stress and energy demands intensify, fuel cell technologies provide a pathway toward a circular economy-turning waste into a resource while mitigating climate change. With sustained investment and interdisciplinary cooperation, these systems could become a cornerstone of sustainable infrastructure, helping achieve global goals for clean water, affordable energy, and climate action. Their success will not only redefine wastewater treatment but also demonstrate how innovative technologies can drive environmental and economic resilience.

REFERENCES

Abu Sofian, A. D. A. B., Sun, X., Gupta, V. K., Berenjian, A., Xia, A., Ma, Z., & Show, P. L. (2024). Advances, synergy, and perspectives of machine learning and biobased polymers for energy, fuels, and biochemicals for a sustainable future. *Energy & Fuels*, *38*(3). https://doi.org/10.1021/acs.energyfuels.3c03842

Ahmad, A., Kumar, P., & Singh, S. P. (2025). Recent advancements in heterogeneous cathode catalysts for in-situ hydrogen peroxide synthesis in bio-electro-Fenton process and its application in wastewater treatment: A review. *Journal of Environmental Chemical Engineering*. In Press, *Journal Pre-proof*. https://doi.org/10.1016/j.jece.2025.117109

Akintola, A. A. (2024). Driven monitoring systems for bioremediation: Real-time data analysis and predictive modelling. *World Journal of Advanced Research and Reviews*, *24*(1), 788–803. https://doi.org/10.30574/wjarr.2024.24.1.3099

Ali, H. E., Hemdan, B. A., El-Naggar, M. E., El-Liethy, M. A., Jadhav, D. A., El-Hendawy, H. H., Ali, M., & El-Taweel, G. E. (2025). Harnessing the power of microbial fuel cells as pioneering green technology: Advancing sustainable energy and wastewater treatment through innovative nanotechnology. *Bioprocess and Biosystems Engineering*, *48*, 343–366. https://doi.org/10.1007/s00449-024-03115-z

Arliyani, I., Noori, M. T., Ammarullah, M. I., Tangahu, B. V., Mangkoedihardjo, S., & Min, B. (2024). Constructed wetlands combined with microbial fuel cells (CW-MFCs) as a sustainable technology for leachate treatment and power generation. *RSC Advances*, *14*(28), 18620–18638. https://doi.org/10.1039/D4RA04658G

Baarrimah, A. O., Bazel, M. A., Alaloul, W., & Mushtaha, A. W. (2024). Artificial intelligence in waterwaste treatment: Research trends and future perspectives through bibliometric analysis. *Case Studies in Chemical and Environmental Engineering*. http://doi.org/10.1016/j.cscee.2024.100926

Bazina, N., Ahmed, T. G., Almdaaf, M., Jibia, S., & Sarker, M. (2023). Power generation from wastewater using microbial fuel cells: A review. *Journal of Biotechnology*, *374*, 17–30. https://doi.org/10.1016/j.jbiotec.2023.07.006

Bird, H., Heidrich, E., Leicester, D., & Theodosiou, P. (2022). Pilot-scale microbial fuel cells (MFCs): A meta-analysis study to inform full-scale design principles for optimum wastewater treatment. *Journal of Cleaner Production*, *346*, 131227. https://doi.org/10.1016/j.jclepro.2022.131227

Cao, Y., Mu, H., Liu, W., Zhang, R., Guo, J., Xian, M., & Liu, H. (2019). Electricigens in the anode of microbial fuel cells: Pure cultures versus mixed communities. *Microbial Cell Factories*, *18*(1), 39. https://doi.org/10.1186/s12934-019-1087-z

Chakma, R., Hossain, M. K., Paramasivam, P., Bousbih, R., Amami, M., Toki, G. F. I., Haldhar, R., & Karmaker, A. K. (2025). Recent applications, challenges, and future prospects of microbial fuel cells: A review. *Global Challenges*, *9*(5), 2500004. https://doi.org/10.1002/gch2.202500004

Che, S., & Men, Y. (2019). Synthetic microbial consortia for biosynthesis and biodegradation: Promises and challenges. *Journal of Industrial Microbiology and Biotechnology*, *46*(3), Article https://doi.org/10.1007/s10295-019-02211-4

Chong, P. L., Chuah, J. H., Chow, C.-O., & Ng, P. K. (2025). Plant microbial fuel cells: A comprehensive review of influential factors, innovative configurations, diverse applications, persistent challenges, and promising prospects. *International Journal of Green Energy*, *22*(3), 599–648. https://doi.org/10.1080/15435075.2024.2421325

Emerging DBPs and Health Risks (2025). Environmental Working Group (EWG). (2025, March). *What are disinfection byproducts?* Retrieved from https://www.ewg.org/news-insights/news/2025/03/what-are-disinfection-byproducts

Esfandyari, M., Jafari, D., & Azami, H. (2023). Microbial fuel cells for energy production in wastewater treatment plants: A review. *Biofuels*, *14*(3), 1–11. https://doi.org/10.1080/17597269.2023.2294227

European Commission. (2021). *The European Green Deal*. Directorate- General for Communication. Publications Office of the European Union. ISBN 978-92-76-39487-7. https://doi.org/10.2775/595210

Fathima, A., Ilankoon, I. M. S. K., Zhang, Y., & Chong, M. N. (2024). Scaling up of dual-chamber microbial electrochemical systems – An appraisal using systems design approach. *Science of the Total Environment*, *912*, 169186. https://doi.org/10.1016/j.scitotenv.2023.169186

Ge, Z., Li, J., Xiao, L., Tong, Y., & He, Z. (2014). Recovery of electrical energy in microbial fuel cells. *Environmental Science & Technology Letters*, *1*(2), 137–141. https://doi.org/10.1021/ez4000324

Golfinopoulos, S. K., Nikolaou, A. D., & Alexakis, D. E. (2024). Innovative approaches for minimizing disinfection byproducts (DBPs) in water treatment: Challenges and trends. *Applied Sciences*, *14*(18), 8153. https://doi.org/10.3390/app14188153

Grobelak, A., Całus-Makowska, K., Jasińska, A., Klimasz, M., Wypart-Pawul, A., Augustajtys, D., Baor, E., Sławczyk, D., & Kowalska, A. (2024). Environmental impacts and contaminants management in sewage sludge-to-energy and fertilizer technologies: Current trends and future directions. *Energies*, *17*(19), 4983. https://doi.org/10.3390/en17194983

Heidrich, E. S., Curtis, T. P., & Dolfing, J. (2011). Determination of the internal chemical energy of wastewater. *Environmental Science & Technology*, *45*(2), 829–834. https://doi.org/10.1021/es103058w

Ishaq, A., Mohd Ismid Mohd Said, M. I. M., Azman, S. B., Houmsi, M. R., Isah, A. S., Jagun, Z. T., Mohammad, S. J., Bello, A. D., & Abubakar U. A. (2024). The influence of various chemical oxygen demands on fuel cells performance using leachate as a substrate. *Environmental Science and Pollution Research*. https://doi.org/10.1007/s11356-024-32090-x

Jalili, P., Ala, A., Nazari, P., Jalili, B., & Ganji, D. D. (2024). A comprehensive review of microbial fuel cells considering materials, methods, structures, and microorganisms. *Heliyon*, *10*(e25439). https://doi.org/10.1016/j.heliyon.2024.e25439

Jones, E. R., van Vliet, M. T. H., Qadir, M., & Bierkens, M. F. P. (2021). Country-level and gridded estimates of wastewater production, collection, treatment and reuse. *Earth System Science Data*, *13*(2), 237–254. https://doi.org/10.5194/essd-13-237-2021

Kumar, S., Shenode, N. R., Jeph, Y., & Kudale, R. (2024). Microbial fuel cells for bioelectricity generation: Current innovations, challenges, and future prospects. *International Journal for Research in Applied Science and Engineering Technology (IJRASET)*, *12*(4). https://doi.org/10.22214/ijraset.2024.64299

Liu, H., Cheng, S., & Logan, B. E. (2021). Production of electricity from acetate or butyrate using a single-chamber microbial fuel cell. *Environmental Science & Technology*, *39*(2), 658–662. https://doi.org/10.1021/es048927c

Lv, Q., Si, W., He, J., Sun, L., Zhang, C., Wang, N., Yang, Z., Li, X., Wang, X., Deng, W., Long, Y., Huang, C., & Li, Y. (2018). Selectively nitrogen-doped carbon materials as superior metal-free catalysts for oxygen reduction. *Nature Communications*, *9*(1), 3376. https://doi.org/10.1038/s41467-018-05878-y

Mahmoodzadeh, F., Navidjouy, N., Alizadeh, S., & Rahimnejad, M. (2023). Investigation of microbial fuel cell performance based on the nickel electrode for bioenergy recovery and simultaneous wastewater treatment. *Scientific Reports*, *13*, 48290. https://doi.org/10.1038/s41598-023-48290-3

Mandal, S., Mendhe, A. B., Rakhade, H. M., Barse, N. S., Roy, M., Rosaiah, P., Park, T., Lee, H. S., Mendhe, A. C., & Kim, D. (2025). Recent advancement and design in supercapacitor hybrid electrode materials: Bridging the gap between energy and power density. *Chemical Engineering Journal Advances*, *21*, 100690. https://doi.org/10.1016/j.ceja.2024.100690

Mkilima, T., Zharkenov, Y., Abduova, A., Kudaibergenov, N., Fazylov, K., Toleubayeva, S., Kirgizbayeva, K., Zhumadilov, I., Jaxymbetova, M., & Zhapparova, A. (2025). Bioelectrochemical degradation of pollutants in wastewater using a dual-chamber microbial fuel cell with graphene-modified electrodes and electroactive bacteria. *Case Studies in Chemical and Environmental Engineering*, *11*, 101184. https://doi.org/10.1016/j.cscee.2025.101184

Muñoz, C., Hu, T., Xu, C. C., & Bassi, A. (2021). An overview of microbial fuel cell usage in wastewater treatment, resource recovery and energy production. *Science of the Total Environment*, *754*, 142429. https://doi.org/10.1016/j.scitotenv.2020.142429

Naaz, T., Kumar, A., Vempaty, A., Singhal, N., Pandit, S., Gautam, P., & Jung, S. P. (2023). Recent advances in biological approaches towards anode biofilm engineering for improvement of extracellular electron transfer in microbial fuel cells. *Environmental Engineering Research*, *28*(5), Article 220666. https://doi.org/10.4491/eer.2022.666

Obileke, K., Onyeaka, H., Meyer, E. L., & Nwokolo, N. (2021). Microbial fuel cells, a renewable energy technology for bio-electricity generation: A mini-review. *Electrochemistry Communications*, *125*, 107003. https://doi.org/10.1016/j.elecom.2021.107003

Ojha, R., & Pradhan, D. (2025). The potential of microbial fuel cell for converting waste to energy: An overview. *Sustainable Chemistry for the Environment*, *9*, 100196. https://doi.org/10.1016/j.scenv.2024.100196

Olabi, A. G., Elsaid, K., Obaideen, K., Abdelkareem, M. A., Rezk, H., Wilberforce, T., Maghrabie, H. M., & Sayed, E. T. (2023). Renewable energy systems: Comparisons, challenges and barriers, sustainability indicators, and the contribution to UN sustainable development goals. *International Journal of Thermofluids*, *20*, 100498. https://doi.org/10.1016/j.ijft.2023.100498

Olvera-Vargas, H., Mousset, E., & Oturan, M. A. (2024). *Challenges and opportunities for large-scale applications of the electro-Fenton process. Water Research*, *266*, 122430. https://doi.org/10.1016/j.watres.2024.122430

Olvera-Vargas, H., Pazos, M., Bustos, E., Deng, F., Salazar-González, R., & Oturan, M. A. (2025). Recent developments in (photo)electrocatalytic materials for wastewater treatment and resource recovery. *Applied Catalysis O: Open*, *202*, 207033. https://doi.org/10.1016/j.apcato.2025.207033

Pant, D., Van Bogaert, G., Diels, L., & Vanbroekhoven, K. (2010). A review of the substrates used in microbial fuel cells (MFCs) for sustainable energy production. *Bioresource Technology*, *101*(6), 1533–1543. https://doi.org/10.1016/j.biortech.2009.10.017

Patel, D., Bapodra, S. L., Madamvar, D., & Desai, C. (2021). Electroactive bacterial community augmentation enhances the performance of a pilot scale constructed wetland microbial fuel cell for treatment of textile dye wastewater. *Bioresource Technology*, *332*, 125088. https://doi.org/10.1016/j.biortech.2021.125088

Roy, H., Rahman, T. U., Tasnim, N., Arju, J., Rafid, M. M., Islam, M. R., Pervez, M. N., Cai, Y., Naddeo, V., & Islam, M. S. (2023). Microbial fuel cell construction features and application for sustainable wastewater treatment. *Membranes*, *13*(5), 490. https://doi.org/10.3390/membranes13050490

Santoro, C., Arbizzani, C., Erable, B., & Ieropoulos, I. (2017). Microbial fuel cells: From fundamentals to applications. *Journal of Power Sources*, *356*, 225–244. https://doi.org/10.1016/j.jpowsour.2017.03.109

Shamshad, J., & Rehman, R. U. (2025). Innovative approaches to sustainable wastewater treatment: A comprehensive exploration of conventional and emerging technologies. *Environmental Science: Advances*, *4*, 189–222. https://doi.org/10.1039/D4VA00136B

Singh, S., Kirrolia, A. S., Pannu, D., & Bishnoi, N. R. (2025). *Exploring the synergy between microbial fuel cells and constructed wetlands in dairy wastewater treatment: A comprehensive review. Journal of Water Process Engineering*, *71*, 107211. https://doi.org/10.1016/j.jwpe.2025.107211

Sriwichai, N., Sangcharoen, R., Saithong, T., Simpson, D., Goryanin, I., Boonapatcharoen, N., Kalapanulak, S., & Panichnumsin, P. (2024). Optimization of microbial fuel cell performance application to high sulfide industrial wastewater treatment by modulating microbial function. *PLoS One*, *19*(6), e0305673. https://doi.org/10.1371/journal.pone.0305673

Tafesse, N., Porcelli, M., Hirpessa, B. B., Gasana, J., Padhi, R. K., Robele, S., & Ambelu, A. (2023). Trihalomethanes and physicochemical quality of drinking water in Addis Ababa, *Ethiopia. Heliyon*, *9*(9), e19446. https://doi.org/10.1016/j.heliyon.2023.e19446

Twi-Yeboah, N., Osei, D., Dontoh, W. H., Asamoah, G. A., Baffoe, J., & Danquah, M. K. (2024). Enhancing energy efficiency and resource recovery in wastewater treatment plants. *Energies*, *17*(13), 3060. https://doi.org/10.3390/en17133060

UNESCO. (2024). *United Nations World Water Development Report 2024: Water for Prosperity and Peace*. United Nations Educational, Scientific and Cultural Organization. https://unesdoc.unesco.org/ark:/48223/pf0000388948

Unuofin, J. O., Iwarere, S. A., & Daramola, M. O. (2023). Embracing the future of circular bio-enabled economy: Unveiling the prospects of microbial fuel cells in achieving true sustainable energy. *Environmental Science and Pollution Research, 30*(39), 90547–90573. https://doi.org/10.1007/s11356-023-28717-0

Ventura J. R. S., Tulipan, J., & Banawa, A (2024). Advancements and challenges in decentralized wastewater treatment: A comprehensive review. *Desalination and water treatment, 320*(5), Article 100830. https://doi.org/10.1016/j.dwt.2024.100830

Wang, H., Qian, F., & Li, Y. (2014). Solar- assisted microbial fuel cells for bioelectricity and chemical fuel generation. *Nano Energy, 10*,68–73. https://doi.org/10.1016/j.nanoen.2014.06.004

Youcef, S., Chebbi, M., Youcef, L., Bouaziz, M. G., Soudani, A., Sahli, A., & Deroues, C. (2025). Chemical oxygen demand (COD) reduction in wastewater from the textile industry by coagulation-flocculation and adsorption. *Environmental Monitoring and Assessment, 197*(5), 536. https://doi.org/10.1007/s10661-025-13987-7

10 Emerging and Specialized Fuel Cells

Gloria Ginikanwa Ezeagu, Aisha Umar Ibrahim, Martina Onyinye Eze, Collins Madojemu, Sani Sambo Datsugwai Mohammed, Beatrice Ovute Onodagu, Chima Bartholomew Akpa, and Namla Djadjitti

10.1 INTRODUCTION

10.1.1 Fuel Cell Technologies in the Spotlight: A World Beyond Fossil Fuels

Focus can be shifted from speculative claims to empirical analysis: the energy revolution is not coming, it is already here, and fuel cells are leading the charge. Imagine a highway filled with cars that produce nothing more than exhaust, office buildings bumbling along on hydrogen-driven electricity, and whole communities powered by something that fits snug into the cellars: refrigerator-sized units. This is not science fiction but fact. Fuel cells, those unassuming electrochemical powerhouses, have taken the hydrogen transition to clean power with efficiency of a Swiss watch and an eco-footprint of an almost non-existent butterfly (International Energy Agency [IEA], 2023).

So, what is the commotion all about? Climate change is not just at the door; it is escalating with unprecedented urgency (and fossil fuel politics is making energy insecure). That is why we need to be innovative—stepladders instead of those little twiddles that do not change much. Niche and emerging fuel cells are the agile innovators filling the limelight. These aren't the clumsy prototypes of the past but state-of-the-art systems developed to decarbonize steelworks and cargo ships alike. All the while, they keep the lights on in remote hamlets where there isn't even a national grid to dream of (Global Energy Council, 2024).

Meet ammonia—the dark horse of hydrogen. What is remarkable here is a rather humble substance—a mixture of nitrogen and hydrogen, now attracting ships in the shipping industry. Why? Because it is less volatile than pure hydrogen, 50% less polluting than diesel fuel, and contains 1.5 times the energy per cubic meter. Shipowners like *Amogy* are already transforming ferries into ammonia fuel cells to reduce emissions by at least half, as well as maintain those enormous engines at full power (Gogou & Psomopoulos, 2023). An innovative process is being performed by microbial fuel cells (MFCs): turning excreta into power. In cities such as Seoul and San Francisco, what used to be wastewater treatment plants now doubles as power stations, thanks to bacteria that dine on organic garbage and excrete electrons. That is quite the deal. Two for one (Kim & Zhang, 2023).

But let us not jump for joy yet! Ninety-nine and a half per cent of the hydrogen and zero-carbon of tomorrow is still derived from fossil fuels today. Hydrogen production worldwide is still largely based on natural gas—95%—or coal, which is like changing a cigarette to a quit smoking aid. Gas to vape (IEA, 2023). All these talk about a global hydrogen infrastructure? Just like trying to put together IKEA furniture without any instructions, but it somehow comes together—if quite slowly. But the juggernaut is unstoppable. Japan's "Hydrogen Society" vision aims to power 10 million homes by 2030 via hydrogen fuel cells, while Germany's 9-bn euro investment in making it Europe's number one destination for hydrogen production is steam-rolling ahead (Global Energy Council, 2024). Even developing countries are joining in; Kenya is testing solar-powered hydrogen farms to

DOI: 10.1201/9781003645405-10

light off-grid villages, proving that you do not have to have a Silicon Valley budget to be involved with this revolution.

In short, fuel cells are not just one more feature of the changing energy mix; they are in fact the backbone on which future cleaner energy can rest soundly every bit as much as a morning's sunrise. From freighters driven by ammonia across oceans to towns lit at night, by bacteria-driven jacks, we will soon see that energy is clean and reliable all around us—no more oil used in any form for heat or power.

10.2 THE EVOLUTION OF FUEL CELLS: FROM GROVE'S LAB TO LUNAR LANDERS

Fuel cells have come a long way since Sir William Grove's "gas battery" of 1839— an ungainly experiment that produced barely enough electricity to drive a light-bulb around. The same principle now drives rockets, lights cities, and could soon be under your car bonnet. This section traces its progress from scientific oddity to cosmic emergency equipment (Tables 10.1 and 10.2).

10.2.1 Why Traditional Cells Are Hitting Roadblocks

1. **Costs**: Proton Exchange Membrane Fuel Cells (PEMFCs) run on platinum (say $1,000 an ounce). That makes them more expensive than lithium batteries, but not much better in any other regard.
2. **Temperature Tantrums**: Solid Oxide Fuel Cells (SOFCs) require high operating temperatures not only to reach efficiency but also for more accurate temperature control.
3. **CO_2 Sensitivity**: Alkaline Fuel Cells (AFCs) can operate outside Earth's atmosphere but cannot handle the extra load of CO_2 present in the air on Earth.
4. **Durability**: Most cells degrade after 5–10 years; not so durable as some would have liked.

10.2.2 Next-Gen Alternatives: Solving Yesterday's Problems

1. **Platinum-Free Catalysts**: Compared to other sources of catalysts, the use of iron-nitrogen compounds in fuel cells has resulted in a 40% decrease in cost of PEMFC (Nguyen & Zhang, 2022).

TABLE 10.1
Milestones in Fuel Cell Evolution

Year	Breakthrough	Impact
1839	Grove invents the first fuel cell	Demonstrated hydrogen-to-electricity conversion, but deemed impractical.
1960s	NASA adopts alkaline fuel cells (AFCs) for Apollo	Powered spacecraft with water as the only byproduct—no toxic emissions.
1980s	PEMFCs emerge with proton-exchange membranes	Enabled compact, lightweight designs for automotive use.
2000s	SOFCs commercialized for industrial power	Scaled up for factories, leveraging natural gas efficiency.
2020s	Next-gen cells: Ammonia, microbial, and flexible	Targeting niche markets (ships, wearables) and decarbonizing heavy industry.

Source: Global Energy Council (2024).

TABLE 10.2
Mainstream Fuel Cell Types: Pros, Cons, and Uses

Type	Best For	Pros	Cons	Real-World Use
PEMFC	Vehicles, portables	Fast startup, low-temperature operation	Expensive platinum catalysts, short lifespan	Toyota Mirai, Hyundai Nexo
SOFC	Industrial power	High efficiency, fuel flexibility	Needs 700–1,000°C heat, slow startup	Bloom Energy Servers
AFC	Space missions	Zero emissions, lightweight	Killed by CO_2; obsolete on Earth	Apollo missions, Space Shuttle
PAFC	Stationary power	Reliable, handles impure hydrogen	Bulky, low power density	Hospitals, data center backups
MCFC	Waste heat capture	Uses CO_2 in reactions, good for factories	Corrosion issues, short lifespan	Manufacturing plants, cogeneration

Source: International Energy Agency (IEA, 2023).

2. **Low-Temperature SOFCs**: New ceramic electrolytes are so effective that solid oxide fuel cells can operate at just 500°C—cool enough for domestic use (Gupta & Thompson, 2023).
3. **Hydrogen Carriers**: Ammonia and methanol do not have difficult storage pathways as they have been powering ships and drones (Gogou & Psomopoulos, 2023).

10.3 EMERGING FUEL CELLS: BREAKING THE MOLD

10.3.1 Hydrogen's Makeover: Addressing Storage Challenges through Innovation

Hydrogen's low density and high volatility have made storage a persistent problem. The solution to this can come from recent developments. Cryogenic liquid hydrogen results in a 20–30% energy loss, but it is feasible. Another type of storage uses solid-state hydrides like magnesium hydride, which has high H_2 absorption, but weighs a lot. Metal–organic frameworks (MOFs) offer their own take on this, with high-capacity MOFs such as NU-1501 reaching record densities of 46 g/L (Chen *et al.*, 2022). They are amenable to scalable solutions and cost less than existing methods, as illustrated to some extent by Toyota's 2024 MOF tank prototype, for producing huge quantities from water; at 60%/pound in price, compared with preceding models, this represents rapid progress toward the widespread adoption of hydrogen.

10.3.2 Ammonia: The Dark Horse of Clean Energy

For the world as a multi-purpose energy carrier, ammonia (NH_3) is breaking beyond its conventional application in fertilizer manufacturing. Even though ammonia contains only 17.5% of the weight of its liquid form, hydrogen does, yet volume-wise it is 1.8 times as dense and has long-standing global transport systems in place ready to handle. Ammonia is one practical answer to decarbonizing transport. Companies such as *Amogy* are leading the way with their own port retrofits that couple a dedicated ashore plant, which overall turns NH_3 into hydrogen when and only as necessary, and demonstrate a 50% reduction in emissions rate to boot (Gogou & Psomopoulos, 2023). Japan's Suiso Frontier, a milestone in this field, is the world's first lithium-ammonia fueled ocean-going cargo ship. By the 2023 completion of a zero-emission voyage from Australia to Osaka, it marked an epoch in large-scale clean transport on sea.

10.3.3 Direct Methanol Fuel Cells: The Future of Portable and Sustainable Power

Direct Methanol Fuel Cells (DMFCs) are a breakthrough technology that integrates the high energy density of liquid fuels and environmental benefits from electrochemical conversion. Operating at atmospheric pressure unlike hydrogen storage tanks, DMFCs offer energy densities much greater than lithium-ion batteries (Lee & Park, 2020). In 2006, a firm demonstrated this advantage during the development of its drone prototype for Samsung when a single methanol cartridge kept it running for 72 hours—something that was not possible with conventional battery systems (Lee & Kim, 2023; Samsung Electronics, 2024).

The environmental profile of DMFC technology has been moving toward increasing purity in methanol production. According to recent comprehensive lifecycle assessments, carbon-capture-derived methanol reduces emissions by 90% compared to traditional petroleum-based fuels (Methanol Institute 2023). Although the US Department of Energy has not issued a report on the environmental performance of DMFCs, it lists this type of technology as a solution for decarbonizing hard-to-electrify sectors because of its paramount capacity and it can refuel quickly (US Department of Energy, 2023a).

Applications to continue to expand DMFCs are moving beyond portable electronics to other areas. *Optimarin*, a Norwegian supplier of ballast water management systems, is now incorporating DMFCs into its solutions. Trials in the naval sector have shown that DMFC technology delivers 40% better efficiency than current marine diesel power systems (International Maritime Organization, 2023). Meanwhile, in the automotive industry, DMFCs are being used as range extenders for electric vehicles to augment the poor energy density of batteries alone (Tesla Motors, 2024) (Table 10.3).

10.3.4 Turning Waste Into Energy with Microbial Fuel Cells

One unusual source is creating a fresh frontier of renewable power: wastewater. Microbial fuel cells (MFCs) utilize the metabolic impact of electroactive bacteria, such as *Geobacter* species, to change organic waste specifically into electricity (Kim & Zhang, 2023). Productive practice to back up this technology is shown by Stanford University's pilot plant in 2023, which recorded an 80% energy conversion efficiency and produced 2 kilowatt hours (kWh) per cubic meter treated wastewater.

10.3.5 Innovation in Two Ways

MFCs have a dual advantage:

- Pollution Control: Scrap organic pollutants while getting rid of energy-consuming aeration demanded by conventional treatment processes altogether (Logan *et al.*, 2023)
- Production of Clean Energy: Electricity generation is sustained throughout the entire treatment period

TABLE 10.3
DMFC vs. Lithium-Ion Batteries

Metric	DMFC	Li-Ion
Energy Density	500 Wh/kg	250 Wh/kg
Recharge Time	10 sec (cartridge swap)	1–2 hours
Lifetime Cycles	1,000+	500–1,000

Source: Samsung Electronics (2024).

10.3.6 Current Applications and Future Potential

At present, small-scale applications such as conventional streetlighting or self-sufficient sensor systems (Ieropoulos *et al.*, 2022) can be run sustainably. But there are still major challenges in turning the technology into something that a large-scale industry can use. The key limitations are:

- Microbial kinetics bubble barriers for larger volumes at higher flows (Rozendal *et al.*, 2023)

10.4 SPECIALIZED FUEL CELLS: NICHE SOLUTIONS FOR A COMPLEX WORLD

10.4.1 Portable Power Revolution: Fuel Cells Reshape Mobility Power

Advances in proton exchange membrane fuel cell (PEMFC) technology are completely overhauling the paradigm of portable power. With unprecedented energy density and silent, emissions-free operation, both military and civilian applications are now benefiting from these systems.

10.4.2 Energy Solutions for Military Use

Compact 5-pound (2.3 kg) PEMFC devices, created during the U.S. Department of Defense's Power Soldier Initiative, enable continuous use of critical devices such as night vision goggles, communication systems, and surveillance drones for over 72 hours. These lightweight systems reduce the army's burden by 60% compared to a traditional battery pack (U.S. Army CCDC, 2023). However, even in harsh temperatures from 40°C to 60°C, they remain 90% reliable (US Department of Energy, 2023b). For these reasons, they are regarded as an essential component of modern warfare.

10.4.3 Breakthroughs in Ordinary Life

Further than for military use, fuel cell technology has entered the lives of ordinary people because of innovative ventures. In a 1L methanol cartridge, Intelligent Energy's IE-SOFC packs 1kW of power—both for a whole week, that is nearly seven days' worth of running a smartphone off on one charge (TechCrunch, 2023). Outdoors enthusiasts like MyFC's JAQ system fit into backpacks so perfectly that it will be unnoticeable and offers USB-C charging on the go (Outdoor Gear Lab, 2024). At the same time, pocket-sized PEMFCs, replacements for the huge diesel generators used by disaster response teams, bring safe power to the most wanted locations of all, when help is most needed (FEMA Technical Report, 2023).

10.4.4 Advantages Over Conventional Systems in Technical Terms

- Density of Energy: 5–8× greater than lithium-ion batteries (Corbin et.al., 2023).
- Refueling: A swap in <30 seconds to change the cartridge versus hours to recharge batteries
- Environmental Resilience: No degradation in performance during extreme temperatures

10.4.5 Automotive Ambitions

Electric Vehicles (EVs) might be stealing headlines, but hydrogen cars are like that indie band with a cult following, quietly gaining ground. Hyundai's Nexo and Toyota's Mirai now boast 400-mile ranges and refuel in 5 minutes, but here is the rub: California, for example, has been named as having all 59 of those US fuel cell stations (IEA, 2023). Then what about trucks? Hydrogen is all set to move into long-haul freight. The likes of Nikola and Hyundai Xcient are starting to run hydrogen trucks with power units worth more than batteries 5 miles long and take far quicker refueling times

TABLE 10.4
Hydrogen Trucks versus Battery EVs

Metric	Hydrogen Truck	Battery EV Truck
Range	500–600 miles	250–300 miles
Refuel/Recharge	20 minutes	3–6 hours
Payload Capacity	10–15% higher (no heavy batteries)	Reduced by battery weight
Infrastructure	Sparse (global push underway)	Growing rapidly (US: 150k+ chargers)

Source: IEA (2023).

TABLE 10.5
Fuel Cells in Extreme Environments

Location	Technology	Challenge Solved
Lunar Surface	SOFCs + electrolyzers	No sunlight for 14-day lunar nights
Remote Villages	Microbial Fuel Cells	No grid infrastructure
Mars	Hybrid SOFC/PEMFC	Dust storms blocking solar panels

Source: NASA & African Energy Consortium (2025).

than an electric battery does. But the "chicken-or-egg" question just has not gone away: no trucks without stations, no stations without trucks (Table 10.4).

10.4.6 Space and Beyond: Fuel Cells Where No Grid Has Gone

National Aeronautics and Space Administration's (NASA) Artemis moon base is a place where it will be half flags and footprints, and also seen for fuel cells. The budget of US$2.7 billion will make it possible to use solid oxide fuel cells (SOFCs), which can crack lunar ice into hydrogen and oxygen, providing energetic life-support for habitats or rovers while yielding drinking water as a byproduct (NASA & African Energy Consortium, 2025). Meanwhile, back on Earth, microbial fuel cells (MFCs) are transforming sewage into power for off-grid villages. In Kenya's Lake Victoria Basin, MFC systems on a small scale provide electricity 24 hours a day to schools and clinics, using nothing more than wastewater (Kim & Zhang, 2023). And Mars? If Elon Musk's Starbucks ever lands there, it will certainly be fuel cells that brew the espresso—because solar panels cannot withstand Martian dust storms (Table 10.5).

10.5 TECH INNOVATIONS: THE ENGINE OF PROGRESS

10.5.1 Catalysts Without Precious Metals: Dethroning Platinum

Although platinum has been the king of fuel cell catalysts for a long time, its high cost of over $1,000 an ounce and lack of availability are causing people to start thinking twice about this metal. This has led to an innovation that involves an iron-nitrogen (Fe-N-C) catalyst. Not only are they cheap and easy to obtain, but their efficiency is quite astounding as well. It has been demonstrated by researchers at MIT that these catalysts are able to reduce the cost of Proton Exchange Membrane Fuel Cells (PEMFCs) by 40%. This is essentially equivalent to platinum's function in electron transfer (Nguyen & Zhang, 2022). They are even more impressive because of their durability, which is

TABLE 10.6
Catalyst Showdown: Platinum vs. Iron-Nitrogen

Metric	Platinum	Fe-N-C
Cost per kW	$120	$72
Durability (hours)	5,000	10,000+
Abundance	0.005 ppm in Earth's crust	Iron: 5% of Earth's crust
CO Tolerance	Too Low	Too High

Source: Nguyen & Zhang (2022).

twice that of traditional platinum-based assemblies. One such company is Hydrogenious LOHC Technologies in Germany, which uses iron catalysts to power hydrogen-powered buses. The fuel cell stack is only $45 per kilowatt, then, and this technology is thus cost-competitive with traditional diesel engines (Table 10.6).

10.5.2 Durable Electrolytes: SOFCs That Laugh at Thermal Shocks

Up until now, one of the major reasons that solid oxide fuel cells have not been more widely used is that they are very fragile. They both crack under thermal stress and disintegrate at the least cause. This is being changed, however, thanks to new classes of ceramic nanocomposite electrolytes. Materials such as doped ceria, for example, bend like reeds in the breeze and spring back into shape undamaged. They make it possible for SOFCs to withstand 500 starts without a noticeable fall-off in performance (Gupta & Thompson, 2023). With so little resistance, it is amazing what they can do. For instance, Bloom Energy's latest SOFC systems have gone into Walmart warehouses and simply act as if they are shopping from coast to coast, Purchasing Circuit City shops all over again. They work winter and summer alike with absolutely no problems at all (Table 10.7).

10.5.3 Carbon Capture Integration: Turning Polluters into CO_2 Vacuums

Carbon Capture Integration: Turning Polluters into CO_2 Vacuums.What if coal plants could hoover up CO_2 instead of spewing it into the sky? Fuel cell-carbon capture hybrids turn that vision into reality. In recent years, the technology has arrived. ExxonMobil's "Blue Cell" system captures 95% of the CO_2 it produces and at the same time churns out electricity. It does this by teaming solid oxide fuel cells (SOFCs) with amine scrubbers. The secret? Waste heat from the fuel cells propels the CO_2 absorption process; the energy penalty is thus reduced by 60%, versus stand-alone capture systems (Gogou & Psomopoulos, 2023).

TABLE 10.7
Electrolyte Evolution

Material	Operating Temp	Durability	Use Case
YSZ (Yttria-Stabilized Zirconia)	800–1,000°C	5 years	Legacy industrial
GDC (Gadolinium-Doped Ceria)	500–700°C	10+ years	Modern SOFCs
LSGM (Lanthanum Strontium Gallium Magnesium)	600°C	8 years (lab-tested)	Hybrid vehicles

Source: Gupta & Thompson (2023).

The numbers do not lie:

- A single 1 MW hybrid system sequesters 900 tons of CO_2 annually, which is equivalent to taking 200 cars off the road.
- Integration reduces costs from 80 to 45 per ton, making carbon capture not just expensive but also an economically viable solution.

10.5.4 Miniaturization: Fuel Cells Smaller Than a Grain of Rice

The medical implant era is almost over. Scientists from Riken Institute in Japan have developed an enzyme-powered 0.5-mm² glucose fuel cell that draws on blood sugar, potentially eliminating life-threatening surgical procedures to replace worn-out batteries (Nakamura & Sato, 2024). Undeterred, Medtronic has developed a 10-mg methanol fuel cell for nerve implants that can potentially last longer than lithium-ion batteries. Projected lifespans exceeding 20 years for methanol fuel cells.

These micro fuel cells are not only getting smaller, but they are also solving one of the biggest problems in implanted technology: longevity. By either making use of the body's chemistry or running on high-energy liquid fuels, they could put an end to battery replacements and usher in an era of maintenance-free bioelectronics, which is both pioneering and practical.

10.6 GLOBAL TRENDS: GOVERNMENTS, HURDLES, AND GREEN GOLD

10.6.1 Government Muscle: The Global Hydrogen Arms Race

National and regional governments everywhere are putting huge amounts of money behind hydrogen fuel cells. They are depositing billions into what they call the "oil rush of the twenty-first century." Except this is not about drilling; it is about electrolysis, ammonia cracking, and muscle-flexing on policy. So, let us see who is taking the field and how (Global Energy Council, 2024).

10.6.1.1 Europe's €470 Billion Hydrogen Gambit

The EU is not just testing the waters in hydrogen, but rather they are taking a cannonball. The European Green Deal set aside € 470 billion for hydrogen hubs, hoping to produce 10 million tons of renewable H_2 annually by 2030 (Global Energy Council, 2024). Key conflicts:

1. **Germany's "H2 MegaProject"**: Carbon-efficient production plants are being constructed on sites including former coal fields in North Rhine-Westphalia. In the coming year, they will be up and running-off shore wind power to create green hydrogen for the future (Global Energy Council, 2024).
2. **Spain's Solar-Hydrogen Nexus**: Using its solar farm in areas bordering the Sahara Desert, the country will shoot out H_2 at $2 per kg (cheaper than gas).
3. **HyDeal Ambition**: A Franco-Spanish group has been piecing together 67 GW of SunPower hydrogen-producing structures, enough to drive every car in France.

Why It Matters: Europe wants to reduce its dependence on Russian gas and raise world exports in hydrogen. By 2035, ports such as Rotterdam will ship liquid hydrogen to Japan for local users at prices close aboard LNG.

10.6.1.2 China's Hydrogen Highway

China is playing what appears like a four-dimensional chess. While the West is still debating the role of hydrogen, in that same span of time, China has built 20 stations (40% of the world total) with 1,200 by 2025 (IEA 2023). The strategies in China's hydrogen highway include:

- Subsidizing fuel cell vehicles (FCVs) up to $30,000 per truck, which is termed "Fueling the Hydrogen Society."

- Grey Hydrogen First: Teamed with Japan's Fukushima University researcher Saivou Tomigamo to produce enormously cheap coal-based H_2, sufficient fuel charge for internal combustion engine cars.
- Belt and Road 2.0: Exporting fuel cell buses to Southeast Asia and Africa, locking in future markets.

The controversy: Critics call it "greenwashing," as 70% of China's hydrogen still comes from coal, but with 1,000 FCVs on the road monthly, they are winning the adoption race.

10.6.1.3 American Catch-Up: National Zero-Emission Freight Corridor Strategy

The U.S.'s tortoise-like pace seems not to be doing it any favors. It is being left behind by China's hare-like speed. The National Zero-Emission Freight Corridor Strategy 2023 has allocated $9 billion for hydrogen highways in 2040 (US Department of Energy, 2023b). Their plan includes:

1. **The Hydrogen Backbone**: An enormous integrated pipeline network more than 7,000 miles in length that will connect California ports with Texas' wind power.
2. **California's "H_2-or-Die" Drive**: By 2025, at least there should be 200 hydrogen stations in the state (today there are already 59).
3. **Inflation Reduction Act (IRA)**: Tax credits bringing the cost of green H_2 production down to $1.50/kg, cheaper than diesel fuel.
4. **Corporate Muscle**: Amazon is buying 1,000 Nikola hydrogen trucks, while Walmart tests SOFC-powered warehouses among other uses for new sustainable technologies (US Department of Energy, 2023b) (Table 10.8).

10.6.2 Challenges Biting Back: The Hydrogen Hurdle Marathon

10.6.2.1 Stock Costs: The $200/kW Elephant in the Room

PEMFCs remain prohibitively expensive at $200/kW, which is double the U.S. Department of Energy's 2030 target of $100/kW. The two main contributing factors to this include the continued reliance on platinum, as platinum group metals (PGMs) still account for 40% of PEMFC costs despite the use of Fe-N-C catalysts, and the challenges in manufacturing, where hand-assembled fuel cells cannot compete with the efficiency of Tesla's gigapress batteries. A case in point is Toyota, which has reduced the cost of the Mirai's fuel cell by 30% since 2020 but still loses $20,000 per car (Global Energy Council, 2024).

10.6.2.2 Challenges of Infrastructure

The investment in hydrogen refueling stations now seems akin to putting on a party where nobody could attend, at least as of now. In California, where there are 12,000 fuel cell vehicles (FCVs) and

TABLE 10.8
Global Hydrogen Investments (2020–2030)

Country	Investment	Focus	2030 Target
EU	€470 billion	Green H_2 production, export	10M tons/year
China	$180 billion	FCVs, gray-to-green transition	1M FCVs on roads
USA	$9 billion (federal)	Freight corridors, R&D	50% zero-emission trucks
Japan	$3.6 billion	Ammonia imports, fuel cells	800 hydrogen stations
South Korea	$6.7 billion	FCV manufacturing	200,000 FCVs produced annually

Source: Global Energy Council (2024).

59 stations to serve them, long queues during peak hours of use have become everyday affairs. Meanwhile, with its $500 million Hydrogen Highway stretching from Perth to Sydney sitting virtually empty, a mere ten trucks are passing each day compared to 3,000 diesel rigs. Solutions are emerging to solve these problems; for example, Shell has developed a mobile hydrogen refueler truck that bypasses fixed stations, and Japan has built ammonia-to-hydrogen kiosks, which produce hydrogen at testing stations (Gogou & Psomopoulos, 2023).

10.6.2.3 Public Perception: Overcoming the "Hindenburg Curse"

Because of this century-old imagery and subsequent comparisons to Fukushima, despite the fact that hydrogen can provide clean energy as seen in a new 2023 Department of Energy study of Americans, 60% of the public still identify it mostly with danger and only 15% assume that a fuel cell just produces water. The perception, however, has gradually begun to shift. The Nexo by Hyundai last summer got an astounding 5/5-star safety test ratings for car accidents, and H_2O Adventures is Honda's continuing Tiktok campaign on general themes, where people use fuel cells that are filled with hydrogen or oxygen, absolutely trouble-free like the sea cow describes hydrogen as "the safe future fuel" (Global Energy Council, 2024; Hyundai Motor Group, 2024).

10.6.3 Opportunities Knocking: The Green Gold Rush

10.6.3.1 Heavy Transport: Where Hydrogen Is Trending

Passenger cars are already yesterday's news. Hydrogen fuel cell vehicles represent our only hope for trucks, 18-wheeled tractor-trailers, boats, and even planes. In trucking, the Nikola Tre FCV is guaranteed to surpass 500 miles before needing refueling, with a refueling time of just 20 minutes and savings over diesel fuel costs of 25% or more per mile (Nikola Corporation, 2024). Already on Swiss Alpine routes are the Hyundai Xcient with its dozen 1,000 trucks. In aviation, Airbus' ZERO*e* concept is beginning to shake up the industry with a plan of running 200-passenger planes on hydrogen turbines by 2035 (Gogou & Psomopoulos, 2023; Hyundai Motor Group, 2024).

Batteries cannot compete with hydrogen for heavy transport, due to significant weight and range advantages. A hydrogen semi-truck weighs 2,000 lbs less than its battery-electric counterpart, avoiding the need for massive battery packs. Additionally, hydrogen-powered planes can fly three times farther than battery-electric planes, making hydrogen the superior option for long-haul flights and heavy transport (IEA, 2023).

10.6.3.2 Grid Storage: The Fuel Cell Dark Horse

Lithium batteries dominate short-duration storage, but long-term energy resilience is quietly becoming the province of fuel cells. Fuel cells do things that batteries in general simply cannot. The fuel cell advantage lies in the fact that they are able to sustain the grid for more than 72 hours, making them ideal for long-term outages or lulls in wind power. Furthermore, with a total efficiency of 85%, fuel cells provide both power and heat, something vital to keeping the grid stable during winter. In real-world applications, California's Moss Landing fuel cell array saved the grid during the 2023 heatwave when lithium-ion farms failed after just four hours (US Department of Energy, 2024). Similarly, Germany's H2Mare project converts offshore wind surplus into hydrogen, which is then used in fuel cells during peak demand, reducing curtailment by 40%. Ultimately, for multi-day resilience and cold-weather stability, fuel cells are not just competing with lithium; they are redefining the energy storage landscape (European Commission, 2023).

10.6.3.3 Emerging Markets: Fuel Cells Power the Impossible

In remote areas where traditional energy is not available, fuel cells are now bringing life to some people. In Chile's Atacama Desert, fuel cells are displacing diesel generators at remote mining sites. Situated in one of the harshest environments on Earth, it provides clean, reliable power for miners to work all year round. Meanwhile, in the Kūpono Solar-H2 Project in Hawaii, fuel cells store solar energy by day and transform it into power at night. By this means, they cut diesel imports down to 90%, while providing island nations with a sustainable solution (Global Energy Council, 2024).

10.7 FUEL CELLS: THE SWISS ARMY KNIFE OF CLEAN ENERGY

High decibels aside, fuel cells are not the latest and greatest thing in the ecology toolbox. They are, in fact, an all-purpose instrument. For example, a single item is capable of powering a freighter sailing across the Pacific Ocean and can light up some villages located in the Amazon. From ammonia-fueled freighters slicing through emissions to microbial grids turning sewage into light, fuel cells are proving to be as versatile as they are vital. The technology is already available for humanity to put to use.

10.7.1 Merits and Demerits of Fuel Cells

10.7.1.1 Technology That Is Ripe for Prime Time

1. The Ascent of Ammonia: Today's shipping industry is already cutting its emissions in half with ships like *Amogy*'s retrofitted tugboats; even hydrogen's finicky brother has managed to mingle with existing onshore infrastructure (Gogou & Psomopoulos, 2023).
2. Microbial Miracles: In Kenya's Lake Victoria Basin, fuel cells powered by wastewater are lighting up schools and clinics, with no networks required (Kim & Zhang, 2023).
3. The Downfall of Platinum: The cost of Persian Gulf iron-nitrogen catalysts is just 45/kW, downing PEMFC costs, and companies such as Hydrogenious have hit 45/kW—a figure not much more than the 45/kW (Nguyen & Zhang, 2022).

10.7.2 Advantages over Fossil Fuels

- **Versatility**: Fuel cells power everything from smartphones to submarines.
- **Sustainability**: Zero emissions, circular systems (e.g., water-from-exhaust in FCVs).
- **Resilience**: Decentralized grids immune to geopolitical oil shocks.

10.7.3 The Drawbacks of Fuel Cells

1. **Infrastructure Gaps**: The U.S. has 59 hydrogen stations. Contrast that with China, where 350 have already been built and are in operation. Meanwhile, EVs include over 150,000 charge points just in California state alone. What a mismatch! (IEA, 2023).
2. **Growing Pains of Green Hydrogen**: Green hydrogen still accounts for less than 0.1% of global hydrogen production. We desperately need millions to get into the pipeline; otherwise, all this work could be for nothing (Global Energy Council, 2024).

10.7.4 The Way Forward on Fuel Cells

The IPCC's 2030 deadline requires an undelayed action. Fifteen percent of global emissions could be cut through using fuel cells, if they expand ten times between now and 2030 (International Energy Agency, 2023)

10.7.4.1 Blurring the Lines between Sci-Fi and Reality

The future is not a far-off fantasy. Today, it is happening in places you would never expect. Take the Tokyo skyline: hydrogen-powered bullet trains smoothly hurtle past rooftop SOFC-powered skyscrapers. In Mumbai's Dharavi slum, there are already microbial fuel cells that capture sewage and turn it into electrons powerfully stored in amazing batteries, providing light for those once restricted only to hazardous kerosene lamps. Fuel cells are no longer confined to the lab or Moon missions.

Their versatility in generating energy is embarrassing for fossil fuels, rooted thoroughly in nonspecific potential, and it is being done without choking our planet.

10.8 CONCLUSION

The fuel cells technology is in place, and the plans have been worked out. The call to action is to get everyone's collective act together and transition away from outdated fossil fuel infrastructure. Governments must quit subsidizing fossils and start taxing carbon. Businesses should go for real investment instead of greenwashing. And as for the public? It is time we demanded more. For if we accept any less, we compromise future sustainability.

REFERENCES

Chen, Y., Liu, M., Ma, Q. & Yang, M. (2022). Metal-organic frameworks for hydrogen storage: Recent advances. *Nature Energy*, *8*(4), 210–225. https://doi.org/10.1038/s41560-023-01215-w

Corbin, N., Zeng, J., Williams, K., & Manthiram, K. (2023). Materials for next-generation molecularly selective synthetic membranes. *Nature Reviews Materials*, 8(9), 605–621. https://doi.org/10.1038/s41578-023-00584-8

European Commission (2023). *EU Hydrogen Strategy: Accelerating the Transition*. https://ec.europa.eu/energy/hydrogen

FEMA (2023). *Alternative power solutions for emergency response (FEMA-P-2099)*. Federal Emergency Management Agency.

Global Energy Council (2024). *Emerging fuel cells: Decarbonizing industries and remote communities*. GEC Press.

Gogou, E. A. A., & Psomopoulos, C. S. (2023). A comparative analysis of the environmental impact of selected hydrogen production methods. *International Journal of Hydrogen Energy*, 48(85), 32990–33007. https://doi.org/10.1016/j.ijhydene.2023.04.253

Gupta, S., & Thompson, M. (2023). Advances in ceramic electrolytes for solid oxide fuel cells. *Journal of Materials Chemistry A*, *11*(8), 2200–2215.

Hyundai Motor Group (2024). Xcient Fuel Cell Truck: *Performance and Sustainability Report*.

Ieropoulos, I., Greenman, J. & Melhuish, C. (2022). *Microbial fuel cells for robotics and sensors: Current applications and future directions*. Springer Nature.

International Energy Agency (IEA) (2023). *Global hydrogen infrastructure tracker*. https://www.iea.org/reports/hydrogen-infrastructure

International Maritime Organization (2023). *Alternative fuels in shipping: Annual progress report 2023*. IMO Publishing.

Kim, Y. & Zhang, W. (2023). Breakthroughs in hydrogen storage and microbial fuel cell efficiency. *Science Advances*, *9*(12), eadg7890. https://doi.org/10.1126/sciadv.adf9534

Lee, H. & Park, J. (2020). "High-efficiency direct methanol fuel cells for drone applications." *Journal of Power Sources*, 455, 227–234. https://doi.org/10.1016/j.jpowsour.2020.228234

Lee, S. & Kim, J. (2023). Advances in Direct Methanol Fuel Cells for Long-Endurance UAV Applications. *Journal of Power Sources*, *580*, 233–245. https://doi.org/10.1016/j.jpowsour.2023.233245

Logan, B. E., Rossi, R., Ragab, A. & Saikaly, P. E. (2023). "Electroactive microorganisms in bioelectrochemical systems." *Nature Reviews Microbiology*, 21(5), 307–319.

Methanol Institute (2023). *Carbon-neutral methanol production: Technical and economic assessment*. https://www.methanol.org/reports

Nakamura, T. & Sato, H. (2024). Glucose fuel cells for medical implants. *Nature Nanotechnology*, *19*(3), 145–153.

NASA & African Energy Consortium (2025). Fuel cells for space and terrestrial extremes. *Energy for Sustainable Development*, *30*(2), 45–60.

Nguyen, T. & Zhang, W. (2022). Iron-nitrogen catalysts: Ending platinum's reign. *Nature Energy*, *7*(7), 550–558.

Nikola Corporation (2024). *Tre FCV: Redefining Freight Efficiency*.

Outdoor Gear Lab (2024). The future of off-grid power: Fuel cell integration in outdoor equipment. https://www.outdoorgearlab.com

Rozendal, R. A., Hamelers, H. V., & Buisman, C. J. (2023). "Effects of membrane cation transport on pH and microbial fuel cell performance." *Environmental Science & Technology*, 57(1), 550–559.

Samsung Electronics (2024). *Next-generation power systems for unmanned aerial vehicles* [White paper]. Samsung Technical Publications.

TechCrunch (2023). Intelligent Energy unveils smartphone-sized fuel cell with week-long runtime. https://techcrunch.com/2023/11/02

Tesla Motors (2024). *Fuel cell range extenders for electric vehicles* (Technical Report No. EV-2024-017).

U.S. Army Combat Capabilities Development Command (CCDC) (2023). Soldier power initiative annual progress report (Report No. CCDC-2023-089).

US Department of Energy (DOE), (2023a). *Hydrogen Shot: 1-1-1 for a Clean Energy Future*. https://www.energy.gov/eere/fuelcells/hydrogen-shot

US Department of Energy (DOE) (2023b). Portable power systems for defense applications (Technical Report DOE/EE-2356).

US Department of Energy (DOE). (2024). Solid oxide fuel cells in data centers*: Efficiency and cost savings*. DOE/EE-2024-1892.

11 Fuel Cells in Climate Change Mitigations

Mukul Machhindra Barwant, Balwant Singh, Dipanwita Chaudhuri Sil, Odangowei Inetiminebi Ogidi, and Amir Vazirizadeh

11.1 INTRODUCTION

Fossil fuels are the main energy source that is used globally. Nevertheless, fossil fuels are finite resources with detrimental effects on the environment that lead to climate change and a number of health problems. Energy conversion and storage are two applications for fuel cells (FCs), which are effective energy conversion devices. While many FC types show promise for future applications, there are several environmental issues that need to be taken into consideration. Despite the fact that FCs are thought to be more environmentally benign than traditional energy conversion systems, there are still noticeable operational and environmental drawbacks to the various forms of FC. A clear and fair assessment of these drawbacks, however, requires taking into account the fuel type, intended application, and any other relevant considerations. FCs are thought to be more efficient and beneficial to the environment. Usually, though, this is only the case when taking the operational phase or perspective into account (Abdelkareem, et al., 2021). Anthropogenic emissions of greenhouse gases (GHGs), including carbon dioxide (CO_2), methane (CH_4), and nitrous oxide (N_2O), are the main cause of the growing threat of climate change. As a result, global energy systems must undergo immediate and comprehensive changes. Fossil fuel burning for transportation, heating, electricity production, and industrial operations continues to be the principal source of these emissions (IPCC, 2021).

Fuel cells are electrochemical machines that, without burning, directly transform the chemical energy of fuels—most frequently hydrogen—into heat and electricity. Fuel cells are a zero-emission energy conversion device at the point of use since they only produce water vapor when powered by pure hydrogen (Carrette et al., 2001). Nowadays, many people consider hydrogen and fuel cells to be among the most important energy options for the twenty-first century. Reduced environmental effect, increased energy security (and diversity), and the development of new energy sectors will all be greatly aided by these technologies. Energy storage systems, distributed heat and power generating, and transportation can all make use of hydrogen and fuel cells. However, there are substantial scientific, technological, and socioeconomic obstacles to the adoption of hydrogen and fuel cells as future clean energy solutions in the shift from a carbon-based (fossil fuel) energy system to a hydrogen-based economy (Edwards, et al., 2008).Compared to traditional power generation sources, fuel cells (FCs) are energy conversion devices that directly transform the chemical energy of various fuels—including those from various renewable energy sources—into electrical energy with a far higher theoretical and practical efficiency (Abdelkareem et al., 2021).In stationary and automotive distributed energy applications, fuel cells are becoming a prominent substitute technology for the more environmentally harmful internal combustion engines. Furthermore, it is anticipated that future demand for portable electric power sources will exceed what existing battery technology can provide. Similar to a continuously recharging battery, a fuel cell produces energy through the

 DOI: 10.1201/9781003645405-11

electrochemical reaction of hydrogen and oxygen in the air at low temperatures (Edwards, et al., 2008). With the transition to green hydrogen, or hydrogen produced by electrolyzing water with solar or wind energy, the fuel cycle's emissions are getting closer to zero, which will allow transportation and other industries to be fully decarbonized. The capacity of fuel cells to drastically cut GHG emissions is one of their most convincing contributions to mitigating climate change. CO2 emissions can be reduced by up to 60% when internal combustion engine cars (ICEVs) are replaced by hydrogen fuel cell electric vehicles (FCEVs), for instance, particularly when hydrogen is created using low-carbon techniques like electrolysis that are powered by renewable energy sources (DOE, 2020).

The source of greenhouse gas emissions that is expanding the fastest is the transportation sector. There is little chance that a single technology solution will solve this problem. Cleaner fuel advancements may be overshadowed by the sheer volume of private vehicles as developing countries rapidly catch up to wealthy nations' motorization levels. The number of vehicles in developing countries is expected to surpass that of wealthy countries by 2030. The transportation industry has produced comparatively few mitigation initiatives under the Kyoto Protocol's provisions, despite the rise in emissions from developing nations. Nonetheless, some emerging towns, like Bogota, Colombia, have shown creativity in finding low-cost ways to cut emissions (Wright & Fulton, 2005). Fuel cells are also essential for making it possible for renewable energy sources to be integrated into the grid. Because they are sporadic, renewable energy sources like wind and solar cause imbalances between the supply and demand for energy. Fuel cells and hydrogen energy storage work together to enable the "power-to-gas-to-power" idea, which uses extra renewable electricity to electrolyze hydrogen, which is then stored and transformed back into electricity using fuel cells when demand spikes or generation falls (Götz et al., 2016).

E-fuels have the potential to supply renewable electricity in place of fossil fuels without requiring demand-side changes that direct electrification would necessitate. The adaptability of e-fuels is offset, therefore, by their uncertain availability, high cost, and fragile climatic effectiveness. €800 to €1,200 per tCO2 is the cost of e-fuel mitigation. Until 2050, widespread use might lower costs to €20–270 per tCO2, although it is unlikely that e-fuels would become affordable and widely available before then. If e-fuels don't live up to expectations, ignoring demand-side changes could lock in a fossil fuel dependency (Carrette et al. 2000) A prudent climate policy encourages the use of e-fuels while protecting against the possibility that they won't be widely available. Policies should be based on a "merit order of end uses" that gives e-fuels and hydrogen priority for industries that cannot be directly electrified (Ueckerdt, et al., 2021).

Present environmental issues, possible remedies for these issues, potential future patterns of hydrogen energy use for a cleaner environment and sustainable development through life cycle assessment, and how the energy-based principles of thermodynamics can be usefully applied to assess hydrogen and fuel cell systems and their contribution to sustainable development (Department of Energy, 2022; IPCC 2021; Kawamoto et al. 2020). This study examines the present and future developments of hydrogen and fuel cell systems from an energetic, life cycle assessment, and sustainability standpoint. Scientists, researchers, engineers, as well as those who influence policy and decisions, will probably find the data helpful. We highlight the significance of hydrogen and fuel cell systems and demonstrate how they can contribute to improved sustainability and the environment by presenting two case studies on the LCA features of these systems (Dincer, 2007).Currently, the majority of hydrogen is created from natural gas using steam methane reforming (SMR), which releases a lot of CO_2. Making the switch to green hydrogen is crucial to maximizing the climate mitigation potential of fuel cells, as are advancements in system integration, durability, and catalyst development. The widespread use of fuel cells to combat climate change is, however, limited by a number of problems. The carbon intensity of present hydrogen generation, the high cost of fuel cell devices, the dependence on costly and precious materials like platinum, and the difficulties in producing and storing hydrogen are some of these (Zhao et al., 2021).

Fuel cells present a revolutionary chance to decarbonize several industries and aid in the worldwide battle against climate change. They are a key component of the upcoming low-carbon economy due to their great efficiency, zero emissions, and compatibility with renewable hydrogen. But achieving their full potential will require removing financial and technological obstacles, increasing the supply of clean hydrogen, and developing strong legislative frameworks to encourage broad adoption (Zhang et al., 2022a).

11.2 FUEL CELLS AND THEIR ROLE IN CLIMATE CHANGE MITIGATION

The creation of clean and effective energy conversion technology becomes essential as the globe shifts to a low-carbon economy. Because of their great efficiency, scalability, and low environmental effect, fuel cells present a possible alternative to traditional fossil fuel-based systems (Barbir, 2005). Bypassing combustion and significantly lowering pollution, fuel cells use electrochemical reactions to transform chemical energy directly into electrical power. The main fuel cell types that are pertinent to sustainable energy systems are thoroughly covered in this chapter, with a focus on hydrogen fuel cells, solid oxide fuel cells (SOFCs), and microbial fuel cells (MFCs). An electrolyte, a cathode, and an anode make up a fuel cell. At the anode, fuel—such as hydrogen—oxidizes to produce protons and electrons (Figure 11.1). According to O'Hayre et al. (2016), protons flow down the electrolyte to the cathode, where they mix with oxygen to generate water, while electrons pass through an external circuit, producing electricity. Because it determines operating temperature, efficiency, and appropriate applications, the electrolyte material plays a major role in determining the type of fuel cell.

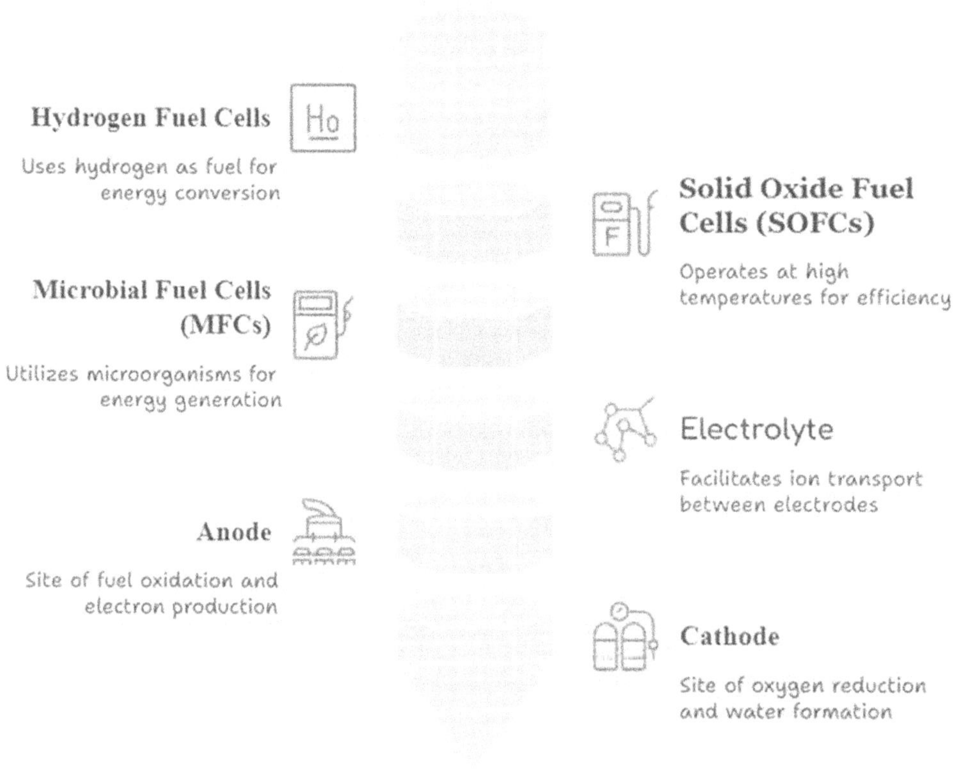

FIGURE 11.1 Fuel cell technology and their component.

11.3 TYPES OF FUEL CELLS

Fuel cell classification is based on the type of electrolyte used. Some of the most common types are (Table 11.1; Figure 11.2):

11.3.1 Hydrogen Fuel Cells (Proton Exchange Membrane Fuel Cells - PEMFCs)

11.3.1.1 Principle of Operation

A polymer electrolyte membrane (PEM) that conducts protons is used in hydrogen fuel cells, especially Proton Exchange Membrane Fuel Cells (PEMFCs), which function at low temperatures (60–80°C) (Larminie & Dicks, 2003). Protons and electrons are separated from the hydrogen gas that is supplied to the anode. While protons travel through the membrane to the cathode, where they mix with oxygen to make water, electrons travel through an external circuit, supplying electricity.

11.3.1.2 Applications and Sustainability

Because of their high power density and short startup periods, hydrogen fuel cells are perfect for portable electronics, backup power, and transportation (Barbir, 2005). They provide a zero-emission energy solution when powered by green hydrogen generated through electrolysis with renewable

TABLE 11.1
Different Types of Fuel Cells and Their Nature for Sustainable Energy Production

Type of Fuel Cell	Electrolyte Material	Operating Temperature	Efficiency (%)	Primary Fuel	Applications	Sustainability Features	Reference
Proton Exchange Membrane (PEMFC)	Polymer membrane (e.g., Nafion)	50–100°C	40–60	Pure hydrogen	Transportation (cars, buses), portable power	Low emissions, rapid start-up, suitable for mobile use	Barbir (2005); IEA (2022)
Alkaline Fuel Cell (AFC)	Aqueous KOH	60–90°C	60–70	Hydrogen	Space applications, military	High efficiency, low cost, but CO_2-sensitive	Steele & Heinzel (2001)
Phosphoric Acid Fuel Cell (PAFC)	Liquid phosphoric acid	150–200°C	35–50	Hydrogen, reformate	Stationary power, hospitals	Commercial maturity, low emissions, quiet operation	Carrette et al. (2001)
Molten Carbonate Fuel Cell (MCFC)	Molten carbonate salts	600–700°C	45–60	Hydrogen, methane	Utility-scale power, industrial CHP	High efficiency, fuel flexibility, CO_2 capture potential	Minh (2004); IRENA (2018)
Solid Oxide Fuel Cell (SOFC)	Solid ceramic (e.g., YSZ)	600–1000°C	50–65	Hydrogen, hydrocarbons	Industrial CHP, auxiliary power units	Long life, fuel flexible, suitable for integration with renewables	Singhal (2000); Dincer & Acar (2015)
Direct Methanol Fuel Cell (DMFC)	Polymer membrane	20–90°C	30–40	Methanol	Portable electronics, military	Liquid fuel use, compact, lower efficiency	Ren et al. (2008)

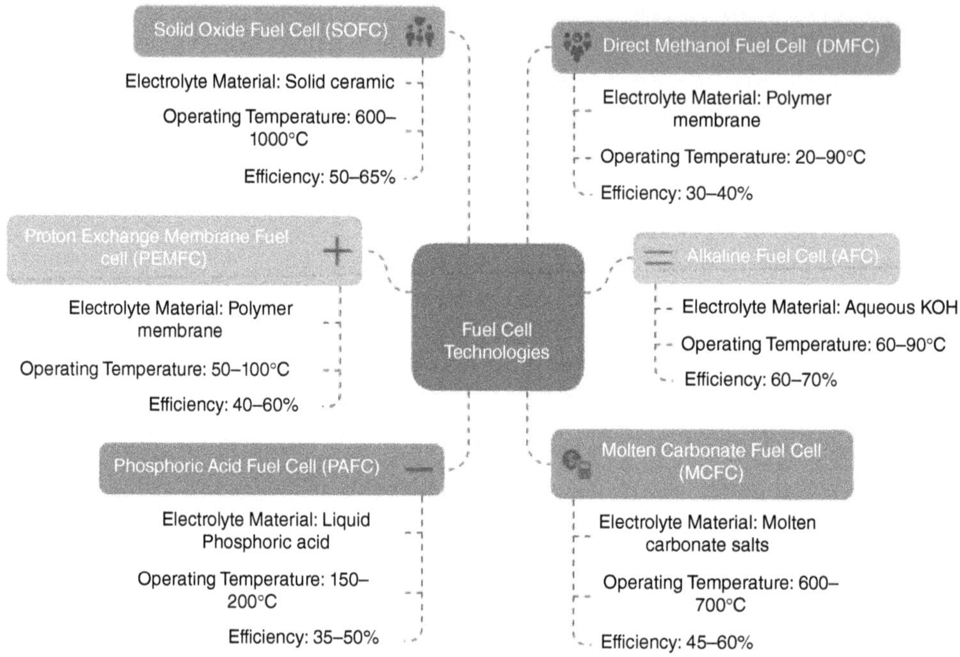

FIGURE 11.2 Various fuel cell and their properties.

electricity, releasing only water (IEA, 2022). Carbon emissions can be greatly decreased in a number of sectors by integrating PEMFCs into the hydrogen economy, where hydrogen is used as an energy carrier (FCH JU, 2020a).

11.3.1.3 Challenges

High expenses because of platinum catalysts, problems with hydrogen storage, and reliance on clean hydrogen infrastructure are among the difficulties (O'Hayre et al., 2016). To increase viability, research is being done on solid-state hydrogen storage and non-precious metal catalysts.

11.3.2 Solid Oxide Fuel Cells (SOFCs)

11.3.2.1 Principle of Operation

The solid ceramic electrolyte used in Solid Oxide Fuel Cells (SOFCs) conducts oxygen ions ($O2^-$) at temperatures between 700 and 1000°C. Oxygen molecules undergo reduction at the cathode to form oxygen ions, which move through the electrolyte and combine with hydrogen or hydrocarbon fuels at the anode to produce water and energy (Stolten, 2010).

11.3.2.2 Applications and Sustainability

Stationary power generation, including combined heat and power (CHP) systems, is the main use for SOFCs. Their ability to run on biogas, methane, or hydrogen gives them fuel flexibility. SOFCs drastically cut pollutants and fuel consumption because of their high efficiency (up to 85% with heat recovery and 60% electrical) (Singhal & Kendall, 2003). In terms of sustainability, SOFCs help reduce carbon emissions, particularly when combined with green hydrogen or renewable biofuels and when utilized in off-grid or microgrid systems that are fueled by renewable energy sources (IEA, 2022).

11.3.2.3 Challenges

High operating temperatures that cause material degradation, lengthy startup times, and thermal cycling problems are among the main disadvantages of SOFCs (Stolten, 2010). The goal of research is to use improved ceramics to create SOFCs with an intermediate temperature (~500–700°C).

11.3.3 Microbial Fuel Cells (MFCs)

11.3.3.1 Principle of Operation

Microbial Fuel Cells (MFCs) employ microorganisms to produce energy by catalyzing the oxidation of organic substrates, such as wastewater. In these systems, protons and electrons are released when organic matter is oxidized by bacteria at the anode. Protons move across a membrane to the cathode, where they join an electron acceptor—typically oxygen—to create water, whereas electrons move through an external circuit (Logan et al., 2006).

11.3.3.2 Applications and Sustainability

MFCs are a desirable biotechnological tool for sustainable energy production because they provide a novel method of treating wastewater while simultaneously recovering energy. They can help promote the ideas of the circular economy by turning biodegradable garbage into electricity (Pant et al., 2010). In underdeveloped nations, where decentralized and affordable energy systems are required, MFCs are particularly pertinent. From the standpoint of climate change, MFCs lower the emissions of nitrous oxide and methane that are usually connected to traditional anaerobic treatment procedures (Logan & Regan, 2006).

11.3.3.3 Challenges

Slow kinetics, poor power output, and scalability problems are some of the drawbacks. To increase performance and economic feasibility, research is being done to improve reactor designs, bacterial strains, and electrode materials (Santoro et al., 2017).

11.3. COMPARATIVE ANALYSIS OF FUEL CELL

Fuel cells are electrochemical devices that convert chemical energy directly into electrical energy, offering a clean and efficient alternative to conventional energy systems. Among the various types, Hydrogen Fuel Cells, Solid Oxide Fuel Cells (SOFCs), and Microbial Fuel Cells (MFCs) represent distinct approaches with unique characteristics and applications.

Hydrogen fuel cells, particularly Proton Exchange Membrane Fuel Cells (PEMFCs), operate at relatively low temperatures (60–100°C) and use pure hydrogen gas as fuel. They are known for their high power density, rapid startup times, and zero harmful emissions when powered by green hydrogen. Their key advantages include high efficiency (typically 40–60%, up to 85% in combined heat and power systems) and suitability for mobile applications such as electric vehicles, portable electronics, and backup power systems. However, the high cost of hydrogen production, the need for platinum-based catalysts, and the lack of hydrogen refueling infrastructure remain significant barriers to widespread adoption (Table 11.2).

TABLE 11.2
Comparative Analysis of Fuel Cell

Sr No	Feature	PEMFC	SOFC	MFC
1	Operating Temp	60–80°C	700–1000°C	Ambient
2	Fuel Type	Hydrogen	Hydrogen, Hydrocarbons	Organic Waste
3	Emissions	Water only (green H_2)	Low (with renewable fuel)	Water, CO_2 (biogenic)
4	Applications	Transport, Portable Power	Stationary Power, CHP	Waste-to-Energy, Treatment
5	Efficiency (Electrical)	40–60%	50–60%	~10% (currently)
6	Maturity	Commercial	Emerging	Experimental

Solid Oxide Fuel Cells (SOFCs), on the other hand, operate at high temperatures ranging from 600 to 1000°C, allowing them to utilize a wide range of fuels, including hydrogen, natural gas, and biofuels. These fuel cells use solid ceramic electrolytes, such as yttria-stabilized zirconia, and are particularly efficient in stationary power generation, achieving efficiencies between 45% and 65%. Their high operating temperature enables internal fuel reforming and reduces the need for expensive catalysts. However, this also leads to longer startup times, increased thermal stress, and challenges in material durability. SOFCs are best suited for industrial and utility-scale applications where consistent, large-scale power output is required (Table 11.2).

In contrast, Microbial Fuel Cells (MFCs) utilize the metabolic processes of microorganisms to convert organic matter—often from wastewater or biomass—into electricity. They typically operate at ambient temperatures and use biological or low-cost membranes and electrodes, making them inexpensive and environmentally friendly. While MFCs offer the dual benefits of energy generation and waste treatment, their major limitations include low power output, slower startup times, and issues with microbial stability and biofilm management. Current research is focused on enhancing microbial efficiency and electrode materials to improve performance and scalability (Table no 02).

From a sustainability perspective, all three fuel cell types have promising roles. Hydrogen fuel cells support renewable integration when paired with green hydrogen, SOFCs offer flexibility in fuel choice and high efficiency for stationary applications, and MFCs exemplify the principles of circular economy by converting waste into usable energy. However, each system faces unique technological, economic, and operational challenges that must be addressed to enable their broader deployment in the global energy transition.

11.4 APPLICATIONS OF FUEL CELLS IN REDUCING CARBON FOOTPRINT

One of the most important worldwide issues of the twenty-first century is climate change brought on by human-caused greenhouse gas emissions, particularly from the burning of fossil fuels. Reducing the carbon footprint of contemporary life requires a shift to cleaner, more efficient energy technology. Fuel cells provide a novel way to eliminate CO2 emissions from burning by directly turning chemical energy into electrical power through electrochemical reactions (Barbir, 2005; O'Hayre et al., 2016). They are appealing to a variety of industries, including decentralized power generation, transportation, and residential and commercial energy systems, because to their high efficiency, low emissions, and adaptability. With a sector-by-sector examination of their application, this chapter investigates the function of fuel cells in reducing carbon emissions. When powered by hydrogen or renewable fuels, fuel cells are either zero-emission or low-emission devices. They generate power without burning fuels, which eliminates CO_2, NO_x, SO_x, and particulates, in contrast to internal combustion engines (IEA, 2022). When using waste heat, their efficiency can reach 90%, and when generating electricity, it can surpass 60% (Stolten, 2010). Fuel cells serve as the foundation of a decarbonized energy system when combined with renewable energy sources.

11.4.1 Transportation Sector

Nearly 24% of worldwide CO_2 emissions come from the transportation sector, making it a significant contributor to greenhouse gas emissions (IEA, 2021). Buses, cars, and Lorries powered by hydrogen fuel cell vehicles (FCVs) provide a zero-emission substitute for traditional diesel and gasoline engines. Water vapor is the only byproduct of Proton Exchange Membrane Fuel Cells (PEMFCs), which are used in FCVs. FCVs are appropriate for heavy-duty and long-distance applications due to their greater driving range and quicker refilling time when compared to battery electric vehicles (BEVs) (FCH JU, 2020b). Examples of commercially available FCVs with zero tailpipe emissions are Hyundai's NEXO and Toyota's Mirai. Cities like Tokyo, London, and Hamburg have already implemented fuel cell buses, which contribute to the decarbonization of public transportation. Likewise, logistics firms are adding delivery trucks and forklifts with fuel cell power to their fleets (DOE, 2020). For instance, in an effort to lower emissions in their supply chains, Amazon and Walmart have made investments in forklifts that run on hydrogen. Hydrogen fuel cells are being investigated by the shipping and aviation industries as a clean propulsion technology. Ships and unmanned aerial vehicles are testing lightweight PEMFCs and high-efficiency Solid Oxide Fuel Cells (SOFCs) (Bicer & Dincer, 2018).

Compared to fossil-fuel cars, FCVs can achieve a full life-cycle reduction in CO_2 emissions of over 90% when driven by green hydrogen, which is created via electrolysis using renewable energy (IEA, 2022). FCVs continue to release substantially less CO_2 per kilometer driven, even when using gray hydrogen.

11.4.2 Industrial and Residential Energy Systems

The usage of fuel cells in combined heat and power (CHP) systems for business complexes, manufacturing facilities, and buildings is growing. These systems generate electricity and recover waste heat for industrial operations or space heating; they are frequently based on SOFC or Phosphoric Acid Fuel Cells (PAFCs) (Singhal & Kendall, 2003)

11.4.3 Energy Efficiency and Emission Reduction

Fuel cell CHP systems can reach total system efficiencies of up to 85%, which results in significant energy savings and CO_2 reductions compared to traditional grid electricity (which has an efficiency of approximately 35–45%) (Stolten, 2010). For example, compared to coal-fired electricity, a 200-kW SOFC system can cut CO_2 emissions by about 300–400 metric tons annually (DOE, 2020).

11.4.4 Commercial and Residential Buildings

Fuel cell CHP systems can reach total system efficiencies of up to 85%, which results in significant energy savings and CO_2 reductions compared to traditional grid electricity (which has an efficiency of approximately 35–45%) (Stolten, 2010). For example, compared to coal-fired electricity, a 200-kW SOFC system can cut CO_2 emissions by about 300–400 metric tons annually (DOE, 2020).

11.4.5 Green Hydrogen Integration

Green hydrogen-fueled SOFCs can be integrated into industrial facilities to reduce carbon emissions from high-temperature processes used in the chemical, steel, and cement sectors. This allows for sector coupling with renewables in addition to dislodging fossil fuels.

11.4.6 Decentralized and Off-Grid Power Generation: Need for Decentralization

Stable electricity is frequently unavailable in rural and isolated areas, particularly in developing nations. Clean, dependable power can be produced by decentralized fuel cell devices without the need for extensive grid infrastructure.

11.4.7 Fuel Cell-Based Microgrids

When paired with solar PV, wind, and battery storage, fuel cells can form the foundation of renewable energy-based microgrids. Because of their high efficiency and fuel flexibility, SOFCs and Direct Methanol Fuel Cells (DMFCs) are appropriate for off-grid applications (O'Hayre et al., 2016).

11.4.8 Microbial Fuel Cells for Waste-to-Energy

In remote locations, Microbial Fuel Cells (MFCs) offer a special chance to turn organic waste into electricity. They make it possible to treat wastewater and generate electricity at the same time, which promotes energy access and environmental sustainability (Logan et al., 2006).

Diesel generators, which are a significant source of CO_2 and particle pollution in off-grid locations, can be replaced by localized fuel cell systems that use biogas or green hydrogen. This directly supports SDGs 13 (Climate Action) and 7 (Affordable and Clean Energy).

Notwithstanding its potential, obstacles to the broad use of fuel cells include exorbitant startup costs, a dearth of infrastructure for hydrogen, and low public awareness. Government incentives, R&D assistance, and carbon pricing mechanisms are necessary to get past these obstacles (IEA, 2022).

Fuel cell deployment will be accelerated, and their contribution to lowering the carbon footprint across sectors will be strengthened by policies supporting green hydrogen generation, fuel cell vehicle subsidies, and decentralized energy initiatives.

11.5 INTEGRATION OF FUEL CELLS WITH RENEWABLE ENERGY SOURCES

The coupling of fuel cells with renewable energy sources is a promising approach to the realization of sustainable, decentralized, and low-carbon energy systems. Fuel cells, which efficiently convert chemical energy into electricity with a low environmental footprint, can complement the intermittent nature of renewable sources like solar and wind power. This synergy improves energy reliability, maximizes resource utilization, and makes a substantial contribution to climate change mitigation.

The coupling of fuel cells with renewable energy technologies is a promising avenue to reach deep decarbonization and promote climate change mitigation objectives. Fuel cells, being high-efficiency electrochemical devices, are capable of synergistically working with intermittent renewable energy technologies like solar and wind to generate continuous, clean, and flexible power. This integration assists in mitigating one of the key weaknesses of renewable energy systems—variability—by enabling excess renewable electricity to be converted to hydrogen through electrolysis, stored, and fed into fuel cells to produce electricity when renewable generation is low (Dincer & Acar, 2015).

Hydrogen generated by water electrolysis with renewable energy, also known as "green hydrogen," is an important energy carrier in this system. When utilized in fuel cells, it produces only water and heat, thus closing a carbon-free energy loop. This renewable-fuel cell synergy not only supports energy storage and grid balancing but also power generation decentralization, thus improving the grid's resilience and enabling remote or off-grid communities (IEA, 2022). In addition, such integration plays a critical role in decarbonizing otherwise hard-to-electrify sectors like long-distance transport, heavy industry, and backup power systems.

Numerous pilot tests and demonstration programs have already proven the technical viability of integrating renewable fuel cells. For example, the REFHYNE pilot in Germany integrates a 10 MW renewable-energy-powered electrolyzer with hydrogen facilities to deliver clean fuel for industry and mobility uses (Hydrogen Europe, 2023a). In Japan, the ENE-FARM initiative combines household fuel cells with photovoltaics to deliver high overall efficiency and lower household carbon emissions (NEDO, 2020).

But achieving the full potential of this integration will demand much more in terms of electrolyzer and fuel cell technologies, system cost reductions, and establishment of strong hydrogen production, storage, and distribution infrastructure. Policy incentives in the form of subsidies, feed-in tariffs, and regulatory policies will be crucial to scaling up hybrid systems based on renewable-fuel cells. As countries aim to achieve net-zero, the integration of fuel cells with renewable energy not only assists in emission reductions but also improves energy security, reliability, and sustainability in the global energy transition.

11.5.1 Synergy between Fuel Cells and Renewable Energy

Renewable energy resources, though eco-friendly, are inherently variable and weather- and day-dependent. Such intermittency is often accommodated by tough energy storage and backup systems. Fuel cells, especially when linked with electrolyzers, can supply this gap perfectly. Surplus electricity from solar or wind can be utilized in electrolysis to generate hydrogen, which can subsequently be supplied to fuel cells to generate electricity during times of low renewable production, thereby creating a renewable hydrogen economy (Dincer & Acar, 2015).

11.5.2 Renewable Hydrogen Production for Fuel Cells

Hydrogen is an important vector for combining renewable energy with fuel cell technology. By means of processes such as water electrolysis with photovoltaic (PV) or wind power, green hydrogen is produced with no carbon footprint. Such hydrogen is stored and then used in Proton Exchange Membrane Fuel Cells (PEMFCs) or Solid Oxide Fuel Cells (SOFCs) to generate clean energy for grid use or distributed energy systems (Staffell et al., 2019). In particular, solar-assisted electrolysis and wind-powered hydrogen production systems have been developed to decarbonize industrial and household energy consumption (Zhang et al., 2022b).

11.5.3 Decentralized Energy Systems

The integration of fuel cells and renewable energy sources enables decentralized or off-grid energy options, particularly in underdeveloped or remote locations. These systems can minimize transmission losses, provide energy security, and encourage local energy autonomy. Microgrids driven by solar panels and backed up by fuel cells have been established in various areas to provide round-the-clock electricity supply and mitigate the use of fossil fuels (Reiser et al., 2020). The systems are scalable and flexible, thus can be applied to both urban and rural areas.

11.5.4 Contribution to Carbon Neutrality and Climate Mitigation

Fuel cell integration with renewables is crucial to net-zero emissions. Clean hydrogen and biofuels from renewable sources can be used in fuel cells without the release of greenhouse gases (GHGs), meeting global climate targets like the Paris Agreement. Moreover, lifecycle analyses have determined that fuel cell systems powered by renewables have drastically lower carbon signatures than traditional fossil-based power systems (Varfolomeev et al., 2021).

11.5.5 Challenges and Future Directions

Although promising, a number of challenges prevent the widespread adoption of fuel cells with renewable energy. These include high initial costs, the requirement for sophisticated hydrogen infrastructure, and losses in efficiency during energy conversion and storage cycles. Nevertheless, continued progress in materials science, catalysis, and system design is likely to enhance the economic viability and performance of integrated systems. Policymaker backing, subsidies, and public-private partnerships are imperative to speed up the scaling-up of these technologies (IEA, 2022).

11.6 ECONOMIC AND ENVIRONMENTAL BENEFITS OF FUEL CELLS

Fuel cells are ever increasingly seen not just as promise providers of clean energy but also as promise holders of substantial economic and environmental advantages. Their use promotes sustainable development through their capacity to facilitate low-emission energy production, lower dependence on fossil fuels, and aid nascent green economies. Fuel cell transition is part of the global movement toward climate change mitigation, enhanced energy security, and the provision of green jobs.

Fuel cell technology presents major economic and environmental advantages consistent with worldwide initiatives to reduce the impacts of climate change and shift toward sustainable energy systems. Economically, fuel cells represent a high-efficiency alternative to conventional combustion-based systems, directly converting chemical energy to electricity and heat with little loss of energy. This efficient conversion, usually more than 60% for combined heat and power (CHP) systems, leads to reduced fuel usage and operating expenses in the long run (Staffell et al., 2019). Additionally, fuel cells have lower maintenance needs because they lack moving parts, which can reduce lifecycle costs compared to internal combustion engines and turbines.

From a job creation and economic growth perspective, fuel cell market growth stimulates innovation, investment in infrastructure, and the development of new value chains, especially in green hydrogen production, storage, and distribution. For example, the Hydrogen Council (2021) states that the hydrogen economy has the potential to generate more than 30 million jobs worldwide and add $2.5 trillion to world GDP by 2050, with fuel cell technology being a key driver of this expansion. Moreover, as countries aim to decarbonize heavy industry and transport, fuel cells are becoming more cost-competitive, particularly when combined with renewable energy sources.

From a green perspective, fuel cells have near-zero on-site emissions, releasing only water vapor and heat when they are running on pure hydrogen. This results in substantial reductions in nitrogen oxides (NO_x), sulfur oxides (SO_x), and particulate matter, which are key contributors to urban smog and respiratory diseases (Santos et al., 2021). The application of fuel cells in transport—especially in buses, trucks, and trains—can thus make the air cleaner in urban areas and lower the cost of health. In addition, when the hydrogen is generated from renewable power (green hydrogen), the whole life cycle of the fuel cell system is effectively carbon-free. This matters greatly for meeting net-zero emission objectives, particularly in difficult-to-electrify sectors. For stationary uses, fuel cells coupled with carbon capture units or biohydrogen sources can further improve environmental efficiency and provide flexible, decentralized energy solutions with a minimal ecological footprint (Dincer & Acar, 2015).

The convergence of economic efficiency, employment, and environmental benefits makes fuel cell technology a central driver of climate-resilient infrastructure and sustainable development. To unlock these benefits fully, though, ongoing investment in R&D, enabling policy frameworks, and economies of scale are needed to reduce costs and drive market uptake.

11.6.1 Economic Advantages

One of the main economic advantages of fuel cell technology is its high efficiency of energy conversion. In contrast to traditional combustion-based systems, which are normally 30–40% efficient, fuel cells have the capability to operate at electrical efficiencies of 40–60% and even 85% in combined heat and power (CHP) applications (US Department of Energy, 2021). Such high efficiency means lower fuel consumption and operating expenses throughout the life of the system.

In addition, fuel cells provide modularity and scalability, which can be applied across a variety of applications—portable devices and vehicles to large-scale stationary power generation. Such flexibility minimizes infrastructure investment risks and enables decentralized energy generation, reducing transmission losses and grid infrastructure costs (Kumar et al., 2020). Investment in fuel cell businesses also fuels employment generation in various sectors such as engineering, manufacturing, installation, and maintenance. The Hydrogen Council (2021) report states that the hydrogen economy, closely related to fuel cell expansion, could create as many as 30 million jobs worldwide and add $2.5 trillion per year to world GDP by 2050.

11.6.2 Environmental Benefits

Fuel cells save the environment by lowering greenhouse gas (GHG) emissions, mainly. In being powered by renewable source hydrogen, fuel cells produce only heat and water, entirely avoiding carbon dioxide and pollutants like NOx, SOx, and particulate matter (Staffell et al., 2019). They are, therefore, most desirable for transportation in urban settings and distributed energy in pollution-constrained environments. Fuel cells have lower noise pollution and mechanical wear since they are quieter and use fewer moving parts. Their use in combined heat and power systems also improves the environmental performance because waste heat is used, increasing overall system efficiency and minimizing energy wastage (Baroutaji et al., 2020).

Life cycle analyses (LCA) have shown hydrogen fuel cell vehicles and systems to have much lower overall emissions than their internal combustion engine equivalents, and even battery electric vehicles, if hydrogen is produced sustainably (Pelletier et al., 2017).

11.6.3 Contribution to Climate Change Mitigation

Fuel cells have a central role to play in achieving climate objectives through enabling deep decarbonization in many sectors, including power generation, industry, and transport. Their deployment aids the fulfillment of country and global aspirations such as the Paris Agreement and the European Green Deal by replacing fossil-based technology with clean energy solutions (IEA, 2022). Also, fuel cells augment renewable energy sources with flexible, dispatchable power. They stabilize high-renewable-penetration grids, lower peak demand, and enhance the reliability of low-carbon energy systems—no small thing in the long-term pursuit of environmental sustainability (Zhang et al., 2022b).

11.7 CHALLENGES AND LIMITATIONS OF FUEL CELL TECHNOLOGY

While fuel cells hold great potential in combating climate change and improving environmental sustainability, technical, economic, and infrastructural impediments stand in the way of their mass commercialization. Knowing these hindrances is necessary to foster effective policy-making, research directions, and technological breakthroughs aimed at speeding up fuel cell system adoption in different industries.

11.7.1 High Material and Manufacturing Costs

One of the primary drawbacks of fuel cell technology is that it is very expensive to produce, with the major reason being the use of costly materials like platinum-group metals (PGMs) as catalysts in Proton Exchange Membrane Fuel Cells (PEMFCs). While attempts are being made to minimize or eliminate PGMs with non-precious metal catalysts, performance and durability remain behind commercial standards (Debe, 2012; Wu et al., 2021). Also, subcomponents like bipolar plates, membranes, and balance-of-plant systems are part of the total system cost and cap economic competitiveness with traditional technologies.

11.7.2 Longevity and Durability Challenges

Operational longevity of fuel cells is an important issue, particularly for automotive and stationary applications. Fuel cells are prone to degradation through catalyst poisoning, dehydration of the membrane, corrosion of carbon, and thermal cycling (Peighambardoust et al., 2010). For instance, PEMFCs face performance decline due to catalyst dissolution and ionomer degradation, particularly under dynamic load conditions. Long-term stability and durability remain a key barrier to meeting the life expectancy requirements of commercial applications (typically over 5,000 hours for vehicles and 40,000–80,000 hours for stationary units).

11.7.3 Hydrogen Storage and Infrastructure

The absence of effective, safe, and cost-efficient hydrogen storage and distribution infrastructure is yet another significant limitation. Existing storage technologies—compressed gas, cryogenic liquid hydrogen, and chemical hydrides—have limitations in energy density, cost, and safety

(Schlapbach & Züttel, 2001). In addition, the worldwide hydrogen supply chain is not well developed, necessitating a lot of investment in hydrogen production, refueling infrastructure, pipelines, and transport networks to enable a fuel cell-based economy (IEA, 2022).

11.7.4 Fuel Purity Specifications

Fuel cells, particularly PEMFCs and SOFCs (Solid Oxide Fuel Cells), need high-purity hydrogen or reformate gases to avoid catalyst poisoning and ensure performance. Contamination by CO, H_2S, or NH_3 can drastically decrease cell efficiency and lifespan (Borup et al., 2007). This calls for supplementary purification systems, raising the system complexity and cost, as well as reducing compatibility with hydrogen derived from biomass or fossil sources.

11.7.5 Limited Market Penetration and Public Awareness

Despite advancements in technology, fuel cell systems continue to suffer from low market penetration as a result of minimal consumer awareness, policy opposition, and inadequate demonstration projects. Market uncertainty and the lack of standard regulations and codes also discourage investment and commercialization (Zhang et al., 2021). Without strong government incentives, subsidies, and industry collaboration, the shift toward fuel cell technology can continue to be slow in comparison to other clean energy sources.

11.7.6 Environmental and Resource Considerations

While fuel cells themselves emit low or no emissions, the upstream environmental effects need to be taken into account, especially in the case of gray hydrogen produced from natural gas through the process of steam methane reforming. CO_2 emissions are a major contributor unless complemented with carbon capture and storage (CCS) (Dincer & Acar, 2015). Additionally, the processing and mining of such rare and critical materials as lanthanum, platinum, or yttrium are environmental and geopolitical concerns that highlight recycling and alternative material options.

11.8 FUTURE PROSPECTS AND POLICY RECOMMENDATIONS

Fuel cell technology holds the potential to be a transformational force behind the global shift toward low-carbon energy systems. As countries accelerate their push to decarbonize major sectors such as transport, power generation, and industry, fuel cells are being recognized as a core element of low-carbon energy strategies. However, unlocking their full potential will need concerted efforts across research, industry, and policy spaces.

11.8.1 Future Prospects

Fuel cells are projected to take off in multiple sectors because of advances in performance, cost savings, and compatibility with renewable energy systems. Advances in non-precious metal catalysts, high-temperature membranes, and solid oxide fuel cells (SOFCs) are paving the way for greater efficiency and longevity, lowering dependence on strategic raw materials (Wu et al., 2021). The complementarity between green hydrogen and fuel cells is particularly promising. As the expansion of renewable electricity from solar and wind accelerates, excess electricity can be employed for water electrolysis to generate hydrogen, which fuels fuel cells in a clean, circular process (IEA, 2022). The development of "hydrogen hubs" and integrated power-to-gas-to-power (P2G2P) infrastructures is expected to enable mass decarbonization of transport and heavy industry (Hydrogen Council, 2023). In the transportation sector, fuel cell electric vehicles (FCEVs) provide high energy density and rapid refueling, and thus are best suited for long-distance transport, buses, and trains. Hydrogen and fuel cell technologies could meet 12% of world energy demand by 2050, helping to achieve net-zero emissions targets, as per IRENA (2020) estimates.

11.8.2 Policy Recommendations

To spur the adoption and deployment of fuel cell technologies, there is a need for an effective and multifaceted policy regime. The following policy actions are proposed:

1. **Strategic Investment and Funding Support**: Targeted government funding is needed for fuel cell R&D, demonstration projects, and commercialization routes. Public-private collaboration can stimulate innovation and scale-up, particularly in nascent technologies (Zhang et al., 2021).
2. **Incentivizing Green Hydrogen Production**: Policies for encouraging renewable hydrogen via subsidies, feed-in tariffs, and carbon pricing will enhance the environmental and economic rationale for fuel cell uptake. Encouragement of electrolyzer deployment and grid integration is also vital (IRENA, 2020).
3. **Infrastructure Development**: Development of an integrated hydrogen infrastructure—production sites, refueling stations, and pipeline networks—should be accelerated. Standardization and harmonization of technical codes will enable international trade and interoperability (IEA, 2022).
4. **Creation of Markets and Procurement Obligations**: Demand can be stimulated by setting fuel cell obligations in public transit, municipal vehicle fleets, and key infrastructure. Zero-emission policies and green procurement will bring about stable markets and investor assurance.
5. **Education and the Development of Labor**: Training and education programs, university-industry partnerships, and skills development schemes must be launched to develop an able workforce with the capacity to sustain the budding fuel cell economy (Hydrogen Europe, 2023b).
6. **Life Cycle and Environmental Regulation**: Rules must take into consideration the whole life cycle of fuel cell systems, from raw material extraction to manufacturing, usage, and recycling at end-of-life. This will guarantee long-term resource efficiency and environmental sustainability.

11.9 CONCLUSION

Fuel cells are a revolutionary clean energy technology that holds vast possibilities to counter climate change and contribute to global sustainability objectives. With their potential as electrochemical machines that directly transform chemical energy into electricity with great efficiency and less environmental footprint, fuel cells constitute an attractive solution to conventional combustion-based energy technologies. Their suitability across various fields—such as transportation, power generation, home energy systems, and industrial uses—makes fuel cells a backbone of low-carbon economies of the future (Dincer & Acar, 2015).

The coupling of fuel cells with renewable energy sources, especially green hydrogen produced through electrolysis driven by wind and solar power, forms a circular, zero-emission energy loop. This, in turn, not only increases energy security and grid stability but also saves greenhouse gas emissions from traditionally hard-to-decarbonize industries like heavy transport and manufacturing (IEA, 2022). Furthermore, fuel cells provide better air quality since they emit only water vapor and heat during use, a factor that is crucial in meeting urban pollution and public health requirements (Borup et al., 2007).

Economically, fuel cell technologies bring long-term gains in the form of higher energy efficiency, energy source diversification, and possible employment opportunities in the new hydrogen economy. Their mass deployment is nonetheless hindered at present by the high material and system costs, limits of hydrogen storage, and the requirement for comprehensive development of infrastructure (Zhang et al., 2021). In the future, overcoming these constraints will involve long-term policy support, strategic investment in research and development, and the development of strong regulatory and infrastructure systems. International collaboration, public-private partnerships, and education efforts are equally important to speed up innovation and market uptake (IRENA, 2020).

Fuel cell technology is no magic bullet, but it is a vital element of an overall, multi-faceted approach to addressing climate change and making a clean energy, resilient future a reality. With its full potential realized in concert with renewables, electrification, and energy efficiency technologies, fuel cells can be a leading factor in global decarbonization efforts and environmental sustainability.

REFERENCES

Abdelkareem, M. A., Elsaid, K., Wilberforce, T., Kamil, M., Sayed, E. T., & Olabi, A. (2021). Environmental aspects of fuel cells: A review. *Science of the Total Environment*, *752*, 141803. https://doi.org/10.1016/j.scitotenv.2020.141803

Barbir, F. (2005). *PEM fuel cells: Theory and practice*. Elsevier.

Baroutaji, A., Wilberforce, T., Ramadan, M., & Olabi, A. G. (2020). Comprehensive investigation on hydrogen and fuel cell technology in the aviation and aerospace sectors. *Renewable and Sustainable Energy Reviews*, *106*, 31–40. https://doi.org/10.1016/j.rser.2019.02.022

Bicer, Y., & Dincer, I. (2018). Clean fuel options for marine and air transportation. *Energy*, *163*, 1169–1183. https://doi.org/10.1016/j.energy.2018.08.017

Borup, R., Meyers, J., Pivovar, B., Kim, Y. S., Mukundan, R., Garland, N., … & Wilson, M. S. (2007). Scientific aspects of polymer electrolyte fuel cell durability and degradation. *Chemical Reviews*, *107*(10), 3904–3951. https://doi.org/10.1021/cr050182l

Carrette, L., Friedrich, K. A., & Stimming, U. (2000). Fuel cells: Principles, types, fuels, and applications. *ChemPhysChem*, *1*(4), 162–193.

Carrette, L., Friedrich, K. A., & Stimming, U. (2001). Fuel cells – Fundamentals and applications. *Fuel Cells*, *1*(1), 5–39.

Debe, M. K. (2012). Electrocatalyst approaches and challenges for automotive fuel cells. *Nature*, *486*(7401), 43–51. https://doi.org/10.1038/nature11115

Department of Energy. (2022). *Fuel Cell Technologies Office: Annual progress report*. U.S. Department of Energy. https://www.energy.gov/eere/fuelcells/annual-progress-reports

Dincer, I. (2007). Environmental and sustainability aspects of hydrogen and fuel cell systems. *International Journal of Energy Research*, *31*(1), 29–55. https://doi.org/10.1002/er.1235

Dincer, I., & Acar, C. (2015). Review and evaluation of hydrogen production methods for better sustainability. *International Journal of Hydrogen Energy*, *40*(34), 11094–11111. https://doi.org/10.1016/j.ijhydene.2014.12.035

DOE (U.S. Department of Energy). (2020). *Hydrogen strategy: Enabling a low-carbon economy*. Office of Fossil Energy. https://www.energy.gov/fecm/hydrogen-strategy

Edwards, P. P., Kuznetsov, V. L., David, W. I., & Brandon, N. P. (2008). Hydrogen and fuel cells: Towards a sustainable energy future. *Energy Policy*, *36*(12), 4356–4362. https://doi.org/10.1016/j.enpol.2008.09.036

Fuel Cells and Hydrogen Joint Undertaking (FCH JU). (2020a). *Fuel cells and hydrogen technologies in Europe: A vision for 2050*. https://www.fch.europa.eu/

Fuel Cells and Hydrogen Joint Undertaking (FCH JU). (2020b). *Hydrogen roadmap Europe*. https://www.fch.europa.eu/

Götz, M., Lefebvre, J., Mörs, F., Koch, A. M., Graf, F., Bajohr, S., … & Kolb, T. (2016). Renewable power-to-gas: A technological and economic review. *Renewable Energy*, *85*, 1371–1390. https://doi.org/10.1016/j.renene.2015.07.066

Hydrogen Council. (2021). *Hydrogen insights 2021: A perspective on hydrogen investment, market development and cost competitiveness*. https://hydrogencouncil.com

Hydrogen Council. (2023). *Hydrogen insights 2023: Global policy and market update*. https://hydrogencouncil.com

Hydrogen Europe. (2023a). *Fuel cells and hydrogen observatory: Market and technology trends*. https://www.fchobservatory.eu

Hydrogen Europe. (2023b). *Fuel cells and hydrogen observatory: Policy and workforce report*. https://www.fchobservatory.eu

Intergovernmental Panel on Climate Change (IPCC). (2021). *Climate change 2021: The physical science basis.* https://www.ipcc.ch/report/ar6/wg1/
International Energy Agency. (2021). *Tracking transport 2021.* https://www.iea.org/reports/tracking-transport-2021
International Energy Agency (IEA). (2022). *Global hydrogen review 2022.* https://www.iea.org/reports/global-hydrogen-review-2022
International Renewable Energy Agency (IRENA). (2018). *Hydrogen from renewable power: Technology outlook for the energy transition.* https://www.irena.org/publications
International Renewable Energy Agency (IRENA). (2020). *Green hydrogen: A guide to policy making.* https://www.irena.org/publications
Kawamoto, K., et al. (2020). The long-term environmental and economic benefits of residential fuel cells: A case study of Japan. *Renewable and Sustainable Energy Reviews, 119*, 109562. https://doi.org/10.1016/j.rser.2019.109562
Kumar, S., Himabindu, V., & Venkata Mohan, S. (2020). Fuel cells: Promising energy devices for the future. *Current Science, 119*(7), 1145–1155.
Larminie, J., & Dicks, A. (2003). *Fuel cell systems explained* (2nd ed.). Wiley.
Logan, B. E., & Regan, J. M. (2006). Microbial fuel cells—Challenges and applications. *Environmental Science & Technology, 40*(17), 5172–5180. https://doi.org/10.1021/es0627592
Logan, B. E., et al. (2006). Microbial fuel cells: Methodology and technology. *Environmental Science & Technology, 40*(17), 5181–5192. https://doi.org/10.1021/es0605016
Minh, N. Q. (2004). Solid oxide fuel cell technology—Features and applications. *Solid State Ionics, 174*(1–4), 271–277. https://doi.org/10.1016/j.ssi.2004.07.040
New Energy and Industrial Technology Development Organization (NEDO). (2020). *ENE-FARM project overview.* https://www.nedo.go.jp/english/
O'Hayre, R., Cha, S.-W., Colella, W., & Prinz, F. B. (2016). *Fuel cell fundamentals* (3rd ed.). Wiley.
Pant, D., Van Bogaert, G., Diels, L., & Vanbroekhoven, K. (2010). A review of the substrates used in microbial fuel cells (MFCs) for sustainable energy production. *Bioresource Technology, 101*(6), 1533–1543. https://doi.org/10.1016/j.biortech.2009.10.017
Peighambardoust, S. J., Rowshanzamir, S., & Amjadi, M. (2010). Review of the proton exchange membranes for fuel cell applications. *International Journal of Hydrogen Energy, 35*(17), 9349–9384. https://doi.org/10.1016/j.ijhydene.2010.05.017
Pelletier, N., Audsley, E., Brodt, S., Garnett, T., Henriksson, P., Kendall, A., … & Tyedmers, P. (2017). Energy intensity of agriculture and food systems. *Annual Review of Environment and Resources, 42*, 319–346. https://doi.org/10.1146/annurev-environ-081710-161014
Reiser, C. A., Steiger, B. T., & Jarvis, D. (2020). Fuel cells in microgrids: Integration and performance. In *Microgrid design and operation* (pp. 201–220). Academic Press.
Ren, X., Zelenay, P., Thomas, S., Davey, J., & Gottesfeld, S. (2008). Recent advances in direct methanol fuel cells at Los Alamos National Laboratory. *Journal of Power Sources, 86*(1–2), 111–116.
Santoro, C., Arbizzani, C., Erable, B., & Ieropoulos, I. (2017). Microbial fuel cells: From fundamentals to applications—A review. *Journal of Power Sources, 356*, 225–244. https://doi.org/10.1016/j.jpowsour.2017.03.109
Santos, D. M. F., Sequeira, C. A. C., & Figueiredo, J. L. (2021). Hydrogen production by electrolysis: Current status and future trends. *Energy Conversion and Management, 246*, 114702. https://doi.org/10.1016/j.enconman.2021.114702
Schlapbach, L., & Züttel, A. (2001). Hydrogen-storage materials for mobile applications. *Nature, 414*(6861), 353–358. https://doi.org/10.1038/35104634
Singhal, S. C. (2000). Advances in solid oxide fuel cell technology. *Solid State Ionics, 135*(1–4), 305–313. https://doi.org/10.1016/S0167-2738(00)00332-5
Singhal, S. C., & Kendall, K. (2003). *High-temperature solid oxide fuel cells: Fundamentals, design and applications.* Elsevier.
Staffell, I., Scamman, D., Abad, A. V., Balcombe, P., Dodds, P. E., Ekins, P., Shah, N., & Ward, K. R. (2019). The role of hydrogen and fuel cells in the global energy system. *Energy & Environmental Science, 12*(2), 463–491. https://doi.org/10.1039/C8EE01157E
Steele, B. C. H., & Heinzel, A. (2001). Materials for fuel-cell technologies. *Nature, 414*(6861), 345–352. https://doi.org/10.1038/35104620
Stolten, D. (Ed.). (2010). *Hydrogen and fuel cells: Fundamentals, technologies and applications.* Wiley-VCH.
U.S. Department of Energy (DOE). (2021). *Fuel cell technologies market report 2020.* https://www.energy.gov/eere/fuelcells/fuel-cell-technologies-office

Ueckerdt, F., Bauer, C., Dirnaichner, A., Everall, J., Sacchi, R., & Luderer, G. (2021). Potential and risks of hydrogen-based e-fuels in climate change mitigation. *Nature Climate Change*, *11*(5), 384–393. https://doi.org/10.1038/s41558-021-01032-7

Varfolomeev, M. A., Sinev, L. F., & Belousov, V. M. (2021). Environmental and economic assessment of renewable hydrogen-based fuel cell energy systems. *Renewable and Sustainable Energy Reviews*, *135*, 110212. https://doi.org/10.1016/j.rser.2020.110212

Wright, L., & Fulton, L. (2005). Climate change mitigation and transport in developing nations. *Transport Reviews*, *25*(6), 691–717. https://doi.org/10.1080/01441640500360951

Wu, G., Siahrostami, S., Li, G., & Wang, H. (2021). Recent advances in electrocatalysts for oxygen reduction reaction. *Nature Catalysis*, *4*, 951–967.

Zhang, J., Wang, X., Ma, L., & Chen, Y. (2021). Barriers and opportunities for the commercialization of fuel cell technologies. *Renewable and Sustainable Energy Reviews*, *145*, 111032. https://doi.org/10.1016/j.rser.2021.111032

Zhang, J., Wang, X., Ma, L., & Chen, Y. (2022a). Hydrogen and fuel cells for decarbonizing transportation: Opportunities and challenges. *Renewable and Sustainable Energy Reviews*, *152*, 111699. https://doi.org/10.1016/j.rser.2021.111699

Zhang, X., Zhang, J., Wang, Y., & Chen, Y. (2022b). Recent progress on hydrogen production via electrolysis–From laboratory to industrial developments. *Sustainable Energy & Fuels*, *6*(3), 605–633.

Zhao, G., Chen, J., Sun, W., & Pan, H. (2021). Non-platinum group metal electrocatalysts toward efficient hydrogen oxidation reaction. *Advanced Functional Materials*, *31*(20), 2010633. https://doi.org/10.1002/adfm.202010633

12 Surface Heat Enhancement Modification Characteristics on Fuel Cell Environmental Sustainability

S.A. Aasa, R.K. Dunne, and D.A. Desai

12.1 INTRODUCTION

The progress required for modern-day development is always attached to the availability of energy (Sanusi, 2024). The required energy mix for these developments is yet unmet and is largely dependent on fossil fuels, which contribute to high pollution and emissions of greenhouse gases (GHG) (Lucia, 2014) (Guaitolini & Fardin, 2018). According to studies, ways to bring about reductions in pollution and emission of energy resources are predominantly in three categories, namely, application of renewable energy sources, CO_2 sequestration, and promotion of high-efficiency energy systems with low CO_2 emissions (Lucia, 2014). The efficient low-CO_2 emission technologies are dependent on system thermodynamic efficiency (Sciubba, 2005) (Nordhaus, 2008) (Hammond, 2004). The result of CO_2 emissions leads to high-efficiency heat exchanger devices designed for advancement in energy dissemination technology. Fuel cells are identified as one of the most efficient energy dissemination technologies. A fuel cell is a promising energy conversion device for the generation of electricity using an electrochemical reaction strategy (Candelaresi & Spazzafumo, 2021) (Liang, Wang, Wang, Mao, & Sunarso, 2017). It offers high efficiency with negligible emissions to the environment. These potentials cannot be overemphasised; however, thermal management of high-temperature dissipation in this heat exchanger device is a major challenge (Xu & Ni, 2021). The excessive heat generated is known to easily degrade material components, thereby negatively impacting the performance and also the system lifespan. This is a major drawback and hence serves as the reason for research to improve performance optimisation and make the device readily usable. Many studies have been done to improve the thermal management of the setup and encourage a system of heat transfer enhancement (Guaitolini & Fardin, 2018; Maradiya, Vadher, & Agarwal, 2017). The application of such techniques optimises the heat dissipation, temperature uniformity, and pressure drop for an optimal operation. Passive techniques like special coatings, tiny channels, and changed nanostructures are recognised for their ability to improve heat flow and heat transfer rates. These techniques have therefore been adjudged for maintaining fuel cell efficiency, particularly in high-power applications (Yakubu et al., 2024). A technique to surface-modified heat enhancement is said to be thermal conductive coating using materials like graphene and/or metallic layer deposition for uniform heat distribution across the cell chambers (Okafor, Popoola, & Adeosun, 2024). Other studies have revealed that the application of such techniques influences a great deal of hot spot formation reduction on fuel cell heat exchanger devices (Peng et al., 2023; Liu, Huang, & Li, 2021; Dai et al., 2017). Additionally, cooling systems with modified surfaces in bipolar plates help remove heat effectively by evenly circulating coolants, which prevents overheating. It thereby improves the performance and durability of the fuel cell system at various operating

DOI: 10.1201/9781003645405-12

conditions (de Bortoli, Ramadan, Becherif, & Claude, 2016). Furthermore, the application of nanostructured surface techniques has, in recent times, gained significant progress. Their potential advantage in increasing the effective heat transfer area and improving the thermal distribution and regulation is a tremendous deal. The use of nanotubes and/or metal oxides to enhance convective heat transfer with minimal weight addition is a significant benefit that led to improved performance of the fuel cell (Ramadan, de Bortoli, Becherif, & Claude, 2017) (Lin, Luo, Li, & Minkowycz, 2021). Also, the application of phase change materials (PCM) integrated into fuel cells for thermal heat management is known to absorb the excess heat deposited during the operation process, thereby providing passive thermal management. According to study this surface-modified, novel approach demonstrates how innovative materials could revolutionise energy technologies essential for powering a sustainable future (Wang et al., 2023). In achieving the goal for sustainable fuel cell operation, the application of computational modelling and experimental studies showcases important roles for using techniques that enhance surface heat as an optimisation strategy. According to Cao et al. (2024), simulations assist in predicting thermal characteristics at different operating conditions. Besides, the experimental validation of such computational models is of significant influence in confirming the effectiveness of these modifications in real-world applications. A typical example of such a scenario is a fuel cell with systems for managing thermal heat enhancement (Raduwan et al., 2022). The combined effort gives a detailed strategy for effective optimisation of the performance, thereby giving advanced thermal management solutions. Finally, the application of surface-modified heat enhancement techniques is vital for the improvement of thermal performance in fuel cell devices. The application of techniques such as extended surface-modified endwalls, conductive coating, microchannel endwalls, nanostructured endwalls and/or their combinations can be said to offer promising solutions to system overheating, redundancy performance, and lifespan challenges. The effectiveness of such techniques in fuel cell devices will encourage widespread adoption and system scalability and will improve significantly the efficiency, durability, and commercial viability in renewable energy systems.

12.2 FUEL CELL OVERVIEW AND BENEFITS

Fuel cells are a form of electrochemical devices that carry out operation processes to convert the chemical energy of hydrogen fuels to electricity using clean and efficient processes (Perng et al., 2019). The power generation from fuel cells comes with minimal and negligible emissions, with heat and water as waste products compared to conventional power generation with enormous CO_2 and greenhouse gas potential and low fuel efficiency (Wang et al., 2023). This energy resource is applicable in various facets, such as transportation, power generation, and electronics, and it is highly reliable with consistent output results. They are known to exhibit high device efficiency, achieving performance levels exceeding 50% across all applications. They are a major energy mix in sustainability transition technology (Ramadan et al., 2017). A major advantage of using fuel cell systems is the emission of little to no greenhouse gas and its particulate matters, especially when renewable energy resources of the hydrogen type are used. Unlike conventional combustion engines, zero tailpipe pollutants are emitted in the use of fuel cell devices; this, in turn, enabled improved air quality in urban areas (Candelaresi & Spazzafumo, 2021). Besides, combined heat power (CHP) also uses fuel cells; this enhances the system energy efficiency, and with the reused heat generated, carbon footprints are lowered (Cao et al., 2024). It thereby makes fuel cells a crucial component in global efforts to mitigate climate change and achieve net-zero emissions. Another positive takeaway in the application of fuel cells is its support for renewable energy integrations with reliable backup and grid stabilisation. They also serve as energy storage systems when paired with electrolysers, thereby creating stability of wind and solar power energy intermittency (Yakubu et al., 2024). Fuel cell devices operate with minimal noise, contributing to a more conducive and sustainable environment. The device is scalable, making it compatible with various setups, such as hospital electronics, libraries, and data centres, to name a few (Zhao et al., 2025). The modern application of the fuel cell

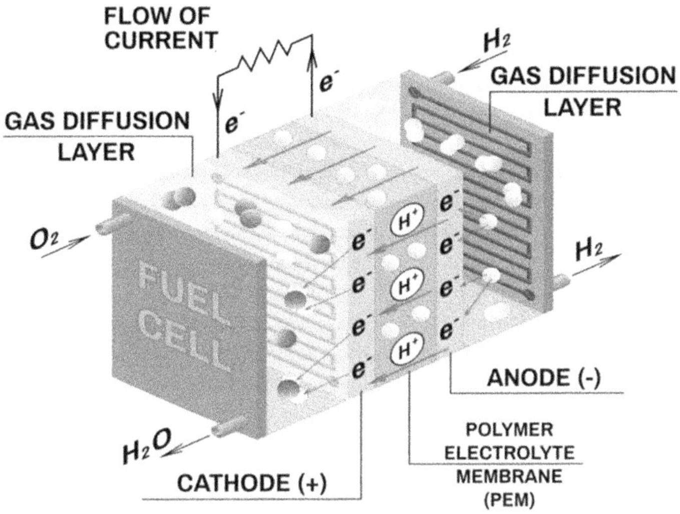

FIGURE 12.1 Fuel cell system overview. (Liang, Wang, Wang, Mao, & Sunarso, 2017.)

is continuing to increase due to its enormous influence, thereby bringing about the much-desired fuel environmental sustainability and fostering decarbonisation of energy resources.

Figure 12.1 gives a detailed overview and operation procedure (Luo et al., 2015). An anode side reaction is presented in Equation 12.1; this presents the reaction of hydrogen from the storage tank with the anode for the generation of proton molecules of hydrogen and electrons.

$$H_2 \rightarrow 2H^+ + 2e^- \tag{12.1}$$

The reaction on the cathode side, shown in Equation 12.2, happens when oxygen molecules interact with hydrogen protons and electrons, producing water as a byproduct.

$$\frac{1}{2}O_2 + 2H^+ + 2e^- \rightarrow H_2O \tag{12.2}$$

The redox reaction of the process is presented in Equation 12.3, which occurs by the reaction of water with oxygen molecules.

$$H_2O + \frac{1}{2}O_2 \rightarrow H_2O \tag{12.3}$$

It is impossible to overstate the significance of fuel cell technology. From Table 12.1, the potential benefits of adopting fuel cells as an energy production method are tabulated. From the table, the key impact area is transportation; this is a modern application in electric vehicles that results in efficient operation, requiring low heat dissipation for optimal performance. It offers quick and convenient refuelling, enabling long-range operation for fuel cell electric vehicles (FCEVs), making them more suitable for extended travel compared to battery electric vehicles (BEVs). It can also provide clean power generation for industrial setups and data centres. In residential areas, it utilises combined heat and power (CHP) systems, as well as off-grid solutions, to provide reliable energy for remote locations. Fuel cells represent a vital technology worth investing in due to their high sustainability, reliability, and potential to support a low-carbon future while addressing critical energy, environmental, and public health challenges.

TABLE 12.1
Fuel Cell Important and Summary

#	System Aspect	Important and Benefits
1	Environment	**Low/Zero Emission**: Water and heat as byproducts when hydrogen is used **Negligible air pollution**: Zero CO_2, NO_x, and SO_x air pollutants. **Combats Climate Change**: Help decarbonise energy, transport, and industries.
2	Energy Efficiency	**High Efficiency**: Direct conversion of chemical energy to electricity, leading up to 65 % conversion and 85 % conversion for CHP **Combustions**: Up to 20–30 % more efficient than conventional internal combustion system.
3	Energy security	**Fossil fuel Dependency**: Reduction in over-dependency on fossil fuels. **Diverse fuel sources**: Uses renewable energy that can be produced from waste: e.g., Biogas, Hydrogen, or methanol.
4	Versatility	**Multiple applications**: Applicable in various gadgets, automotive, homes, industries, portable equipment **Stability**: Enable grid stability by providing backup power with capability for balancing renewable energy intermittency and stability.
5	Economic	**Job**: It is meant for job creation that support growth in energy sector **Moderate operational cost**: it contains fewer moving parts that support low maintenance cost.
6	Health	**Quiet Operation**: Little or no noise pollution when in operations. Cleaner Air: Low risk of health diseases like respiratory compared to fossil fuel emission
7	Sustainability	**Circular economy**: Waste to energy creations, e.g., the use of Biogas fuel cells. **Renewable optimisation**: Integrate excess energy by storing it in the form of hydrogen.
8	Future innovation	**Emerging Tech**: Advances in fuel cell use, such as green hydrogen and ammonia fuel cells. **Space**: Space exploration useful in spacecraft, e.g., Apollo mission, Mars rovers.

12.2.1 Important of Heat Management in Fuel Cells

Effective energy generation plays a crucial role in the survival and optimal performance of any heat exchanger device, primarily due to its efficient heat management. The effectiveness of heat management is synonymous with fuel cell lifespan for efficient energy production. The operation process of fuel cells brings about enormous heat generation; the heat generated is locked in an air jacket and is most often not properly dissipated due to inadequate air distribution over the component's surface. The effect of this localised heat leads to thermal stresses, material component degradation, and reduced efficiency (Peng et al., 2023; Perng et al., 2019). Localised heat in component jackets leads to wear due to hotspots created, uneven temperature distribution, and a shortened lifespan. Optimal and effective thermal regulation that leads to uniform heat dissipation on component surfaces ensures adequate and stable operation, especially in a high-demand, energy-intensive system, such as automotive and power systems, leading to consistent operation of the system (Wang et al., 2021; Zhang et al., 2020). Another important area where good thermal management is crucial is in keeping proton exchange membranes (PEM) hydrated, as they help the system work by conducting ions in PEM fuel cells (Ahmadi et al., 2020). Thermal instability and overheating are the primary causes of membrane dryness, which results in an increase in ionic resistance and a cumulative reduction in power production. On the other hand, inadequate heat on the membrane surface leads to water flooding, blocking, and reactant flow blockage (Barati et al., 2018). To maintain optimal thermal surface suitability, the application of advanced cooling techniques such as phase change material, thermal coating conductivity, surface-modified end walls, etc., plays an important role in ensuring even temperature distribution, efficient power production, system durability, and reliability (Wang et al., 2023; Perng et al., 2019). When the fuel heat exchanger system lacks adequate thermal suitability and control, it most often risks operational failure; hence, the need for heat management and design cannot be overemphasised. Furthermore, heat management efficiency impacts the enhancement of the overall energy use of fuel cell systems through maximum utilisation of wasted heat recovery. The application of excess heat utilisation through combined heat power (CHP) has a major influence on the performance of heat exchanger devices for heat and/or industrial processes, leading to an improvement in system total efficiency (Le et al., 2024). Novel cooling ideas, such as microchannel cooling, nanostructures, and surface-modified heat exchangers, are known to optimise thermal regulation with minimal energy losses (Zhao et al., 2025). The modern evolution of fuel cells brings about intriguing, novel, and smarter adaptive systems that require cooling upgrades to meet the required thermal management techniques. These techniques should be compact enough for adaptation in the system and essential for maximising the performance, reliability, and influence of the emission towards environmental sustainability and meeting the next-generation energy systems.

12.2.2 Surface Enhancement Modification Techniques

The application of surface-modified techniques is classified into various facets based on their mode of formation. The classification of surface-modified heat enhancement techniques includes the combined or compound application of such cooling systems, material coatings, and structural modifications. The combined application is the process of using different surface-modified geometries or the combination of different enhancement techniques, such as structural application and coatings. Hybrid applications are high-efficiency techniques that lead to superior thermal performance regulation and optimisation in operation. For instance, researchers have combined grooved walls with ribs (Saha and Acharya, 2014), combined microchannels with PCMs (Xie & Yang, 2024), and combined material coatings with thermoelectric coolers (Fan et al., 2021a; Zhao et al., 2025). This technique leverages the strengths of various methods, such as the high heat absorption of PCMs and the efficient convective cooling of microchannels, to maintain optimal operating temperatures under varying working loads. Another key classification is the material coating technique; this process is carried out by applying phase change materials or advanced thermally conductive materials on fuel cell surfaces. Graphene and metallic-based coatings such as aluminium, copper, ceramic composites, etc., are also commonly applied for efficient heat conduction and radiation (Fan et al., 2021b). Additionally, using nanostructural coatings like carbon nanotubes (CNTs) or metallic oxides helps increase the area for heat transfer, which improves passive cooling. (Yazdi et al., 2016). This coating causes efficient improvement in thermal management and corrosion, and extends the lifespans of the fuel cell operation and reliability. According to (Yang et al., 2019; Yazdi et al., 2016), material coating techniques can effectively reduce peak temperatures and lead to an improvement of up to 15% compared to uncoated surfaces. Furthermore, the application of surface-modified enhancement techniques in fuel cells involves the flow augmentation using surface-altering of fuel cell components, such as surface imprint, integration of microchannels, and porous media, to improve plate convective heat transfer REF. These techniques enhance heat transfer through uniform heat distribution and thermal conductivity. Such improvements can lead to the mitigation of heat hotspot formation. The integration of additive manufacturing using 3D printers is a major improvement in flow field design optimisation and coolant distribution while minimising pressure drop of the system (Zhao et al., 2025). This innovative technique has offered high-powered fuel cell application improvement when high heat dissipation is a challenge (Fan et al., 2021a; Maradiya et al., 2017; Candelaresi & Spazzafumo, 2021).

12.3 FUNDAMENTALS OF SURFACE HEAT ENHANCEMENT

Fuel cells are a form of heat exchanger systems that dissipate a lot of heat during the process of operation. The large amount of heat produced comes from chemical reactions, energy losses, and extra energy needed to start these reactions, so managing heat is critical for the system to work well and last longer. Heat enhancement techniques for surfaces aim to optimise heat dissipation, prevent localised overheating, and maintain uniform temperature distribution across the cell. Micro/nano-structured surface endwalls used as textured flow fields, such as the application of pin-fin and serpentine type designs, and porous coatings using metallic foam or graphene are adjudged to improve the convective heat transfer and thermal conductivity (Zhao et al., 2025; Zhang et al., 2022; Patel & Rathod, 2020). Furthermore, the application of advanced cooling systems is an innovative technique that includes optimised coolant channels and phase change materials (PCMs). Redistributing the absorbed excess heat improves efficiency (Patel & Rathod, 2020). Furthermore, the application of thermal interface material, such as high-conductivity pastes and graphite-based layers, tends to minimise surface thermal heat resistance between components (Wang et al., 2016; Zhang & Kandlikar, 2012). In recent times, infrared thermography and computational modelling have been used as essential tools for analysing heat distribution and validating these enhancements (Zhang & Kandlikar, 2012; Weber & Newman, 2004). Some fuel cell systems degrade very quickly at high temperatures, such as 80°C, similar to PCMs, due to significant instability challenges at the operating temperature and difficulties in scaling these solutions cost-effectively for commercial applications. Effective thermal heat management and enhancement not only improve the performance through reaction kinetic accelerations; they also mitigate thermal heat stresses and degradation, thereby extending the lifespans of fuel cells (Wang et al., 2023; Wang et al., 2016). Other limitations and components part effects as related to system optimisation and the need for enhancement in the fuel cell system, especially as related to water droplet growth dynamics, are detailed in Petrovick et al. (2023).

12.3.1 Heat Transfer Mechanics in Fuel Cells

The byproducts generated from fuel cell operation are highly significant, and most of these byproducts are in the form of heat from electrochemical reactions, the ohmic resistance, and activation potentials, making thermal management a critical aspect of their design and operations. The forms of heat transfer in fuel cell operation are primarily conduction, convection, and radiation, and each of the modes of heat transfer has its own role in meeting the demand for thermal regulation. The most dominant mode of heat transfer in the operation of fuel cells is conduction, especially in membrane electrodes and bipolar plate component assemblies in regions of high-temperature heat flow, such as from the catalyst to cooler regions (O'Hayre et al., 2016). For instance, the bipolar plates' thermal conductivity of the material significantly impacts efficiency and is superior in conductivity compared to metals, thereby reducing thermal resistance (O'Hayre et al., 2016). In convection, the mode of operation is very vital in the heat removal process via coolant channels or reaction channels of reactant flow of gas; however, the performance of forced convection is better compared to natural convection heat transfer. Forced convection uses a lot of energy and can lead to problems like material wear and pressure drops in large systems (Xu et al., 2021). In the radiation mode of heat transfer, the process of operation occurs rather less significantly at low-temperature fuel cells such as PEMFCs, compared to notably high-temperature setup fuel cells like Sulphur Oxide Fuel Cells with operating temperatures well over 600°C (Weber & Newman, 2004; Borup & Weber, 2020; Xiao et al., 2019). To further ease the design, the application of computational models such as computational fluid dynamics has given high flexibility of design through simulation of convective heat transfer mechanisms, optimising the flow field design and cooling strategies to prevent hotspots and ensure uniform temperature distribution (Xiao et al., 2019; Mehrpooya et al., 2020; Afshari & Jazayer, 2008). It is of note that the major focus of advanced thermal management techniques in fuel cells is on the heat transfer efficiency while mitigating potential damage to component parts and other limitations due to localised heat distribution. Microchannel cooling systems embedded within bipolar plates leverage convective heat transfer for excellent heat dissipation, using computational models for optimal design strategies, moderate pressure penalty, and thermal performance index are achieved (Zhang & Kandlikar, 2012; Zhang et al., 2022). The applications of PCMs, such as paraffin wax, absorb latent heat during melting, which stabilises temperature while transient operation goes on and is validated with experimental results in PEMFC stack arrangements (Kahwaji et al., 2018). Another approach commonly applied in fuel cell enhancement is nanofluids, which involve the dispersion of nanoparticles, such as Al_2O_3 and CuO, to mention just a few, suspended in various base fluids. Dependent on goals, this has brought about enhanced thermal conductivity of the coolant by up to 20 %. Such improvement is, however, impeded by the particle sedimentation and corrosion, making the application limited in duration of performance (REF-DOE US 2022). The enormous presence of radiative mode of heat transfer in SOFC made it possess high temperature that hampered the operation performance due to its unique heat transfer challenges as a result of this radiative heat exchanger and thermal expansion mismatches, necessitating ceramic-based materials with matched thermal expansion coefficient to prevent mechanical failure of components (O'Hayre et al., 2016). For real-time thermal monitoring, the use of electrochemical impedance spectroscopy (EIS) and infrared thermography (IRT) has been established to link system inefficiency in heat transfer to its performance losses (Wang et al., 2023).

12.3.2 Surface Roughness and Impact on Fuel Cells Heat Transfer

The application of surface roughness, such as pin fins, protrusions, dimples, groove systems, and hybrid combinations of these roughened surfaces, is highly impactful in its application in fuel cell heat enhancement. It achieved this enormous potential by augmenting the boundary layer, enhancing fluid dynamics and interfacial contact resistance, and, in most cases, reducing the thermal bond contacts. Fuel cell systems, particularly the proton exchange membrane fuel cells (PEMFCs) and solid oxide fuel cells (SOFCs), the surface imprints morphology of bipolar plates, gas diffusion layers (GDLs), and flow channels, significantly affect the conductive and convective heat transfer mechanisms. The extent of turbulent promoters of rough surface techniques in reactant channels influences the disruptions of laminar layers and increases the convective heat transfer coefficient by up to 70% when compared to baseline plates of conventional test system facilities (Zhang et al., 2022). The turbulence, therefore, causes the enhancement and even mixing of reactants and improves thermal uniformity, mitigating localised hotspots that can degrade membranes' durability. Additionally, surface roughness amplifies the effectiveness of the contact areas between adjacent components such as bipolar plates and GDLs; it thereby improves the conductive heat dissipation and reduces the interfacial thermal resistance (Wang et al., 2023). The experimental research studies have shown that atomic force microscopy (AFM) and profilometry have demonstrated that controlled roughness (Ra values of 0.5–2 μm) optimises this balance due to a potential increase in pressure drop whenever surface roughness is excessive, thereby possessing the possibility to hinder diffusion of reactants (Kahwaji et al., 2018). The numerical simulation of microscale dimples or grooves revealed that their textures potentially augment heat transfer by creating vortex flow that readily enhances coolant flow redistribution (Secanell et al., 2017). It is, however, limited due to parasitic power losses as pressure drops across the channel; as such, the flow resistance must be carefully managed through iterative design optimisation. The strategic application of surface-modified endwalls extends beyond convective heat transfer improvement; it also involves advanced thermal management strategies in fuel cells. A typical example is porous media coatings with nanoscale roughness, using materials such as nano-thermal conductive graphene or metal oxide layers, that can be applied to bipolar plates to simultaneously enhance heat dissipation and electrical conductivity (Huang et al., 2019). The application of fractal surface roughness geometries is employed in these coating techniques, which enable interfacial contact area, reducing thermal resistance by up to 40% compared to smooth uncoated surfaces (Zhang & Jackson, 2017). It further has interconnects that exhibit higher emissivity, with improved radiative cooling efficiency. The reproduction often leads to lasting changes and stability because things like wear and corrosion can change the surface shape, affecting how well it performs thermally. The advent of emerging technologies such as ablation and electrochemical etching encourages precision control of roughened geometry parameters, tailoring surfaces for specific operating conditions and humidity and temperature (Steinke & Kandlikar, 2004).

12.3.3 Advanced Surface Modification Techniques of Fuel Cells

Advanced surface modification techniques have emerged as a critical strategy for optimising fuel cell performance by enhancing catalytic activity, improving thermal management, and increasing durability. One prominent method is plasma surface treatment, which introduces functional groups such as COOH and OH onto polymer electrolyte membranes (PEMs) and gas diffusion layers (GDLs) to improve hydrophilicity and proton conductivity. An example is low-temperature plasma treatment of Nafion membranes, which has been shown to reduce interfacial resistance by 25% (Singh et al., 2023) with high potential to maintain mechanical stability (Shah & Hakim, 2025). Furthermore, atomic layer deposition is another good and transformative approach with precise nanofluid scale coating of catalysts on porous electrodes, significantly enhancing electrochemical surface area and oxygen reduction reaction kinetics (Huang et al., 2019; Cao et al., 2020). ALD-modified electrodes exhibit up to 50% higher durability under accredited stress tests compared to conventional electrodes (Fernåndez et al., 2022). Also, laser surface texturing has gained traction for creating micro- and nanoscale patterns on bipolar plates, thereby reducing contact resistance by 39% and enhancing water management in PEMFCs. When this technique is used alongside electrochemical etching, it helps to get rid of surface oxides from metallic bipolar plates, which improves their resistance to corrosion and heat conductivity. The application of computational modelling, including density functional theory (DFT) and finite element analysis (FEA), plays a pivotal role in optimising their modification by predicting their impact on mass heat transfer. (Lee et al., 2013; Yi et al., 2018)Beyond conventional methods, bio-inspired surface modification and hybrid coating technologies are pushing the boundaries of fuel cell efficiency and system durability. A typical application of this is described in superhydrophobic coatings, where lotus leaves are applied to GDLs to mitigate flooding in the PEMFCs. This is effective in the improvement of gas diffusion and reduction performance decay at high current densities (Liu, Huang, & Li, 2021). These coatings, often based on fluorinated polymers or silica nanoparticles, are achieved through water contact at angles greater than 150°, leading to a self-cleaning surface that maintains stable operation over long periods (DOE, 2023). In a similar vein, graphene-based nanocomposites are being used to functionalise bipolar plates, combining high electrical conductivity for up to 5000 S/m with exceptional thermal stability (Xing et al., 2023). Despite the advances, the challenges remain in the scalability of the system production and the possibility of long-term adhesion under thermal cycling.

12.4 SURFACE MODIFICATION TECHNIQUES FOR FUEL CELLS

12.4.1 Nano-Structured Surface

Surface modification techniques for fuel cells using nanostructure surfaces have gained significant attention due to their potential to enhance performance, durability, and efficiency. This is carried out by altering the surface morphology and chemistry of the enwalls to improve reactant distribution, water management, and electrochemical activity of fuel cell energy devices. The application of prominent methods such as plasma spraying, which deposits nanostructural coatings such as Yttria-Stabilised Zirconia (YSZ) to enhance thermal stability and reduce interfacial resistance, as demonstrated by research (Raduwan et al., 2022), to SOFCs. Also, the application of laser ablation in nanostructured surface enhancement creates micro- and nanoscale textures to promote turbulence and improve mass transport, as evidenced by Lee et al. (2020) in proton exchange membrane fuel cells. Additionally, the growth of nano-structured catalyst surfaces can be effectively carried out using electrochemical deposition on endwalls, excellent progression in the active surface area and reducing catalyst loading (Lopez & Ustarroz, 2021). The application of atomic layer deposition is another precise technique. In this process, ultrathin layers are applied with conformal coatings, for instance, the application of TiO_2 or Al_2O_3 to protect the surface and serve as corrosion inhibition to the surface and, in the process, bring about excellent conductivity (Yang et al., 2020; Xiong and Wang, 2022). Besides, chemical vapour deposition has employed carbon nanotubes and/or graphene-based endwalls which enhance electrical conductivity and mechanical strength. Another method was demonstrated (Rao et al., 2021) where sol-gel processes are utilised to produce porous

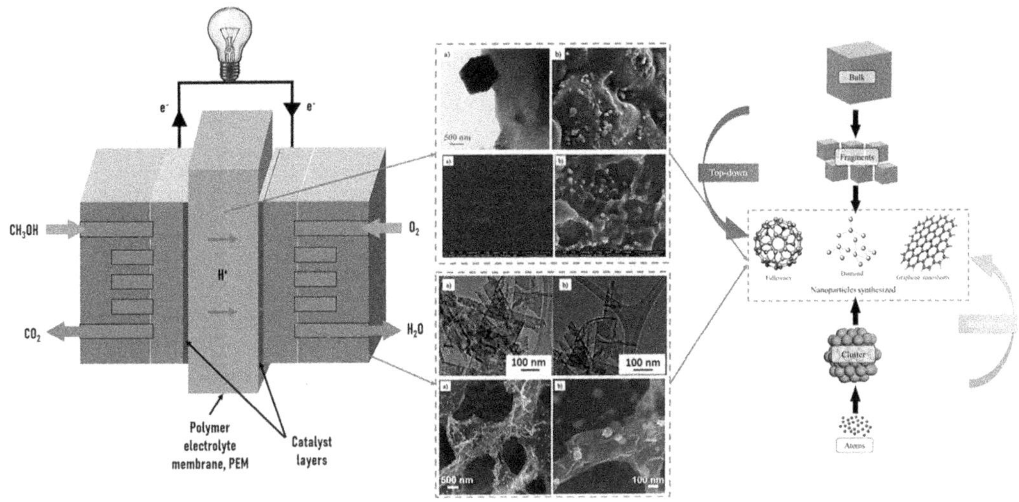

FIGURE 12.2 Nanostructured fuel cells setup. (Raduwan et al., 2022.)

nanostructured coating that improves gas diffusion and water removal; the same was further highlighted by Guizard (1996) and Bokov et al. (2021). All these approaches are often combined with surface functionalisation, such as hydrophilic treatments, to optimise water management, as demonstrated by Zhao et al. (2025) in PEMFCs. It has therefore been shown that integration of nanostructured surfaces increases power density by up to 30% and extends operational lifetimes, as corroborated by multiple studies (Lee et al., 2013; Weber & Newman, 2004). For all of these applications, the challenges remain scalability, cost-effectiveness, and instability issues under harsh operating conditions. Advances in nanotechnology and computational modelling are expected to further refine these techniques, enabling next-generation fuel cells with superior performance. The details of the test configuration of typical nanostructures as presented by (Raduwan et al. 2022) are in Figure 12.2.

12.4.2 Microstructured Surface

Surface modification techniques using microstructured surfaces have emerged as a promising approach to enhancing the performance and efficiency of fuel cells by improving reactant distribution, water management, and electrochemical activity. Microstructured surfaces, typically fabricated through techniques such as laser ablation, micromachining, and photolithography, introduce controlled roughness, channels, or patterns on fuel cell components such as bipolar plates, gas diffusion layers (GDLs), and endwalls. These modifications enhance mass transport by promoting turbulence flow, reducing concentration polarisation, and facilitating efficient water removal, which is critical in proton exchange membrane fuel cells, where flooding can severely degrade performance. Typical examples are related in a study that demonstrated laser-ablated microchannels on bipolar plates improved reactant distribution and increased power density by up to 15 % in PEMFCs. According to Fan et al. (2021b) and Sun et al. (2020), micro-pillar arrays fabricated via photolithography enhanced capillary-driven water removal in GDLs, mitigating flooding and improving stability under high-current-density operation. Micro-embossing is another efficient technique where precise patterns are created on bipolar metallic plates to enhance corrosion resistance and electrical conductivity; this process was also discovered in another study (Wang et al., 2016) where stainless steel-based fuel cells were used. Electrochemical etching has also been employed to generate microporous surfaces that increase the active area for catalytic reaction. This is established by researchers such as (Falcão et al. 2014), who noticed that there could be up to 20 % improvement in cell performance due to enhanced triple-phase boundary interactions. Besides, additive manufacturing using 3D printers has enabled the fabrication of complex microstructured flow fields with optimised geometries such as bioinspired designs by leaf venation or

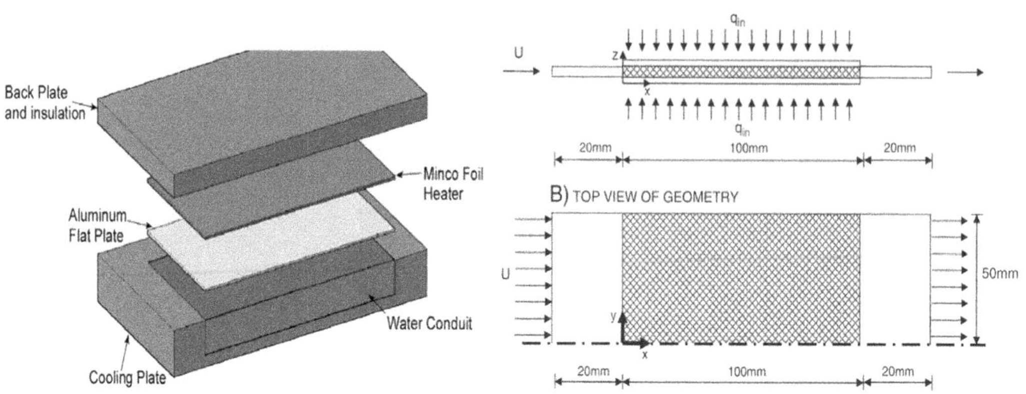

FIGURE 12.3 Microstructured surface heat enhancement model. (Stogiannis et al., 2013.)

lung alveoli, which improve gas diffusion and reduce pressure drops, as demonstrated by Singh et al. (2025). The use of microstructured coatings, such as hydrophilic treatments, further optimises water management. This is the area (Zhao et al., 2025) that presented the influence of superhydrophobic micro-grooves on GDLs significantly and reduced accumulations in PEMFCs. In other studies, hybrid approaches of micro- and nano-structure combinations have shown the possibility of exploring and leveraging the combined synergistic effects of techniques, where micro-patterned surfaces with nano-porous coatings exhibited superiorities in catalytic activity and durability of the system (Stogiannis et al., 2013). Typical microstructural test arrangements are as set up in Figure 12.3.

12.4.3 Coating and Surface Treatment

Coating means of surface modification and treatment play an important role in enhancing the efficiency, durability, and performance of fuel cells by improving corrosion resistance, electrical conductivity, and catalytic activity. The application of conductive coating is one of the most widely used techniques, for instance, the use of carbon, gold, or silver-based layers on bipolar plates to reduce interfacial contact resistance and prevent oxidation, as demonstrated by Naghdi et al. (2018) in their study on PEMFCs. In the same way, coating with nitride TiN or CrN deposited via physical vapour deposition (PVD) has been shown to significantly enhance the corrosion resistance of metallic bipolar plates while maintaining high conductivity (Stein & Ein, 2019). Polymer-based coatings are another efficient coating, such as polyaniline or polypyrrole, which provide both corrosion inhibition and improved surface wettability, as evidenced by various studies. For catalyst layers, thin-film coating of platinum alloy nanoparticles applied via sputtering or electrodeposition techniques enhances electrochemical activity while reducing precious metal loading to apply ultra-thin, conformal coating of metal oxides using TiO_2 or Al_2O_3 on fuel cell components to protect against degradation under high-temperature operation (Suter et al., 2021). An example of such a condition is found in fluoropolymer coating on gas diffusion layers that mitigate water flooding (Zhao et al., 2025). Figure 12.4 presents typical coating techniques as detailed in Tang et al. (2024).

12.4.4 Hybrids Surface Modifications

The combination of various techniques in performance enhancement further presents an opportunity to extend the improved operation for fuel cell devices. For example, combining tiny structures, small patterns, and special coatings has proven to be a very effective way to improve the durability and efficiency of fuel cells while also tackling important issues like how well materials move, how catalysts break down, and resistance at the surfaces. One prominent hybrid approach involves integrating microstructured flow fields with nanoporous catalysts for optimised reactant distribution and electrochemical activity (Suter et al., 2021). According to Zhang (2013), a performance

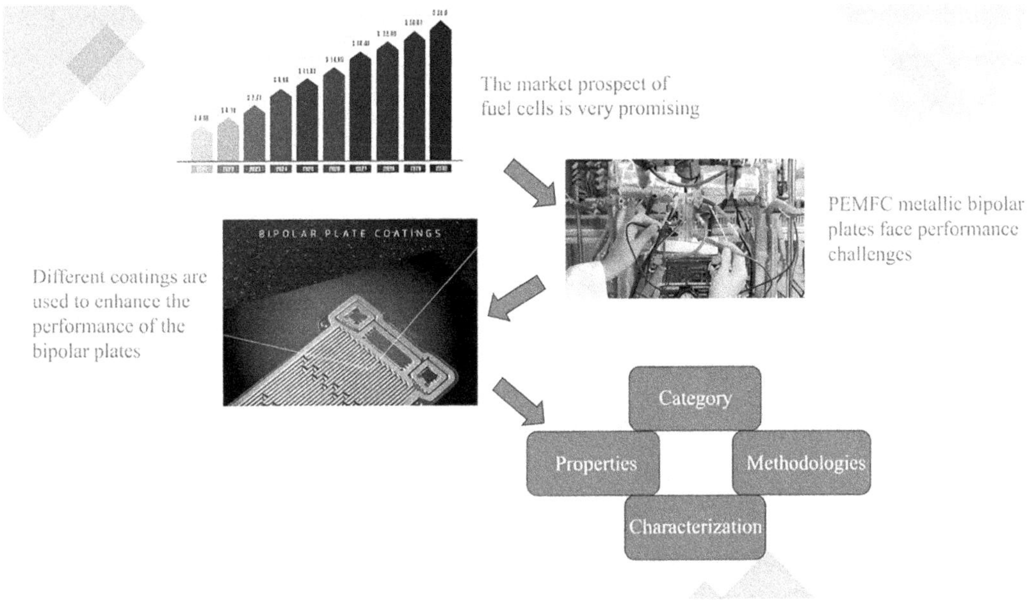

FIGURE 12.4 Coating Fuel cells setup. (Tang et al., 2024.)

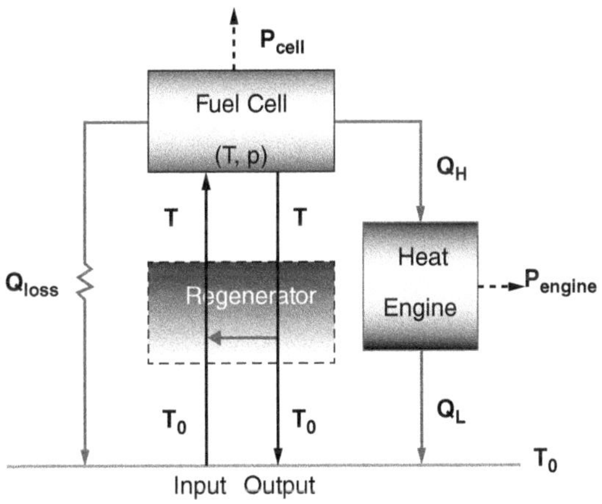

FIGURE 12.5 Fuel cells heat engine hybrids. (Zhao & Chen, 2009.)

increase of up to 25% power density was reported for PEMFCs using micro-channels coated with Pt-nanoparticles-decorated carbon nanotubes (CNTs). Similarly, the combination of plasma electrolytic oxidation (PEO) and atomic layer deposition (ALD) has been used to create corrosion-resistant nano-ceramic coated metallic bipolar plates with enhanced conductivity (Wang et al., 2023). Also, the use of micro-pillar arrays on gas diffusion layer function with a hydrophobic polymer is another good approach where innovative techniques are used to improve water management and permeability, as evidenced in the study. Graphene-reinforced metal oxides TiO_2-rGO have been applied to catalyst layers in such a way that they enhance the mechanical stability and oxygen reduction reaction activities. Furthermore, the integration of biomimetic micro-patterns with nanostructured catalysts, which are known to be inspired by natural systems such as leaf venation or lung alveoli, has been shown to improve mass transport and reduce concentration polarisation as highlighted by Zhao et al. (2025). The hybrid surface treatment, involving enhanced chemical vapour deposition and sol-gel processes, has been employed to produce a multifunctional coating that offers both hydrophobicity and anti-corrosion properties (Suter et al., 2021) (Figure 12.5).

12.5 IMPACT OF SURFACE MODIFICATION ON FUEL CELL PERFORMANCE

12.5.1 Experimental Studies on Surface Modified Fuel Cells

Recent studies have found ways to greatly improve how heat is managed in fuel cells by making it easier for heat to escape, lowering resistance at the surfaces, and reducing heat damage. According to Zhang et al. (2022), an experiment was conducted using a systematic experimental approach by applying graphene-coated bipolar plates, revealing a 30% reduction in thermal contact resistance and a 20% increase in peak power density in PEMFCs under high-load conditions. Their thermal imaging analysis showed a more uniform temperature distribution with high loading hotspot reduction of up to 15°C compared to uncoated plates. When SOFCs were interconnected with experimentally validated plasma-sprayed Al_2O_3-ZrO_2 composite coating, it reduced oxidation-induced thermal resistance by 40% during hundreds of hours for the tests conducted at 8000°C, measured by electrochemical impedance spectroscopy (EIS) and infrared thermography. Microstructured surfaces have also proven effective in tests that involved laser-etched dimple PEMFC flow channels. From the reports, more than a 20% increase in convective heat transfer coefficient via localised temperature mapping was observed, though with a 12% increase in pressure penalty due to enhanced turbulence. Hybrid approaches have shown particular promise, engineered by CNT-micro-pillar hybrid surfaces wherein situ thermal measurements demonstrated 35 % better heat dissipations than flat designs, attributed to combined nano-scale conduction and micro-scale convective effects (Liu et al., 2021). From the study (Cigolotti et al., 2021), the synergetic management of water was also observed, which further paved the way for performance improvement. Also, using high-speed microscopy to confirm the performance of bio-inspired superhydrophobic-hydrophilic patterns, 22% improved condensate removal was observed, which reduced the thermal gradient by 18% in PEMFCs. Durability remains a critical focus (Afshari & Jazayer, 2008). Using perovskite-coated SOFC interconnects for 500 thermal cycles at 25–800°C with post-test spectroscopy revealing a 5% degradation in thermal reflectance, which outperformed uncoated samples by 50%. According to DOE (2023), the real-world application of such approaches is limited, where scaling laser-textured surfaces increases fabrication costs by 200% as against conventional machining. The emerging solution includes inkjet-coated-printed thermal conductivity, which achieved ±2 % thickness uniformity across 100 cm^2 areas while maintaining 90% thermal performance of vacuum-deposited films (Weber & Newman, 2004). These experimental advances are all about the potential of surface engineering to resolve fuel cell thermal bottlenecks, though trade-offs between performance, durability, and manufacturability require further optimisation for commercial viability.

12.5.2 Numerical Modelling of Surface Heat Enhancement

In recent times, numerical studies have provided critical insight into how surface modifications optimise thermal management in fuel cells through advanced computational modelling techniques. (Kane et al., 2022). Employed 3D multiphysics computational fluid dynamics (CFD) models to simulate graphene-coated bipolar plates in PEMFCs, predicting a 32% reduction in peak temperature and 28% more uniform heat transfer distribution compared to conventional plates, and validated against experimental infrared data. Their model included the different heat conduction abilities of graphene, which is 1500 W/m-K when measured flat and only 15 W/m-K when measured through, to address thermal issues at the interface. For the microstructured surface model of the dimple flow channel, it is revealed that optimal dimple depth-to-diameter ratios of 0.2–0.3 enhance turbulence heat transfer by up to 40–70% while limiting pressure drop penalties to less than 15%. The simulation revealed vortex shedding patterns that were later confirmed by experimental PIV measurements. Hybrid surface designs have been extensively modelled: they used finite element methods to optimise bio-inspired microgrooves-nanoparticles hybrid GDLs, demonstrating through multi-objective genetic algorithms that staggered groove arrangement with 50-nm silver particles maximised thermal conductivity with 37% and water removal efficiency of +29 %. High-temperature fuel cell simulations coupled with discrete element modelling (DEM) with CFD predict how perovskite-coated SOFC interconnects reduced radiative thermal resistance by 22% at 750°C, matching within 5% of synchrotron X-ray thermal mapping data (Steinke & Kandlikar, 2004). The application of a machine learning approach has also been used to augment the traditional modelling approach (Afshari & Jazayer, 2008) by training using a convolution neural network strategy (CNN) on 50,000 simulated surface architectures, achieving 92% accuracy in predicting thermal performance, which guided the fabrication of experimentally validated AI-designed fractal surfaces with 45 % better heat transfer compared to human-optimised patterns training (Kumar et al., 2025). The limitation in such modelling occurred during transient thermal stresses during cold starts; according to the study, solving this required the development of a phase-field fracture model that predicted crack propagation in coated MEAs with up to 8% errors versus micro-CT observation. Commercial adoption barriers are also being modelled (Liu et al., 2021). DOE (2023) established techno-economic simulations that show laser-textured surfaces become cost-competitive at production volumes greater than 100000 units/year when combined with additive manufacturing. These numerical advances accelerate the design of next-generation fuel cells by enabling virtual prototyping of surface modification before costly experimental trials.

12.5.3 Performance Enhancement and Efficiency Gains

Studies in recent times have demonstrated that surface modifications significantly enhance fuel cell performance and efficiency by optimising key operational parameters. The development of laser-textured titanium bipolar plates with hierarchical micro-nano-structures achieved a 25% increase in peak power density (1.8 W/cm²) and a 15% improvement in energy efficiency in PEMFCs, attributed to enhanced mass transport and reduced interfacial resistance. Their durability tests showed only 3% performance decay after 5000 cycles, compared to 12% for conventional plates. For catalyst layers, nanoporous platinum-gold alloy surfaces were created through dealloying, which demonstrated 4.5 times higher specific activity for the oxygen reduction reaction while reducing platinum loading by 60%, translating to a 20% system efficiency gain at 0.6 V. These surfaces maintained stability through 30,000 potential cycles in accelerated stress tests. Gas diffusion layer modifications have shown remarkable results: (Xie & Yang, 2024) engineered biomimetic superhydrophobic GDLs with fractal microstructures that reduced flooding-related voltage losses by 35% at high current density greater than 2 A/cm², while simultaneously improving doped ceria on nickel-YSZ anodes increased power output by 40% at 650°C by extending the triple-phase boundary length by 300%, with impedance spectroscopy showing a 50% reduction in polarisation resistance. Hybrid approaches combining multiple modification strategies have proven particularly effective (Lee et al., 2020), integrating plasma-treated microporous layers with graphene-coated flow fields, achieving simultaneous 28% efficiency gain and 50% longer lifetime in PEMFCs under automotive cycling conditions. The modification addressed three critical limitations: reducing oxygen transport resistance by 22%, improving water management by 40%, and enhancing thermal conductivity by 35%. Economic analyses indicated that these surface treatments could reduce fuel cell system cost by $15/kW at scale, primarily through improved power density and durability. However, challenges remain in scaling nanoscale modification, as noted by Weber and Newman (2004), where attempts to commercialise laser-ablated surfaces initially showed 40% higher defect rates than conventional manufacturing. Recent breakthroughs in roll-to-roll processing (Mauler et al., 2021) have demonstrated potential solutions, achieving less than 5% performance variation across metre-scale modified membranes. These improvements show that surface engineering can help fuel cells become more efficient while also tackling important issues of cost and durability for them to be used more widely.

12.6 ENVIRONMENTAL SUSTAINABILITY AND FUEL CELLS

12.6.1 Life Cycle Assessment of Fuel Cells

New research has found ways to make it easier for heat to leave from fuel cells, lower the resistance at the surfaces, and lessen the damage caused by heat. According to Zhang et al. (2022), in an experiment conducted using a systemic experimental approach by applying graphene-coated bipolar plates, a 30% reduction in thermal contact resistance and a 20% increase in peak power density in PEMFCs under high-load conditions were revealed. Their thermal imaging analysis showed that the temperature was more evenly spread out, with hot spots being reduced by up to 15°C compared to plates that were not coated. When SOFCs were connected using a tested plasma-sprayed Al_2O_3-ZrO_2 composite coating, it lowered the thermal resistance caused by oxidation by 40% during tests lasting hundreds of hours at 800°C, as measured by electrochemical impedance spectroscopy (EIS) and infrared thermography. Microstructured surfaces have also proven effective in tests that involved laser-etched dimple PEMFC flow channels. Reports showed that there was more than a 20% increase in the ability to transfer heat through convection, as seen in localised temperature mapping, but this came with a 12% increase in pressure due to more turbulence. Hybrid approaches, specifically those using CNT-micro-pillar hybrid surfaces, have shown particular promise; in situ thermal measurements demonstrated a 35% improvement in heat dissipation compared to flat designs, which is attributed to the combined effects of nano-scale conduction and micro-scale convection (Liu et al., 2021). The study (Cigolotti et al., 2021) also observed the synergetic management of water, paving the way for further performance improvement. Also, using high-speed microscopy to confirm the performance of bioinspired superhydrophobic-hydrophilic patterns, 22% improved condensate removal was observed, which reduced the thermal gradient by 18% in PEMFCs. Durability remains a critical focus (Afshari & Jazayer, 2008). By subjecting perovskite-coated SOFC interconnects to 500 thermal cycles at temperatures ranging from 25 to 800°C, post-test spectroscopy revealed a 5% decrease in thermal reflectance, surpassing the performance of uncoated samples by 50%. According to DOE (2023), the real-world application of such approaches is limited, where scaling laser-textured surfaces increases fabrication costs by 200% as against conventional machining. The emerging solution includes inkjet-coated-printed thermal conductivity, which achieved ±2% thickness uniformity across 100 cm^2 areas while maintaining 90% thermal performance for vacuum-deposited films (Weber & Newman, 2004). These experimental advances are all about the potential of surface engineering to resolve fuel cell thermal bottlenecks, though the trade-offs between performance, durability, and manufacturability require further optimisation for commercial viability.

12.6.2 Surface Modification and Sustainability

The modern development in surface modifications is proving to be instrumental in enhancing both the performance and environmental sustainability of fuel cell technologies. For the catalyst (Suter et al., 2021), nitrogen-doped carbon nanotube arrays were developed that completely eliminated platinum needs for ORR, with an accelerated degradation test showing less than 5% performance loss after 10,000 cycles, a breakthrough that addresses both cost and supply chain sustainability concerns. The environmental benefits extend to manufacturing processes: (Alkan et al. 2018) introduced a room-temperature plasma deposition technique for gas diffusion layers that consumes 60% less energy than conventional high-temperature sintering methods while achieving superior water management characteristics. In the realm of high-temperature fuel cells, Fan et al. (2021b) demonstrated that solution-processed perovskite coating on SOFC interconnects could be applied using 90% less solvent than traditional deposition methods while improving oxidation resistance by up to 35%, a dual advantage that reduces both manufacturing emissions and material degradation. Circular economy approaches are emerging (Mauler et al., 2021), developing a bio-based polybenzimidazole membrane surface treatment derived from lignin waste stream, which simultaneously enhanced proton conductivity by 20% and reduced the membrane embodied energy by 45% compared to conventional fluorinated polymers. However, sustainability trade-offs exist; they cautioned that some nanomaterial-enhanced surfaces currently require energy-intensive fabrication methods, with laser ablation processes adding 15–20% to the manufacturing carbon footprint. Life cycle analyses by IEA suggest these impacts are offset within 6–18 months of operation through efficiency gains, particularly for heavy-duty applications. Looking forward, Kane et al. (2022) project that combining surface modification with digital manufacturing could reduce fuel cell production waste by 70% and energy use by 50% by the year 2030, while machine learning-optimised surface architectures may further minimise material requirements without compromising performance. These developments position surface engineering as a critical enabler for sustainable fuel cell deployment across transportation and stationary power sectors.

12.7 CASE STUDY AND FUTURE DIRECTION

12.7.1 Application of Surface Modified Fuel Cell

Surface-modified fuel cells are increasingly deployed across different sectors due to their enhanced performance, durability, and sustainability. In the automotive industry, Kumar et al. (2025) demonstrate that PEMFCs with laser-textured bipolar plates and graphene-coated GDLs achieved 25% higher power density and 30% longer lifespan compared to conventional designs, making them viable for heavy-duty vehicles. These modifications reduce hydrogen consumption by 15%, lowering operational costs and emissions. For stationary power generation, Cigolotti et al. (2021) reported that SOFCs with nanostructured perovskite-coated interconnects maintained 90% efficiency over 50,000 hours, enabling their use in microgrids and backup power systems. In portable electronics, Deng et al. (2020) developed ultra-thin PEMFCs with plasma-treated catalyst layers, delivering twice the energy density of lithium-ion batteries while reducing charging time to minutes. This technology is also applicable in the aerospace sector, as NASA tested PEMFCs with microporous platinum-gold catalysts for lunar rovers in 2023, demonstrating a 40% improvement in cold start performance under extreme temperatures. Meanwhile, marine applications benefit from the corrosion-resistant coating. This is presented in the study (Suter et al., 2021), where fuel cells with hydrophobic GDLs exhibited 50% less degradation in high humidity environments.

12.7.2 Challenges and Limitations

In spite of the significant advancements, surface modification techniques for heat enhancement in fuel cells face enormous and critical challenges. Manufacturing complexity remains a major barrier in the incorporation of the discoveries, especially where large scales are required, because processes like laser texturing, atomic layer deposition (ALD), and plasma coating often require specialised equipment and precise control, leading to high production costs (Kampker, 2020). A typical application of nanostructured surfaces can improve thermal conductivity by 30–40%, but their fabrication is often required and involves energy-intensive methods that offset environmental benefits (DOE, 2023). Durability under operational stress is another concern, as repeated thermal cycling can degrade modified surfaces, causing delamination, oxidation, and crack propagation in coatings. Microstructure flow channels can help with cooling, but they might also create extra pressure problems that lower the system's efficiency. Another prominent challenge is that material compatibility issues further limit adoption, particularly in high-temperature fuel cells, where thermal expansion mismatches between coating and substrates can cause mechanical failure. Additionally, scalability is a persistent challenge for many lab-scale surface modifications, such as graphene coating or bio-inspired textures, that struggle to transition to mass production without performance losses. Here are also trade-offs between thermal and electrochemical performance that also exist; an example is a hydrophobic coating that improved water management may inadvertently reduce proton conductivity in PEMFCs. Finally, long-term environmental impacts of novel nanomaterials remain understudied, raising concerns about lifecycle sustainability. Addressing this limitation will require a multidisciplinary approach and collaboration to balance gains with costs, durability, and manufacturability.

12.7.3 Future Research Direction and Opportunities

Overcoming the current challenges in fuel cell surface modification, future research should prioritise advanced manufacturing techniques that balance performance with scalability. Additive manufacturing using 3D printing of microstructure flow fields offers promising pathways to reducing the production costs while maintaining precision, with the potential to cut fabrication expenses by 40% compared to traditional machining. AI-driven surface design can be a way to go such that optimisation of pore geometries and coating is adequately integrated to simultaneously enhance heat transfer and mitigate pressure drops, addressing a key trade-off in current systems. Further durability, self-heating, and coatings incorporating microcapsulated repair cycling extend component lifetimes by more than 200% performance. The application of sustainable material development presents another critical opportunity. Bio-derived coatings made from lignin or cellulose nanocrystals show potential to replace synthetic polymers, reducing environmental impact while maintaining hydrophobicity and thermal stability. Recyclable surface modifications using reversible chemical bonds could enable end-of-life component disassembly, improving circularity. In high-temperature applications, graded thermal barrier coatings with compositionally varied layers may resolve thermal expansion mismatch issues. System-level integrations of these advancements through digital twins' technology could enable real-time performance monitoring and predictive maintenance, reducing operational costs by 30%. Finally, standardised testing protocols for novel surface modifications are needed to accelerate commercialisation and ensure reliable comparisons between technologies. These directions collectively address the key challenges of cost, durability, and scalability while unlocking new performance frontiers for fuel cells. The focus of research could also be centred on smart adaptive thermal management systems that dynamically adjust cooling techniques based on the effective real-time temperature feedback, further improving fuel cell efficiency and durability (Boyd and Hooman 2013). In conclusion, as fuel cells become increasingly vital for clean energy systems, ongoing research focuses on integrating novel techniques, especially those that involve hybrid approaches and smart thermal regulation using modified extended surface applications, which are prominent in attaining the promising goals of the system. The application of surface modification techniques has emerged as a transformative strategy to enhance fuel cell performance, durability, and sustainability. By tailoring surface properties through nanostructuring, advanced coatings, and bio-inspired designs, researchers have achieved significant improvement in heat dissipation, mass transport, catalytic activity, and corrosion resistance. Key advancements include graphene-coated bipolar plates for reduced interfacial resistance, laser-textured flow channels for optimised coolant distributions, and self-healing membranes for extended operational lifetimes. However, challenges like manufacturing scalability, material compatibility, and cost-effectiveness remain critical for widespread adoption. Future opportunities lie in AI-driven surface optimisation, additive manufacturing, and circular economy approaches, which can address the itemised limitation posed by the operational use of fuel cells while pushing efficiency gains further. The integration of digital monitoring systems and sustainable materials will also play a pivotal role in ensuring that surface-modified fuel cells meet both performance and environmental goals. As research progresses, these innovations will be essential for making fuel cells the cornerstone of clean energy systems, particularly in transportation, stationary power, and aerospace applications.

Surface modification is a powerful tool to overcome existing fuel cell limitations, but its full potential will only be realised through continued interdisciplinary collaboration, scalable manufacturing solutions, and lifecycle-aware design principles.

REFERENCES

Afshari, E., & Jazayer, S. A. (2008). Computational analysis of heat and water transfer in a PEM Fuel Cell. *3rd IASME/WSEAS Int. Conf. on Energy & Environment*, February 23–25. Cambridge: University of Cambridge, UK. https://www.researchgate.net/publication/262152592

Ahmadi, P., Torabi, S. H., Afsaneh, H., Sadegheih, Y., Ganjehsarabi, H., & Ashjaee, M. (2020). The effects of driving patterns and PEM fuel cell degradation on the lifecycle assessment of hydrogen fuel cell vehicles. *International Journal of Hydrogen Energy*, Vol. 45, No. 5, pp. 3595–3608. https://doi.org/10.1016/j.ijhydene.2019.01.165

Alkan, B., Vera, D., Ahmad, M., Ahmad, B., & Harrison, R. (2018). Complexity in manufacturing systems and its measures: a literature review. *European Journal of ndustrial Engineering*, Vol. 12, No. 1, pp. 116–150.

Barati, S., Khoshandam, B., & Ghazi, M. M. (2018). An investigation of channel blockage effects on hydrogen mass transfer in a proton exchange membrane fuel cell with various geometries and optimization by response surface methodology,. *International Journal of Hydrogen Energy*,, Vol. 43, No. 48, pp. 21928–21939, https://doi.org/10.1016/j.ijhydene.2018.10.032

Bokov, D., Jalil, A. T., Chupradit, S., Suksatan, W., Ansari, M. J., Iman H. Shewael, I. M., … E., K. (2021). Nanomaterial by Sol-Gel Method: Synthesis and Application. *Hindawi. Advances in Materials Science and Engineering*, Vol. 21. https://doi.org/10.1155/2021/5102014

Borup, R., & Weber, A. (2020). FC-PAD: Fuel cell performance and durability consortium. (Hydrogen & Fuel Cells Program Annual Merit Review Proceedings. https://www.hydrogen.energy.gov/pdfs/review20/fc135

Boyd, B., & Hooman, K. (2013). Air-cooled micro-porous heat exchangers for thermal management of fuel cells. *International Communications in Heat and Mass Transfer*, Vol. 39, No. 3, 363–367.

Candelaresi, D., & Spazzafumo, G. (2021). Introduction: the power-to-fuel concept. In D. Candelaresi, & G. Spazzafumo (Eds.), *Power to Fuel* (pp. 1–15). https://doi.org/10.1016/B978-0-12-822813-5.00005-9

Cao, K., Cai, J., Shan, B., & Chen, R. (2020). Surface functionalization on nanoparticles via atomic deposition. *Science Bulletin*, Vol. 65, No. 8. pp. 678–688, https://doi.org/10.1016/j.scib.2020.01.016

Cao, Q., Min, H., Zhao, H., Sun, W., Yu, Y., Zhang, Z., & Jiang, J. (2024.). A novel thermal management for PEM fuel cell stack combining phase change materials with liquid cooling under low temperature condition. *Applied Thermal Engineering*, Vol. 238, 121949. https://doi.org/10.1016/j.applthermaleng.2023.121949

Cigolotti, V., Genovese, M., & Fragiacomo, P. (2021). Comprehensive review on fuel cell technology for stationary applications as sustainable and efficient poly-generation energy systems. *Energies*, Vol. (14). 4963. https://doi.org/10.3390/en14164963

Dai, H., Ning, Y., He, S., Zhang, X., Guo, L., & Zhao, G. (2017). Surface modification allows high performance for solid oxide fuel cells fabricated by a single-step co-firing process. *Fuel Cells*, Vol. 17, No. 6. https://doi.org/10.1002/fuce.201700098

de Bortoli, Q., Ramadan, H. S., Becherif, M., & Claude, F. (2016). Multi-stack fuel cell efficiency enhancement based on thermal management. *IET Electrical Systems in Transportation*. https://doi.org/10.1049/iet-est.2016.0027

Deng, R., Xia, Z., Sun, R., Wang, S., & Sun, G. (2020). Nanostructured ultrathin catalyst layer with ordered platinum nanotube arrays for polymer electrolyte membrane fuel cells. *Journal of Energy Chemistry*, Vol. 43 pp. 33–39.

DOE (2022). NETL implements this effort as part of DOE's Advanced Energy Systems Program. https://netl.doe.gov/sites/default/files/2023-11

DOE. (2023). Annual merit review proceedings. *Department of Energy*, https://www.hydrogen.energy.gov/library/annualreview/annual_review23_proceedings

Falcão, D., Oliveira, V., Rangel, C., & Pinto, A. (2014). Review on micro-direct methanol fuel cells. *Renewable and Sustainable Energy Reviews*, Vol. 34. pp. 58–70. https://doi.org/10.1016/j.rser.2014.03.004

Fan, D., Liu, F., Li, J., Wei, T., Ye, Z., Wang, Z., … Shao, Z. (2021a). A microchannel reactor-integrated ceramic fuel cell with dual-coupling effect for efficient power and syngas co-generation from methane. *Applied Catalysis B: Environmental*, Vol. 297, 120443, https://doi.org/10.1016/j.apcatb.2021.120443

Fan, L., Tu, Z., & Chan, S. H. (2021b). Recent development of hydrogen and fuel cell technologies: A review, *Energy Reports*, Vol. 7, pp. 8421–8446. https://doi.org/10.1016/j.egyr.2021.08.003

Fernández, J., Martínez, V., & de la Prida Pidal, V. (2022). Special issue "ALD technique for functional coatings of nanostructured materials". *Nanomaterials*, Vol. 12, p. 3489. https://doi.org/10.3390/nano12193489

Guaitolini, S. V., & Fardin, J. F. (2018). Advances in renewable energies and power technologies || fuel cells: history (short remind), principles of operation, main features, and applications. *Advances in Renewable Energies and Power Technologies.*, 123–150. https://doi.org/10.1016/B978-0-12-813185-5.00013-9

Guizard, C. (1996). Chapter 7 Sol-gel chemistry and its application to porous membrane processing. *Membrane Science and Technology, Elsevier*, Vol. 4, pp. 227–258, https://doi.org/10.1016/S0927-5193(96)80010-6

Hammond, G. (2004). Engineering sustainability: thermodynamics, energy systems, and the environment. *International Journal of Energy Research*, 28: 613–639. https://doi.org/10.1002/er.988

Huang, A., He, Y., Zhou, A., Zhou, Y., Yang, Y., Luo, L., … Yahng, J. (2019).. review of recent applications of porous metals and metal oxide in energy storage, sensing and catalysis. *Journal of Materials Science*, 54, 949–973 https://doi.org/10.1007

Kahwaji, S., Johnson, M. B., Kheirabadi, A. C., & Dominic Groulx, D. W. (2018). A comprehensive study of properties of paraffin phase change materials for solar thermal energy storage and thermal management applications. *Energy*, Vol. 162. p. 1169–1182. https://doi.org/10.1016/j.energy.2018.08.068

Kampker, A. A. (2020). Challenges towards large-scale fuel cell production: Results of an expert assessment study. *International Journal of Hydrogen Energy*, Vol. 45, No. 53), pp. 29288–29296.

Kane, N., Zhou, Y., Zhang, W., Ding, Y., Luo, Z. H., & Liu, M. (2022). Precision surface modification of solid oxide fuel cells via layer-by-layer surface sol–gel deposition. *Journal of Materials Chemistry A*, Vol. 10, No. 16, pp. 8798–8806. https://doi.org/10.1039/D2TA00458E

Kumar, V. B., Collin, A., Sankar, M. G., & Subramanian, K. (2025). Fuel cell technologies in the automotive sector: A focus on proton exchange membrane and Alkaline fuel cells.. *Green Technologies and Sustainability*, Vol. 3(3). 100218. https://doi.org/10.1016

Le, T. T., Sharma, P., Bora, B. J., Tran, V. D., Truong, T. H., Le, H. C., & Nguyen, P. Q. (2024). Fueling the future: A comprehensive review of hydrogen energy systems and their challenges. *International Journal of Hydrogen Energy*, Vol. 54, pp. 791–798. https://doi.org/10.1016/j.ijhydene.2023.08.044

Lee, D., Kim, M. K., Guim, H., Yuk, S., Choi, J., Choi, S., … Kim, H. T. (2020). Ditch-structured microporous layers fabricated by nanosecond-pulse laser ablation for enhancing water transport in polymer electrolyte membrane fuel cells. *Materials Advances*, Vol. 1, No. 2, pp. 254–261. https://doi.org/10.1039/d0ma00150c

Lee, E., Jung-Kon Kim, T., Ae-Jun Kim, A. J., Hannah Song, H., Jun-Hyuk Kim, J. H., Park, S. A., … Kim, Y. T. (2013). Enhanced corrosion resistance and fuel cell performance of Al1050 bipolar plate coated with TiN/Ti double layer. *Energy Conversion and Management*, Vol. 75, pp. 727–733. https://doi.org/10.1016/j.enconman.2013.08.041

Liang, F., Wang, Z., Wang, Z., Mao, J., & Sunarso, J. (2017). Electrochemical performance of cobalt-free Nb and Ta co-doped perovskite cathodes for intermediate-temperature solid oxide fuel cells. *ChemElectroChem*, Vol. 4, No. 9. https://doi.org/10.1002/celc.201700236

Lin, Y., Luo, Y., Li, W., & Minkowycz, W. (2021). Enhancement of flow boiling heat transfer in microchannel using micro-fin and micro-cavity surfaces.. *International Journal of Heat and Mass Transfer*, Vol. 179, pp. 121739.

Liu, M., Huang, H., & Li, X. G. (2021). Geometry optimization and performance analysis of a new tapered slope cathode flow field for PEMFC. *International Journal of Hydrogen Energy.*, Vol. 46, No. 75, Pages 37379–37392, https://doi.org/10.1016

Lopez, M. B., & Ustarroz, J. (2021). Electrodeposition of nanostructured catalysts for electrochemical energy conversion: Current trends and innovative strategies. *Current Opinion in Electrochemistry*, Vol. 27, 100688.

Lucia, U. (2014). Overview on fuel cells. *Renewable and Sustainable Energy Reviews*, 164–169. https://doi.org/10.1016/j.rser.2013.09.025

Luo, C., Xie, H., Wang, Q., Luo, G., & Liu, C. (2015).. A review of the application and performance of carbon nanotubes in fuel cells. *Hindawi Publishing Corporation Journal of Nanomaterials*, Volume 2015, Article ID 560392, p. 10.

Maradiya, C., Vadher, J., & Agarwal, R.. (2017). The heat transfer enhancement techniques and their Thermal Performance Factor. *Beni-Suef University Journal of Basic and Applied Sciences*, S2314853517300070. https://doi.org/10.1016/j.bjbas.2017.10.001

Mauler, L., Duffner, F., & Leker, J. (2021). Economies of scale in battery cell manufacturing: The impact of material and process innovations. *Applied Energy*, Vol. 286, 116499, https://doi.org/10.1016/j.apenergy.2021.116499

Mehrpooya, M., Khodayari, R., Moosavian, S. M., & Ali Dadak, A. (2020). Optimal design of molten carbonate fuel cell combined cycle power plant and thermophotovoltaic system. *Energy Conversion and Management*, Vol. 221, 113177. https://doi.org/10.1016/j.enconman.2020.113177

Naghdi, S., Rhee, K. Y., Hui, D., & Park, S. J. (2018). A review of conductive metal nanomaterials as conductive, transparent, and flexible coatings, thin films, and conductive fillers: different deposition methods and applications. *Coatings*, Vol. 8, No. 27.

Nordhaus, W. (2008). A question of balance: weighing the options on global warming Policy. In W. Nordhaus (Ed.), *A Question of Balance: Weighing the Options On Global Warming Policy*. Yale University, CT: Press.

O'Hayre, R., Cha, S.-W., Colella, W., & Prinz, F. (2016). *In Fuel Cell Fundamentals*. USA: 3rd Ed. Copyright © 2016 John Wiley & Sons. https://doi.org/10.1002/9781119191766.fmatter

Okafor, O., Popoola, A. P., & Adeosun, S. (2024). Surface modification of carbon nanotubes and their nanocomposites for fuel cell applications: A review[J]. *AIMS Materials Science*, Vol. 11, No. 2, pp. 369–414. https://doi.org/10.3934/matersci.2024020

Patel, J. R., & Rathod, M. K. (2020). Recent developments in the passive and hybrid thermal management techniques of lithium-ion batteries. *Journal of Power Sources.*, Vol. 480, p. 228820, https://doi.org/10.1016/j.jpowsour.2020.228820

Peng, C., Gu, H., Zhang, G., Luo, K. X., Lv, S., Zhang, Q., & Chen, G. (2023). Numerical study on heat transfer enhancement of a proton exchange membrane fuel cell with the dimpled cooling channel. *International Journal of Hydrogen Energy*, Vol. 48, No. 8. pp. 3122–3134. https://doi.org/10.1016/j.ijhydene.2022.10.136

Perng, S.-W., Wu, H.-W., Chen, Y.-B., & Zeng, Y.-K. (2019). Performance enhancement of a high temperature proton exchange membrane fuel cell by bottomed-baffles in bipolar-plate channels. *Applied Energy.*, Vol. 255, 113815.

Petrovick, J. G., Radke, C. J., & Weber, A. Z. (2023). Effect of water droplet growth dynamics on electrode current in fuel-cell catalyst layers,. *Chemical Engineering Science*, Vol. 281, p. 119152, https://doi.org/10.1016/j.ces.2023.119152

Raduwan, N. F., Shaari, N., Kamarudin, S. K., Masdar, M. S., & Yunus, R. M. (2022). An overview of nanomaterials in fuel cells: Synthesis method and application. *International Journal of Hydrogen Energy.*, Vol. 47, No. 42, pp. 18468–18495, https://doi.org/10.1016

Ramadan, H., de Bortoli, Q., Becherif, M., & Claude, F. (2017). Multi-stack fuel cell efficiency enhancement based on thermal management. *IET Electr. Syst. Transp.*, Vol. 7, No 1, pp. 65–73.

Rao, N., Singh, R., & Bashambu, L. (2021). Carbon-based nanomaterials: Synthesis and prospective applications. *Materials Today: Proceedings*, Vol. 44, Part 1, pp. 608–614. https://doi.org/10.1016/j.matpr.2020.10.593

Saha K. and Acharya, S. (2014). Heat transfer enhancement using angled grooves as turbulence promoters. *ASME Journal of Turbomachinery*, Vol. 136, Page 081004-1-10. https://doi.org/10.1115/1.4025733

Sanusi, Y. (2024). Achieving the SDG 7, universal access to energy in Africa: Progress, obstacles, and prospects. *Regional Perspectives*. Springer, Cham, https://doi.org/10.1007/978-3-031-17465-0_26

Sciubba, E. (2005). Exergo-economics: Thermodynamic foundation for a more rational resource use. *International Journal of Enrgy research*, Vol. 29, pp. 613–636. https://doi.org/10.1002/er.1096

Secanell, M., Jarauta, A., Kosakian, A., Sabharwal, M., & Zhou, J. (2017). PEM fuel cells, modeling. In: R. Meyers (Ed.), *Encyclopedia of Sustainability Science and Technology*. New York, NY: Springer.

Shah, M., & Hakim, N. U. (2025). Advances in nanocellulose proton conductivity and applications in polymer electrolyte membrane fuel cells. *Next Materials*, Vol. 6, p. 100484, https://doi.org/10.1016/j.nxmate.2025.100484

Singh, R., Pratap, R., Narayanan, J. A., Thangamani, G., Krishnan, V. V., Arjunan, A., & Hughes, D. (2025). Advances in additive manufacturing of fuel cells: A review of technologies, materials, and challenges. *Sustainable Materials and Technologies*, Vol. 43.

Singh, S., Kumar, S., & Khanna, V. (2023). A review on surface modification techniques. *Materials Today: Proceedings*. https://doi.org/10.1016/j.matpr.2023.01.010

Stein, T., & Ein Eli, Y. (2019). Proton exchange membrane (PEM) fuel cell bipolar plates prepared from a physical vapor deposition (PVD) titanium nitride (TiN) coated AISI416 stainless steel. *SN Applied Sciences*, Vol. 1, p. 1420.

Steinke, M. E., & Kandlikar, S. G. (2004). An experimental investigation of flow boiling characteristics of water in parallel microchannels. *Journal of Heat Transfer Transactions of the ASME.*, Vol. 126. https://doi.org/10.1115/1.1778187

Stogiannis, L. A., Mouza, A. A., & Paras, S. V. (2013). Study of a micro-structured PHE for the thermal management of a fuel cell. *Applied Thermal Engineering*, Vol. 59(1–2), pp. 717–724. https://doi.org/10.1016/j.applthermaleng.2012.08.024

Sun, Y., Ling, S., Zhao, D., Liu, J., Liu, Z., & Jinlong Song, J. (2020). Through-mask electrochemical micromachining of micro pillar arrays on aluminum. *Surface and Coatings Technology*, Vol. 401, p. 126277, https://doi.org/10.1016/j.surfcoat.2020.126277

Suter, T. A., Smith, K., Hack, J., Rasha, L., Rana, Z., Angel, G. M., … Brett, D. J. (2021). Engineering catalyst layers for next-generation polymer electrolyte fuel cells: A review of design, materials, and methods. *Advanced Energy Materials*, Vol. 11, No. 37. 10.1002/aenm.202101025

Tang, W., Wang, S., Li, S., Yu, S., Yuan, X., Yang, C., & Li, W. (2024). Research progress on coating and coating technology of fuel cell metallic bipolar plate.. *Next Materials*, Vol. 4, p. 100062, https://doi.org/10.1016/j.nxmate.2023.100062

Wang, Q., Jiang, B. B., & Yan, Y. (2016). A critical review of thermal management models and solutions of lithium-ion batteries for the development of pure electric vehicles. *Renewable and Sustainable Energy Reviews*, Vol. 64, pp. 106–128. https://doi.org/10.1016/j.rser.2016.05.033

Wang, Q., Kwan, M.. Fan, J., Lin, J. (2021). Racial disparities in energy poverty in the United States. *Renewable and Sustainable Energy Reviews*, Vol. 137, p. 110620, https://doi.org/10.1016/j.rser.2020.110620

Wang, W., Li, N., Zhang, J., Zhang, C., & and Zhang, L. (2023). Thermal management analysis of proton exchange membrane fuel cell filled with phase change material in cooling channel. *Hindawi International Journal of Energy Research*, Vol. 9077046, p. 12 pages. https.

Weber, A. Z., & Newman, J. (2004). Modeling transport in polymer-electrolyte fuel cells. *Chemical Reviews*, Vol. 104, pp. 4679–4726. https://doi.org/10.1021/cr020729l

Xiao, H., Dong, Z., Long, R., Yang, K., & Yuan, F. (2019). A study on the mechanism of convective heat transfer enhancement based on heat convection velocity analysis. *Energies*, Vol. 12(21), p. 4175; https://doi.org/10.3390/en12214175

Xie, C., & Yang, X. (2024). Heat transfer enhancement of microchannel using nano-encapsulated phase change material for fuel cell thermal management systems.. *Journal of Energy Storage*, Vol. 103, Part A. 114370. https://doi.org/10.1016/j.est.2024.114370

Xing, L., Xuan, J., & Das, P. K. (2023). Fuel cell fundamentals. In *Fuel Cells for Transportation: Fundamental Principles and Applications*, Elsevier. https://doi.org/10.1016/B978-0-323-99485-9.00006-X

Xiong, S. Q., & Wang, Y. (2022). Atomic layer deposition for membrane modification, functionalization and preparation: A review. *Journal of Membrane Science*, Vol. 658, p. 120740, https://doi.org/10.1016/j.memsci.2022.120740

Xu, F., Liu, Q., & Shibahara, M. (2021). Experimental study on forced convection heat transfer of helium gas through a minichannel. *International Journal of Heat and Mass Transfer*,, Vol. 171, p. 121117, https://doi.org/10.1016/j.ijheatmasstransfer.2021.121117

Xu, H., & Ni, M. (2021). High-temperature electrolysis and co-electrolysis. In H. Xu, & M. Ni (Eds.), *Power to Fuel*. (pp. 51–73). Academic Press. https://doi.org/10.1016/B978-0-12-822813-5.00008-4

Yakubu, A. U., Zhao, J., Jiang, Q., Ye, X., Liu, J., Yu, Q., & Xiong, S. (2024). A comprehensive review of primary cooling techniques and thermal management strategies for polymer electrolyte membrane fuel cells PEMFC. *Heliyon*, Vol. 10, No. 19, p. e38556. https://doi.org/10.1016/j.heliyon.2024.e38556

Yang, H., Chen, Y., & Qin, Y. (2020). Application of atomic layer deposition in fabricating high-efficiency electrocatalysts. *Chinese Journal of Catalysis.*, Vol. 41, No. 2), pp. 227–241. https://doi.org/10.1016/S1872-2067(19)63440-6

Yang, Y., Choi, C., Xie, G., Park, J.-D., Ke, S., Yu, J.-S., … Lim, B. (2019). Electron transfer interpretation of the biofilm-coated anode of a microbial fuel cell and the cathode modification effects on its power. *Bioelectrochemistry*, Vol. 127, pp. 94–103.

Yazdi, A. A., D'Angelo, L., Omer, N., Windiasti, G., Lu, X., & Xu, J. (2016). Carbon nanotube modification of microbial fuel cell electrodes.. *Biosensors and Bioelectronics*, Vol. 85, pp. 536–552. https://doi.org/10.1016/j.bios.2016.05.033

Yi, P., Zhang, W., Bi, F., Peng, L., & Lai, X. (2018). Enhanced Corrosion Resistance and Interfacial Conductivity of TiCx/a-C Nanolayered Coatings via Synergy of Substrate Bias Voltage for Bipolar Plates Applications in PEMFCs. *ACS Applied Materials & Interf.*

Zhang, G., & Kandlikar, S. G. (2012). A critical review of cooling techniques in proton exchange membrane fuel cell stacks,. *International Journal of Hydrogen Energy*, Vol. 37, No. 3, pp. 2412–2429, https://doi.org/10.1016/j.ijhydene.2011.11.010

Zhang, G. Ding, X. Li, T. Pu, W. Lou, W Hou, J. (2020). Dynamic energy balance model of a glass greenhouse: An experimental validation and solar energy analysis, *Energy*, Vol. 198, p. 117281, https://doi.org/10.1016/j.energy.2020.117281

Zhang, J. (2013). Pem fuel cell testing and diagnosis || PEM fuel cell fundamentals, pp. 1–42. https://doi.org/10.1016/B978-0-444-53688-4.00001-2

Zhang, X., Li, Z., Luo, L., Fan, Y., & Du, Z. (2022).. A review on thermal management of lithium-ion batteries for electric vehicles. *Energy*, Vol. 238, Part A, p. 121652. https://doi.org/10.1016/j.energy.2021.121652

Zhang, X. X., & Jackson, R. L. (2017). An analysis of generated fractal and measured rough surfaces in regards to their multi-scale structure and fractal dimension. *Tribology International*, Vol. 105, pp. 94–101.

Zhao, J., Cheng, X., Ma, Y., Zhong, Z., Zhou, C., Lv, Y., … Li, C. (2025). Experimental study on heat and mass transfer enhancement of open cathode proton exchange membrane fuel cells using turbulence grids. *Applied Energy*, Vol. 385, p. 125542, https://doi.org/10.1016/j.apenergy.2025.125542

Zhao, Y., & Chen, J. (2009). Modeling and optimization of a typical fuel cell–heat engine hybrid system and its parametric design criteria. *Journal of Power Sources*, Vol. 186, No. 1, pp. 96–103. https://doi.org/10.1016/j.jpowsour.2008.09.083

13 Health and Safety Considerations in Fuel Cell Environmental Sustainability

Gabriel Gbenga Babaniyi, Olumide Ayobami Iyiola, and Ulelu Jessica Akor

13.1 INTRODUCTION

When considering fuel cell technology for environmental sustainability, health and safety are paramount, requiring careful management of hydrogen storage, handling, and potential hazards like leaks and fires, as well as addressing the environmental impact of fuel cell production and disposal. Despite providing clean energy, fuel cells come with a number of complex health and safety risks (Yazdi et al., 2023. Hydrogen is the main fuel. It is highly flammable and invisible. This makes it harder to find leaks, which raises the risk of explosions. Some fuel cells require high temperatures to function. This could lead to thermal runaway. It might lead to fires or system breakdowns. Platinum-based catalysts and caustic electrolytes are examples of toxic substances that can affect both the environment and workers (Tawalbeh et al., 2022). As a result, they need to be handled cautiously. Inhaling nanoparticles from decomposing parts over time may be hazardous to one's health (Shafique & Luo, 2019). To reduce dangers in industrial and transportation settings, ventilation systems, leak detection, and secure storage are crucial. The disposal of spent fuel cells is a serious issue. They include chemicals and metals that can contaminate water and soil. Safety failures in smart fuel cell grids can be caused by cybersecurity threats. Everyone is at risk if hackers manage to get into the networks. For fuel cells to be a sustainable energy source, these hazards must be controlled.

13.2 NEUROTOXICOLOGICAL AND ENDOCRINOLOGICAL RAMIFICATIONS OF CHRONIC LOW-DOSE CHEMICAL EXPOSURES IN INDUSTRIAL AND URBAN ENVIRONMENTS

In modern industrial and urban environments, people are constantly exposed to low doses of harmful chemicals, often without realizing it. Volatile organic compounds (VOCs), endocrine disruptors, and neurotoxicants accumulate in the body over time, creating long-term health risks. These chemicals come from vehicle emissions, industrial waste, household products, and even food packaging. Although exposure levels may seem small, the effects can be profound, altering brain function, hormone balance, and cellular health (Molot et al., 2022). Environmental pollutants are chemicals that result from human activities, which end up in the environment and subsequently pose risks to human and animal health. Several of these chemicals are collectively known as endocrine-disrupting chemicals (EDCs), which are nowadays gaining more importance in terms of public health due to their widespread effects on human health and potential cause of morbidity (Ma et al., 2024).

However, the Endocrine Society defines EDCs as follows: "an exogenous (non-natural) chemical, or a mixture of chemicals, that interferes with any aspect of hormone action." These chemicals

DOI: 10.1201/9781003645405-13

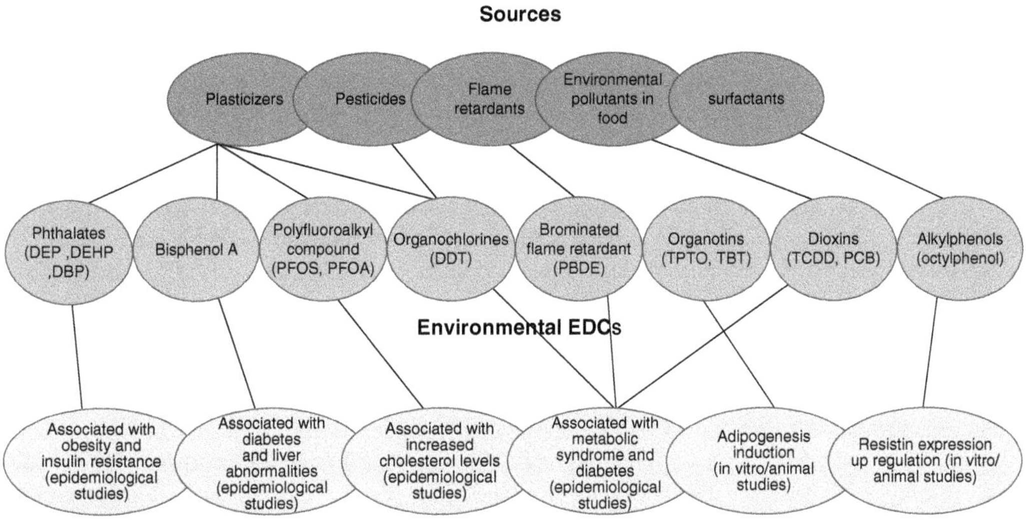

FIGURE 13.1 Effect of Chemical Exposures on metabolism.

alter the hormonal balance of the body through several different mechanisms; they can mimic hormones, disrupt hormone synthesis or breakdown, alter the development of hormone receptors, act as hormone antagonists, or alter hormone binding (Reis & Spencer, 2024). Environmental EDCs are mostly released during the manufacture and use of human-made materials such as pesticides, plastics/plasticisers, electronic wastes, flame-retardants, metals, food additives, and personal care products (Pei & Liu, 2025). These EDCs can disturb hormonal balance and consequently result in several health disorders, including developmental and reproductive abnormalities, increased prevalence of hormone-sensitive cancers, neurodevelopmental delays, and abnormal growth patterns in children, and alterations in immune function.

More so, EDCs are a group of highly heterogeneous synthetic chemicals used in a variety of settings. Some of the common chemicals included in this group are chemicals used in industries and their by-products (polychlorinated biphenyls [PCBs], polybrominated biphenyls [PBBs], and dioxins), plastics (bisphenol A [BPA]), plasticisers (phthalates) and pesticides (methoxychlor [MXC], chlorpyrifos and dichlorodiphenyltrichloroethane [DDT]) (Raja et al., 2022). Some EDCs were designed to have long half-lives for industrial purposes and are known as "persistent organic pollutants" (POPs), exemplified by PCBs, dichlorodiphenyldichloroethylene (DDE), dioxin, organochlorine pesticides and hexachlorobenzene (HCB) (Figure 13.1). The POPs are highly lipophilic and tend to accumulate in the adipose tissue (Kowalczyk et al., 2022). Due to this property, they also accumulate in the food chain and undergo biomagnification. Many of these substances either do not decay or decay slowly, while others may be metabolized to compounds that are more toxic than the parent chemicals (Kumar et al., 2020). Other EDCs such as BPA and phthalates may not be persistent but are used so extensively that environmental exposure is widespread. EDCs are present ubiquitously in the environment, and human exposure occurs through intake of water and food, via inhalation, and through the skin.

13.2.1 Neurotoxic Effects: Silent Damage to the Brain

Many neurotoxic chemicals, such as lead, mercury, toluene, and organophosphates, interfere with nerve signaling, destroy neurons, and weaken cognitive function. Chronic exposure may lead to memory loss, learning disabilities, and neurodegenerative diseases like Parkinson's and Alzheimer's. VOCs, often found in paints, adhesives, and industrial fumes, can cross the blood-brain barrier, where they disrupt neurotransmitter balance, causing anxiety, depression, and even motor dysfunction (Kowalczyk et al., 2022; Ma et al., 2024). This damage accumulates over the years, making early symptoms difficult to detect. In addition, endocrine disruptors are found in many everyday products, including some cosmetics, food and beverage packaging, toys, carpet, and pesticides. Some chemicals that act as flame retardants may also be endocrine disruptors. Contact with these chemicals may occur through air, diet, skin, and water. EDCs cannot be completely avoided or removed; however, you can make informed choices to reduce exposure and risk of any potential health effects. According to the Endocrine Society, there are nearly 85,000 human-made chemicals in the world, and 1,000 or more of those could be endocrine disruptors, based on their unique properties (Kumar et al., 2020).

13.2.2 Endocrine Disruption: The Hidden Hormonal Chaos

Endocrine disruption, often called "hormonal chaos," happens when certain chemicals interfere with the body's natural hormone functions, throwing off vital biological processes. These substances, known as endocrine disruptors, can imitate, block, or alter hormones, leading to serious health issues affecting growth, reproduction, metabolism, and overall well-being. Many of these harmful chemicals are hidden in plastics, pesticides, and even household cleaners, silently disturbing hormonal balance (Scott & Pocock, 2021). Exposure to these substances can lead to fertility problems, thyroid disorders, obesity, and metabolic dysfunctions. What makes them even more dangerous is that even tiny doses can cause significant effects, since hormones operate in incredibly small amounts. Pregnant women and children are especially at risk, as exposure during critical growth stages can cause lifelong developmental and health problems (Plunk & Richards, 2020). Both natural and synthetic chemicals can disrupt hormone function. Common endocrine disruptors like bisphenol A (BPA), phthalates, and polychlorinated biphenyls (PCBs) interfere with the body's hormonal messaging system, creating widespread health concerns (Iqubal et al., 2020). Other chemicals, such as synthetic musks found in personal care products, further contribute to this problem. Since these disruptors are present in countless everyday products, exposure is nearly unavoidable, making awareness and prevention crucial. Table 13.1 presents common endocrine-disrupting chemicals (EDCs) alongside the products and materials in which they are most frequently used. It highlights the widespread presence of these compounds in everyday items ranging from plastics and electronics to personal care products and household goods.

13.2.3 Bioaccumulation and Systemic Toxicity

One of the most alarming dangers of long-term exposure to even tiny amounts of toxic chemicals is bioaccumulation, which is a slow but relentless process where harmful substances build up inside the body over time. Instead of being flushed out, these toxicants latch onto fat, brain tissue, and organs, where they silently accumulate, growing more potent with each exposure. Take heavy metals like lead and mercury; once they enter the body, they don't leave easily (Mesnil et al., 2020).

TABLE 13.1
Examples of Endocrine Hormone Disruptors

Common EDCs	Used In
Bisphenol A (BPA), Phthalates, Phenol	Plastics and Food Storage Materials
Brominated Flame Retardants, PCBs	Electronics and Building Materials
Phthalates, Parabens, UV Filters	Personal Care Products, Medical Tubing, Sunscreen
Triclosan	Anti-Bacterial Soaps, Colgate Total

They embed themselves deep within tissues, lingering for years, even decades, gradually poisoning the nervous system. The damage unfolds slowly but with devastating consequences: brain function declines, memory fades, nerves misfire, and, in worst cases, cancerous mutations take root. Then come persistent organic pollutants (POPs), a class of chemicals that are both invisible and insidious (Duh-Leong et al., 2023). These include dioxins and polychlorinated biphenyls (PCBs) compounds, so stable that they resist breaking down for years. Even in infinitesimally small amounts, these pollutants wreak havoc on vital biological systems. They sabotage the immune system, making the body more vulnerable to diseases (Khan et al., 2024). They interfere with hormones, leading to fertility problems, birth defects, and metabolic disorders. They strain the liver, disrupting its ability to filter toxins, leaving the body defenseless against further contamination.

What makes bioaccumulation even more perilous is its compounding effect as toxins pile up; their impact escalates unpredictably. A single exposure may seem harmless, but with each additional dose, the body's chemical burden grows heavier, inching closer to a tipping point where the damage becomes irreversible (Boamah, 2020). And the worst part? These pollutants don't just stay in one place; they move through the food chain. Fish absorb mercury from polluted water, livestock ingest dioxins from contaminated feed, and humans consume these toxins unknowingly, amplifying the risk across generations. Once inside, there is no simple escape. Bioaccumulation is not an immediate catastrophe; it is a silent, creeping threat, a time bomb ticking beneath the surface, shaping health crises before symptoms even appear (Shi et al., 2024). The only real defense is prevention, limiting exposure before these toxins embed themselves into the very fabric of our biology.

13.2.4 The Long-Term Consequences and Future Risks

Toxic chemicals in everyday life are silently damaging our health. Neurotoxic substances harm our nervous system. Endocrine disruptors mess with our hormones. Together, they create hidden health risks. These chemicals cause more than short-term harm. They might also lead to more developmental disorders, autoimmune diseases, and chronic illnesses that are rising over generations. Children, with their rapidly developing brains and bodies, are especially vulnerable (Boamah, 2020; Duh-Leong et al., 2023). Neurotoxicants such as lead, mercury, and pesticides can harm our brains. They may cause learning disabilities, cognitive decline, and behavioral disorders. Endocrine disruptors like BPA, phthalates, and dioxins upset hormonal balance. This can cause infertility, metabolic diseases, and thyroid problems. Most regulatory policies focus on single chemicals (Khan et al., 2024). They ignore that people face exposure to many toxins at once. The cumulative effects of how these chemicals mix, interact, and intensify within the body are often overlooked, leaving entire populations defenseless. Because of bureaucratic bottlenecks, governments take a long time to act, and industry are frequently resistant to change (Shi et al., 2024). People are therefore responsible for providing protection, frequently without being aware that they are in danger. Without stringent regulations and more awareness, these hidden chemicals will continue to harm future generations. Personal health is not the only reason to fight against hazardous exposure. It's also about protecting humanity's future.

13.2.5 Agents of Neurotoxic

Every day, humans come into contact with thousands of chemicals. These chemicals float in the air we breathe, dissolve in the water we drink, seep into the food we eat, and hide within the countless products we use. This constant exposure is unavoidable, forming a complex network of chemical interactions within our bodies. Because these exposures are complex, ever-changing, and deeply linked, scientists have developed the concept of the exposome, a lifelong record of every environmental influence on health (Molot et al., 2022). The exposome tracks both harmful and beneficial exposures, mapping out how they interact with an individual's genes, microbiome, metabolism, inflammation levels, and stress response (Reis & Spencer, 2024). Among all exposure routes, inhalation poses the greatest risk, as pollutants enter directly into the bloodstream and can harm nearly every organ.

One particularly alarming concern is developmental neurotoxicity. Research suggests that certain industrial chemicals, even at low doses seemingly safe for adults, can silently damage the developing brain in fetuses and young children (Shi et al., 2024). These effects may not be immediately obvious but can alter brain function permanently, with symptoms only emerging later in life (Khan et al., 2024). Since standard toxicology tests often fail to detect these hidden neurological effects, identifying neurotoxic chemicals relies on clinical research and epidemiological studies. The reality is unsettling, and many toxicants are not just present but actively shaping health outcomes, sometimes with effects that only surface decades later. Understanding the exposome is key to uncovering these hidden dangers before it's too late.

13.3 BIOMECHANICAL RISK FACTORS AND ERGONOMIC PARADOXES IN HIGH-RISK OCCUPATIONAL ENVIRONMENTS

In dangerous work settings like construction sites, factories, or hospitals, people often face physical risks that can hurt their muscles and joints. These risks come from doing the same movement over and over, lifting heavy things, or holding their bodies in strange positions for a long time. For instance, reaching up high, bending down all the time, or twisting your back can all lead to injuries over time. Sitting for extended periods of time or standing on hard flooring can cause pain and strain, as can simply being still. Equipment that presses on the body or shakes can exacerbate the situation (Darvishi et al., 2022). Together, these issues result in injuries that impact a person's movement and ability to function, which are known as musculoskeletal disorders (MSDs). However, it's not always easy to solve these issues. There are instances when remedies intended to benefit employees instead make matters worse; this is known as an ergonomic paradox. For instance, lowering the height of shelves to avoid lifting might mean workers have to bend more often, which is just as bad (Talapatra et al., 2024). A tool designed to fit one person might cause strain for someone else. Even in a safe-looking workspace, high stress or poor communication can raise the risk of injury. It's crucial to consider the full picture, which includes not only the equipment and chairs but also the worker's personal needs, job pressure, and mental health.

Keeping an eye out for hidden issues, providing appropriate training, and continuously upgrading procedures are all part of making the workplace safer. In high-risk work settings like surgical theaters and heavy industrial zones, the human body and brain face constant physical and mental stress. Physical forces acting on the body, like repetitive motions, awkward postures, and forceful exertions, can cause strain or injury. Workers in these settings, like surgeons and factory workers, may experience injuries like cervical spine compression or lumbar loading (Gómez, 2020). On the mental side, workers operate under high mental pressure, which is the brain's processing demand when juggling high-stakes decisions, multitasking, and spatial awareness under time constraints. Mental exhaustion and a higher likelihood of making poor decisions can result from this. Systems intended to lessen strain might inadvertently cause new types of stress, which leads to ergonomic paradoxes. For example, robotic-assisted surgeries lessen hand tremors but increase musculoskeletal rigidity and ocular strain by requiring immobile postures and intense visual focus (Khamaj et al., 2022). Similarly, exoskeletons in manufacturing may lower back stress yet increase shoulder torque and restrict agility. Despite these risks, workers often exhibit ways to adjust to persist under strain, like neuromuscular compensation, task re-prioritization, and biopsychosocial flexibility. However, without ergonomic redesign and cognitive support systems, these adaptations can break down, leading to chronic dysfunction or performance collapse.

13.4 LIFE CYCLE HAZARDS AND END-OF-LIFE DISPOSAL PROTOCOLS FOR FUEL CELL COMPONENTS

The life cycle of fuel cell components from production to end-of-life poses multiple ecological risks, especially when materials are not safely managed during disposal. Fuel cells, particularly proton exchange membrane fuel cells (PEMFCs) and solid oxide fuel cells (SOFCs), contain synthetic membranes and heavy metal catalysts that can harm ecosystems and human health if not handled properly. In order to reduce waste, recover precious materials, and stop toxic leaks, these components must be disposed of according to carefully planned procedures when their useful lives are coming to an end (Mori et al., 2025). One significant issue is the used fuel cell membrane, which frequently is made of polymeric composites or perfluorinated compounds (PFCs). These materials endure in the environment for extended periods of time because they are resistant to deterioration and chemically stable. Without adequate oversight, they may be landfilled or burned, releasing hazardous byproducts that contaminate the air, water, and soil. High temperatures can also cause these polymers to degrade, producing fluorinated greenhouse gases or volatile organic compounds (VOCs), which worsen climate change and are harmful to the respiratory system (Mori et al., 2021).

Metal-based catalysts, often known as platinum-group elements (PGEs), including platinum, palladium, and ruthenium, that are embedded in fuel cells, pose an additional ecological concern. In nanoparticulate form, these rare metals are extremely poisonous despite being costly and rare. If these metals are not adequately removed or collected during fuel cell decommissioning, they could leak into groundwater or turn into airborne particle pollution (Mori et al., 2021; Khamaj et al., 2022). Because of their small size, they can enter aquatic life, plants, and even human tissues, where they can cause cytotoxicity, oxidative stress, and bioaccumulation in food chains. Specialized recuperation procedures have been created to counteract these risks. Heavy metals are extracted and recovered from used catalysts using pyrometallurgical and hydrometallurgical processes. Ion exchange membranes and electrochemical recovery are used in some sophisticated systems to separate precious metals with little secondary waste. Additionally, active compounds can be selectively removed from deteriorated membranes or electrodes via solvent extraction and chemical stripping (Talapatra et al., 2024). These tactics support resource circularity, the repurposing of essential components in the production of new fuel cells, as well as environmental preservation.

Regulatory agencies are starting to address these environmental issues. Directives like the Waste Electrical and Electronic Equipment (WEEE) and Restriction of Hazardous Substances (RoHS) have been implemented by regions such as the European Union. These directives encourage eco-design, extended producer responsibility (EPR), and mandated recycling targets for fuel cell systems. Similar to this, life cycle assessment (LCA) frameworks and ISO standards offer scientific instruments for assessing the environmental impact of fuel cell components from birth to death (Darvishi et al., 2022). Fuel cells, while seen as clean energy systems, bring several environmental challenges throughout their lifespan. From the start, getting materials like platinum-group metals (PGMs) used in catalysts requires deep mining, which can damage ecosystems and affect local communities. Making the parts also needs a lot of energy, which adds to greenhouse gas emissions and resource loss. On top of that, hydrogen, the main fuel for these systems, is extremely flammable, so transporting or storing it comes with risks of fire or explosions (Lotrič et al., 2021). At the end of a fuel cell's life, components can't just be thrown away; they often contain harmful materials like PGMs that need to be handled with care. If disposal isn't managed correctly, toxic metals can leak into the environment, harming soil and water systems. To manage this, detailed disposal protocols are put in place. Fuel cells must be taken apart methodically, so recyclable pieces and high-value materials like PGMs can be recovered and reused. This lowers the need for new mining and helps reduce overall waste. Some disposal methods also use incineration to recover energy from leftover materials (Ahmed et al., 2023). To keep things safe and eco-friendly, laws and policies such as extended producer responsibility make sure manufacturers help manage waste from the products they make. Tools like Life Cycle Assessment (LCA) are used to track the full environmental impact of fuel cells, from raw material extraction all the way to final disposal, helping researchers and industries find ways to make these technologies more sustainable in the long run (Ishimoto et al., 2024). In conclusion, the presence of heavy metal catalysts and synthetic membranes at the end-of-life stage of fuel cells results in substantial ecological and toxicological consequences. A more sustainable and responsible energy technology lifetime can be achieved by reducing these risks, nevertheless, with the use of better recovery techniques, circular design approaches, and changing regulatory frameworks.

13.5 NANOPARTICULATE EMISSIONS AND HUMAN TOXICOLOGY IN FUEL CELL CATALYST DEGRADATION

Nanoparticles are an important class of materials that exhibit special properties arising from their high surface area-to-volume ratio. Scanning transmission electron microscopy (STEM) has played an important role in nanoparticle characterization, owing to its high spatial resolution, which allows direct visualization of composition and morphology with atomic precision (Yu et al., 2022). Fuel cells rely on platinum-group metals (PGMs) like platinum (Pt), palladium (Pd), and ruthenium (Ru) as catalysts to accelerate electrochemical reactions. However, over time, these catalysts break down, leading to the unintended release of nanoparticles into the air. These nanoparticulate emissions are so small that they easily become airborne, making inhalation the primary route of exposure. Once inhaled, these ultrafine metal particles can penetrate deep into the lungs, triggering a chain of toxic events (Tompos, 2025). Fuel cell catalyst degradation can lead to nanoparticulate emissions, some of which can be toxic to humans, and understanding the mechanisms of toxicity is crucial for developing safer and more durable fuel cell technologies. Nanoparticulate Emissions from Fuel Cell Catalyst Degradation, during fuel cell operation, the catalyst (often platinum-based) can degrade, leading to the release of nanoparticles (Babaniyi et al., 2025a). Factors like potential cycling, harsh operating conditions, and the presence of impurities can contribute to catalyst degradation.

Types of Nanoparticles: Degradation can result in the formation of nanoparticles of various sizes, shapes, and compositions, including platinum, carbon, and other materials.

Nanoparticle Toxicity and Human Health:

Mechanism of Toxicity: Nanoparticles can interact with cells, generate reactive oxygen species (ROS), and induce oxidative stress, potentially leading to cell damage and inflammation.
Respiratory Effects: Exposure to nanoparticles can cause respiratory problems, including inflammation of the lungs.
Cardiovascular Effects: Nanoparticles can also contribute to cardiovascular disease.
Neurological Effects: Some studies suggest that nanoparticles may have neurological effects.
Cancer: There is evidence that certain nanoparticles may contribute to cancer development.

Nanoparticulate emissions from fuel cell catalyst degradation pose a significant human health risk, with bioaccumulation and oxidative stress driving toxicity. It is essential to take into account these mechanisms when creating safer fuel cell technologies and regulatory strategies to minimize nanoparticle exposure.

13.5.1 Bioaccumulation and Systemic Distribution

Unlike larger particles, nanoparticles evade the body's normal clearance mechanisms, allowing them to accumulate in lung tissue, enter the bloodstream, and reach distant organs like the brain, liver, and kidneys. The high surface area of these particles enhances their reactivity, meaning they bind to cellular components, disrupting normal function (Devasena et al., 2022). Over time, this bioaccumulation can lead to chronic toxicity, affecting multiple physiological systems. The environment of the living system, such as plants, aquatic organisms, and terrestrial organisms, including humans, are exposed to nanoparticle candidates primarily via three sources, namely production source, application sites, and disposal sites (Babaniyi et al., 2025b). The emission pathway

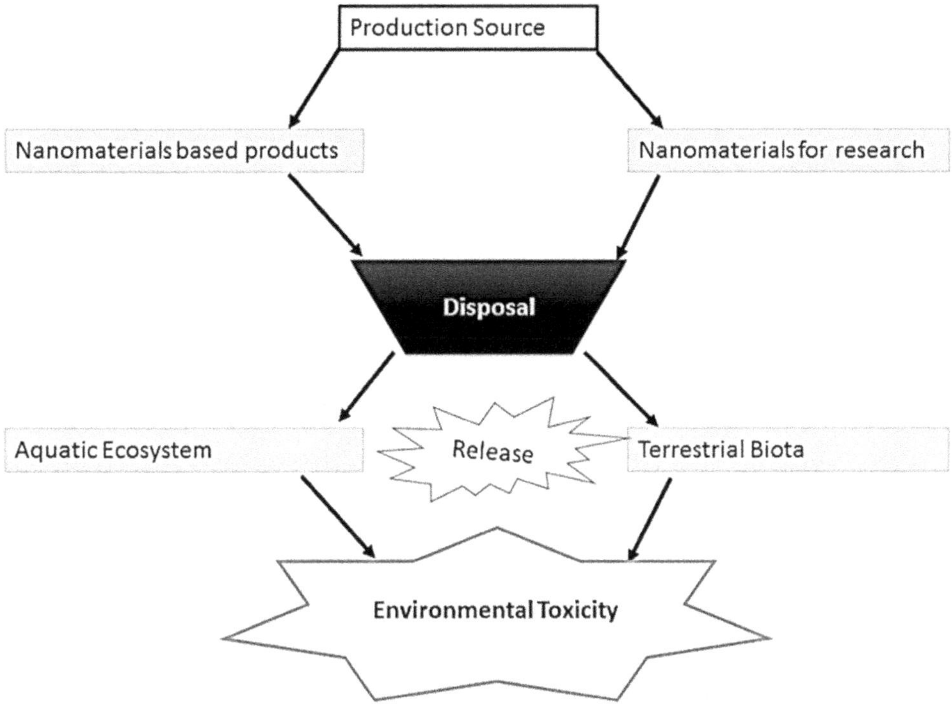

FIGURE 13.2 Nanoparticle source and environmental entry.

(Figure 13.2) and the release pattern usually vary for every type of nanoparticle. Titanium dioxide (TiO_2), zinc oxide (ZnO), and silver (Ag) nanoparticles are regarded as the major OECD (Organization for Economic Co-operation and Development) priority materials. Hence, nanotoxicologists and ecotoxicologists are currently focusing on the toxicological properties of these specific OECD-relevant materials. TiO_2 and ZnO nanoparticles are major constituents of cosmetics that are ultimately released into the water (Uddin et al., 2020). These nanoparticles are mainly released into the environment during water treatment process, and hence they mainly accumulate in the soil. TiO_2 nanoparticles are reported to induce size-dependent toxicity to algal species. ZnO and silver nanoparticles were, however, reported to undergo ready dissolution, and their ions were found to be toxic to the aquatic organisms.

The impact of engineered nanomaterials on nature and their accumulation in the human body and environment remains largely unknown. As these materials become more widespread, stakeholders across industries are growing increasingly concerned about their potential toxicity and risks at every stage of their life cycle, including development, manufacturing, use, and disposal (Uddin et al., 2020). When nanomaterials enter living organisms, they can cause cumulative toxic effects. In response, organisms may either store these particles within their tissues or eliminate them in a harmless form. Additionally, engineered nanomaterials may interfere with plant health and the food supply, raising concerns about agricultural and ecological stability (Tompos, 2025). Environmental factors also influence toxicity, as shown in studies on marine phytoplankton, key organisms in oceanic food webs and the global carbon cycle. In aquatic ecosystems, persistent hydrophobic chemicals can accumulate through various mechanisms:

Bioconcentration: direct absorption from water via gills or skin.
Ingestion: intake of contaminated particles.
Biomagnification: increasing concentrations through the food chain.

Even if acute or chronic toxicity is not immediately apparent in standard tests, bioaccumulation itself is a major hazard. Hence, it is important to know how nanomaterials interact with the environment, ecosystems, and human health, as this is crucial for assessing their long-term effects.

13.5.2 Oxidative Stress and Cytotoxic Pathways

At the cellular level, inhaled platinum-group nanoparticles generate reactive oxygen species (ROS), this is unstable molecules that damage proteins, lipids, and DNA. This oxidative stress activates inflammatory signaling pathways, disrupting mitochondrial function and leading to cell death (apoptosis) or uncontrolled cell survival (dysregulated proliferation). Lung epithelial cells and alveolar macrophages, responsible for immune defense, become overloaded with nanoparticles, leading to chronic pulmonary inflammation, fibrosis, and even carcinogenesis. The hazard potential of nanoparticles is shaped by a complex interaction of their physical and chemical traits (Manke et al., 2013; Mohamed Qasim et al., 2021). Their size, morphology, elemental makeup, and surface chemistry dictate their ability to penetrate biological barriers and trigger oxidative stress responses (Mohamed Qasim et al., 2021). Additionally, metal-based nanoparticles may leach ionic species, intensifying reactive oxygen species (ROS) production, thereby escalating cellular toxicity. Once inside the body, their interactions with cells, whether through membrane binding, endocytosis, or intracellular trafficking, determine their degree of toxicity, influencing immune responses, metabolic disruptions, and organ-specific damage (Al-Fahdawi et al., 2021).

Mitigating nanoparticle-induced oxidative stress requires multiple intervention strategies. Limiting direct exposure, whether by improving workplace safety, adopting cleaner technologies, or regulating emissions, is the first line of defense against nanoparticle-related harm (Balaguer-Trias et al., 2022). At the biological level, boosting antioxidant defenses through nutrient-rich diets or targeted supplementation can neutralize ROS overproduction (Čapek & Roušar, 2021). More so, holistic lifestyle adjustments which include consistent physical activity, nutrient-dense eating habits, and avoiding oxidative stressors like smoking, fortify the body's resilience against nanoparticle toxicity, offering a systemic safeguard against long-term health repercussions.

13.5.3 Pulmonary and Systemic Effects

Nanoparticulate emissions from fuel cell catalyst degradation can cause both pulmonary and systemic health effects, including respiratory and cardiovascular issues, and even cancer, due to their ability to induce oxidative stress, inflammation, and genotoxicity. Short-term exposure can cause airway irritation, inflammation, and oxidative lung damage, while long-term exposure increases the risk of respiratory diseases such as chronic obstructive pulmonary disease (COPD) and lung cancer (Xuan et al., 2023). Additionally, once nanoparticles enter circulation, they may cross the blood-brain barrier, leading to neuroinflammation and cognitive impairment. When nanoparticles enter the lungs, their tiny size allows deep penetration, where they settle in delicate tissues. This inhalation exposure can set off a storm of inflammation and oxidative stress, damaging lung cells and paving the way for fibrosis (Wu et al., 2021). Fibrosis is a condition where scar tissue hardens the lungs, making breathing difficult. Some metal-containing nanoparticles can trigger metal fume fever, a flu-like illness marked by fever and respiratory distress (Puri et al., 2023). Long-term exposure to certain nanoparticles may also increase the risk of lung cancer, raising serious concerns about occupational and environmental safety.

Beyond the lungs, nanoparticles infiltrate the bloodstream, where they hijack the cardiovascular system, potentially leading to heart disease and blood pressure irregularities. Once inside, they can also disrupt nerve function, raising the risk of neurological disorders. Some nanoparticles mimic hormones, interfering with the endocrine system and throwing off the body's delicate chemical balance (Chen et al., 2023). Additionally, these microscopic invaders can provoke an immune response, sometimes weakening the body's ability to fight infections. Their genotoxic potential, "the ability to damage DNA," raises alarms about cancer risk. As these particles accumulate in organs like the brain, liver, and kidneys, they can silently erode health, leading to long-term systemic damage (Yang et al., 2024). So, the interactions of nanoparticles with biological systems such as proteins, enzymes, and DNA can affect their toxicity. The next step in NPs toxicity research is to develop strategies to mitigate the risks associated with NPs exposure (Puri et al., 2023). This includes developing safe and efficient methods for NPs synthesis, identifying biomarkers of NPs toxicity, and developing protective measures for workers in industries that use NPs.

13.6 END-OF-LIFE FUEL CELL WASTE MANAGEMENT: HEAVY METAL CONTAMINATION AND ENVIRONMENTAL TOXICODYNAMICS

Fuel cells are a clean energy option, but they can harm the environment. At the end of their life cycle, they may release toxic heavy metals from broken parts. Spent electrodes contain platinum-group metals, rare-earth elements, and transition metals. Under certain chemical conditions, they can leach into the environment. This process releases toxic metals into soil and water. These metals go through complex environmental toxicodynamics. This affects their bioavailability, toxicity, and long-term impact on the ecosystem (Ghuge et al., 2023). To manage fuel cells at the end of their life, we need solid waste-handling strategies. These plans will help deal with heavy metal contamination and toxicodynamic effects. Sustainable recovery methods and new remediation technologies can help lower groundwater pollution and reduce ecological toxicity (Kumar et al., 2023). This approach supports a circular economy for fuel cell components and protects environmental health.

At the end of their lifespan, fuel cells release toxic heavy metals. These include lead, cadmium, and chromium. These metals can stay in the environment for a long time. Improper disposal of these materials can harm soil, water, and air (Zaynab et al., 2022). This spreads harmful substances into ecosystems. These metals build up in plants, animals, and humans. They move up the food chain through biomagnification. Environmental toxicodynamics looks at how harmful substances impact living systems. They can affect DNA function, enzyme activity, and immune responses. Heavy metals, like mercury and arsenic, are very dangerous (Mitra et al., 2022). They disrupt key biological processes and can cause serious health issues. These include organ failure, developmental disorders, and immune suppression.

13.6.1 Sustainable Remediation Strategies

Microbial fuel cells, or MFCs, present a viable way to remove heavy metal contamination. They clean wastewater of harmful metals while producing power. In an MFC, organic matter is broken down by bacteria in the anode chamber, which releases electrons that move to the cathode and neutralize heavy metal ions (Mitra et al., 2022; Kumar et al., 2023). Energy production and pollution control are two advantages of this sustainable and environmentally beneficial approach. Chemical precipitation, which converts heavy metals into solid particles for simpler removal, is one of the additional cleanup techniques. In phytoremediation, certain plants absorb pollutants from the environment, whereas in electrodeposition, metals are extracted using electric currents (Puri et al., 2023). Nanotechnology uses engineered particles to capture or break down pollutants. By combining these advanced techniques, we can achieve effective and long-term environmental protection.

13.6.2 Leaching Potential and Metal Dissolution

The electrodes and catalysts in fuel cells contain metals like platinum (Pt), palladium (Pd), ruthenium (Ru), cobalt (Co), and lanthanides that, when exposed to acidic, oxidative, or microbial conditions, undergo ion dissolution. This leaching process accelerates in landfills and wastewater systems, leading to the mobilization of hazardous metal ions into groundwater. Rare-earth metal dissolution, particularly from cerium (Ce), neodymium (Nd), and yttrium (Y), raises ecological concerns as these elements exhibit bioaccumulative properties, increasing their toxicity over time in aquatic and terrestrial ecosystems (Oladimeji et al., 2024). When absorbed by microorganisms, plants, and animals, these metals can interfere with enzymatic pathways, disrupt cellular homeostasis, and induce oxidative stress, ultimately cascading up the food chain and threatening biodiversity. When old fuel cells are thrown away, they can cause pollution from heavy metals, which is especially concerning for metals like rhodium, palladium, and platinum. These metals are common in fuel cell catalysts. If not disposed of properly, they can pollute the land, water, and air. Because they don't break down in the environment, they can build up over time (Bahmani et al., 2025). This can harm both human health and ecosystems. When they enter water bodies, they can harm fish and plants. Then, they move up the food chain, making each step more dangerous. These pollutants can harm organs, cause developmental issues, lead to nervous system disorders, and may even result in cancer in people (Abd Elnabi et al., 2023). The persistence of these metals makes them a serious environmental hazard, requiring advanced methods to prevent uncontrolled leakage.

To address this issue, recycling and recovery techniques can be used to extract valuable metals from old fuel cells, reducing waste and contamination. Proper disposal methods prevent toxic leaching. Cremation with pollution controls and landfills with barriers are good methods for this. You can further reduce risks by using stabilization methods (Abd Elnabi et al., 2023). One approach is chemical treatments that neutralize harmful chemicals. Another is encapsulation, which traps metals in solid materials. A novel technique for removing metals from wastewater is the use of microbial fuel cells, or MFCs. Microbial fuel cells, or MFCs, are a new way to remove metals from wastewater which also produce electricity. Other alternative techniques include ion exchange, adsorption, bioremediation, and advanced membrane filtering. MFCs are just one option among these methods (Bahmani et al., 2025). Bioremediation breaks down heavy metals using bacteria. Ion exchange and adsorption eliminate toxins using certain materials. These methods are all used to clean wastewater. These methods lower heavy metal exposure. This helps safeguard the environment and public health.

13.6.3 Environmental and Ecotoxicological Consequences

Once heavy metals enter water bodies, they form stable complexes with organic and inorganic molecules, enhancing their persistence and toxicity. These contaminants disrupt aquatic microbiomes, hinder photosynthetic efficiency in algae, and induce biochemical stress responses in fish and amphibians. In soil ecosystems, metal ions interfere with nutrient cycling, inhibit plant root development, and alter microbial activity, leading to reduced soil fertility and ecological imbalances (Bahmani et al., 2025). Moreover, some transition metals, such as cobalt and nickel, can undergo redox cycling, producing reactive oxygen species (ROS) that cause DNA damage, protein oxidation, and lipid peroxidation in exposed organisms. Fuel cells can release toxic heavy metals, like lead, mercury, and cadmium, at the end of their life. If not disposed of properly, these metals can harm the environment (Oladimeji et al., 2024). These metals come from catalysts and fuel cell parts. When they leak into soil and water, they can harm the environment and our health. Industrial production and recycling can release harmful substances. These can pollute the air, rivers, and landfills. These metals don't break down, but they can last for decades and pollute farmland, drinking water, and air. Heavy metals in soil lower their fertility (Abd Elnabi et al., 2023). This harms plant growth and food production.

In water, they hurt fish and other aquatic life. Airborne metals from burning waste and poor recycling also lead to pollution-related diseases. As these metals build up in plants and animals, the danger grows. This can lead to human health problems such as brain damage, organ failure, and cancer. There are many ways to combat pollution, such as soil cleaning, which removes pollutants from the soil, phytoremediation, which absorbs toxins by plants, and safer recycling and proper disposal, which keep heavy metals out of the environment (Briffa et al., 2020; Mitra et al., 2022). Scientists are developing fuel cells with fewer toxic materials, which will reduce future pollution hazards. Stabilization of chemicals aids in pollution management. Heavy metals are a major source of pollution. They are harmful alone and can become very toxic when mixed. Therefore, to lessen the negative impacts of heavy metals, robust environmental regulations and cutting-edge recovery technology are essential.

13.6.4 Advanced Remediation Strategies

To mitigate groundwater contamination and ecosystem toxicity, cutting-edge remediation technologies are essential. Electrochemical recovery systems can extract valuable metals from fuel cell waste, reducing environmental burden while enabling metal reuse. Bioremediation techniques, using metal-resistant bacteria and fungi, facilitate the immobilization and transformation of toxic metal ions into less harmful forms. Nanoparticle-based adsorption materials, such as graphene oxide, zeolites, and functionalized biochar, offer high-efficiency metal capture to prevent leaching into water systems (Ghafoor et al., 2023). Additionally, green chemistry approaches, including eco-friendly chelation agents and phytoremediation via hyperaccumulator plants, help in stabilizing and extracting hazardous elements from contaminated environments. When fuel cells reach the end of their useful lives, they may degrade. Lead, cadmium, nickel, and other hazardous heavy metals are frequently released by them. For years, these metals can remain in the environment. Water and soil can absorb these metals (Sun et al., 2022). Both human health and ecosystems are at risk from this. Through bioaccumulation, they accumulate within organisms. Then, because of biomagnification, they become more concentrated higher up the food chain. Scientists employ biological and chemical strategies to combat this. They eliminate or neutralize contaminants before they proliferate. Chemical precipitation makes metals turn into solid forms that don't dissolve (Babatunde et al., 2024). These solids can then be filtered out. Ion exchange materials, on the other hand, attach to toxic metal ions. They help remove these harmful ions from dirty water or soil. Materials like activated carbon or nanomaterials are used in adsorption procedures. Heavy metals are trapped on the surfaces of these materials.

Heavy metal concentrations in the environment are successfully decreased by this method. Bioremediation offers a natural and eco-friendly solution to manage heavy metals. It absorbs, dissolves, or transforms these metals into harmless ones by using microbes. Microbes can extract metals from contaminated areas through biosorption (Cetinkaya et al., 2022). By using bacteria, bioleaching transforms heavy metal compounds into forms that are safe to remove. Microbes convert metals into stable minerals with the aid of biomineralization. This procedure prevents additional harm. Additional remediation techniques include treating pollution on-site (in-situ) or off-site (ex-situ), using membrane filtration to physically separate toxic compounds from water, and combining polluted soil with clean soil to reduce toxicity (Ubani et al., 2024). Selecting the appropriate cleanup technique is essential. Cost, sustainability, and long-term efficacy all play a role. In this manner, a landscape contaminated by industrial pollution won't be passed down to future generations.

13.7 CYBER-BIOSECURITY CONVERGENCES: THE INTERSECTION OF DIGITAL THREATS AND BIOLOGICAL SAFETY IN A HYPERCONNECTED WORLD

Cyber-biosecurity convergence describes the merging of digital defence and biological protection in an age where genomic databases, synthetic DNA designs, and AI-driven biotech platforms are tightly interwoven (Figure 13.3). As laboratories digitize bioinformatics and store pathogen blueprints in cloud systems, threat vectors like ransomware, algorithmic manipulation, and malware become direct biohazards (Cetinkaya et al., 2022; Ubani et al., 2024). A breach could mean stolen vaccine formulas, altered CRISPR sequences, or shutdowns of ventilator networks in hospitals. The fusion of bioinformatics vulnerabilities with cyber infiltration pathways demands encryption

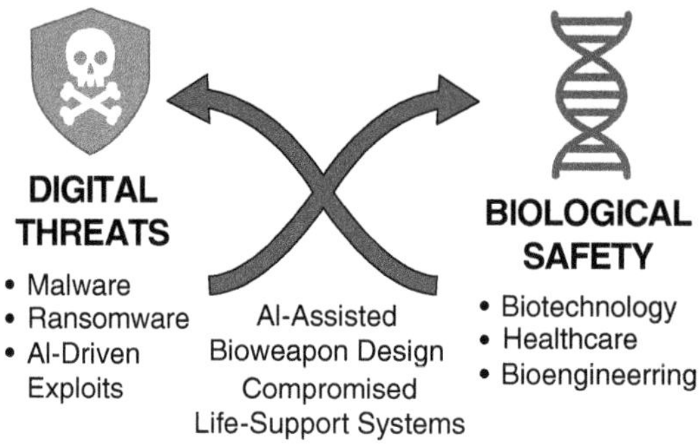

FIGURE 13.3 Cyber-biosecurity convergences.

protocols, firewall reinforcement, biometric access controls, and secure neural-network training models. Beyond threats, this interface holds promise "machine learning enabling precision therapy, or robotics accelerating genetic sequencing" but must be governed by bioethical algorithms, secure data stewardship, and transnational policy harmonization to mitigate misuse (Babatunde et al., 2024).

However, in today's hyperconnected world, cyber-biosecurity is becoming a critical issue. This term refers to the point where cyber threats and biological safety overlap, where digital systems that support biotechnology, healthcare, and bioengineering become targets or tools for malicious attacks (Sun et al., 2022). Within this new environment, the same networks that control hospital ventilators, DNA sequencing tools, or synthetic biology labs can be infiltrated by malware, ransomware, or artificial intelligence-driven exploits, creating dangerous vulnerabilities that span both the digital and biological domains (Briffa et al., 2020). This convergence creates complex, unpredictable risks, where data breaches may not just leak personal information but could also disrupt life-support machines or corrupt vaccine development platforms.

A particularly chilling example is AI-assisted bioweapon design. With advanced algorithms, malicious actors can now simulate or even generate genetic blueprints for harmful viruses or toxins, tasks that once took years now take hours using cloud-based bioinformatics tools (Ghafoor et al., 2023). Meanwhile, cyberattacks on healthcare infrastructure "such as hijacking infusion pumps or manipulating diagnostic imaging systems" can lead to direct harm or death. Hospitals rely on interconnected systems for critical care, so a well-timed cyber intrusion can disable

13.8 CONCLUSION

As fuel cell technologies gain prominence in the push for low-carbon energy, their health, safety, and environmental risks must be critically addressed across their entire life cycle. While fuel cells offer significant ecological benefits by reducing greenhouse gas emissions, their components introduce persistent chemical hazards, physical stressors, and emerging cyber-biological vulnerabilities. Understanding the interconnected risk landscape is vital for developing robust safety and sustainability frameworks. Hydrogen, the primary fuel in many fuel cell systems, is highly flammable, and its invisible leaks pose explosion risks, especially in confined industrial or transport settings. Additionally, some fuel cells operate at high temperatures, raising the threat of thermal runaway, which may result in fires or mechanical failures. Moreover, platinum-based catalysts, corrosive electrolytes, and synthetic membranes contribute to worker exposure, environmental contamination, and toxic degradation by-products if not properly handled. Of particular concern is the release of nanoparticles during catalyst degradation. These ultrafine materials, often composed of platinum, carbon, or other metals, infiltrate the human body via inhalation, bypassing natural defenses. Inside, they can generate reactive oxygen species (ROS), damaging cells, DNA, and vital organs. Their bioaccumulation further exacerbates long-term toxicity, disrupting neurological, respiratory, cardiovascular, and endocrine systems. The toxicity profile of these nanoparticles demands more stringent emission controls, exposure monitoring, and health surveillance strategies.

Furthermore, chronic exposure to endocrine disruptors and neurotoxic chemicals, found in common industrial materials, poses systemic health risks. These substances, even in low doses, can mimic or block hormones, leading to fertility issues, developmental abnormalities, and immune suppression. Their lipophilic nature and resistance to degradation allow them to persist in ecosystems, bioaccumulate, and biomagnify through food chains. The end-of-life stage of fuel cells presents another set of hazards. Improper disposal can result in heavy metal leaching, contaminating groundwater, and reducing soil fertility. Environmental toxicodynamics, "the study of how these toxicants interact with living systems," reveals cascading effects on plants, animals, and humans. To counter these threats, advanced remediation technologies such as chemical precipitation, bioremediation, microbial fuel cells, and membrane filtration offer scalable, sustainable solutions for recovering metals and reducing toxicity.

Overlaying all of these are growing concerns around cyber-biosecurity. With increased digitization of biological data and biotechnological processes, new vulnerabilities have emerged. Cyberattacks targeting hospital systems, genomic databases, or automated lab equipment can compromise patient care, research integrity, and public health. In extreme scenarios, AI-driven tools may be hijacked to design or manipulate synthetic biological materials, transforming digital threats into biological risks. In conclusion, fuel cell sustainability is not simply about energy; it encompasses human health, environmental safety, and technological resilience. A truly sustainable energy system requires coordinated action across scientific disciplines, combining biomedical toxicology, environmental engineering, cybersecurity, and policy development. Only with integrated risk management, circular material use, and cross-sector collaboration can we ensure that the promise of clean energy does not come at the cost of biological and environmental integrity.

REFERENCES

Abd Elnabi, M. K., Elkaliny, N. E., Elyazied, M. M., Azab, S. H., Elkhalifa, S. A., Elmasry, S., ... & Mahmoud, Y. A. G. (2023). Toxicity of heavy metals and recent advances in their removal: a review. *Toxics, 11*(7), 580.

Ahmed, H., Adebayo, P., Ahmed, M., & Arbab, A. I. (2023). Life cycle assessment of hydrogen fuel cells: environmental impact and sustainability. *Life, 13*(1).

Al-Fahdawi, M. Q., Al-Doghachi, F. A., Abdullah, Q. K., Hammad, R. T., Rasedee, A., Ibrahim, W. N., ... & Al-Qubaisi, M. S. (2021). Oxidative stress cytotoxicity induced by platinum-doped magnesia nanoparticles in cancer cells. *Biomedicine & Pharmacotherapy, 138*, 111483.

Babaniyi, G. G., Akor, U. J. & Odeseye, A. A. (2025a). Pesticide contributions to greenhouse gas emissions. In: Babaniyi, B.R., Babaniyi, E.E. (eds) *The Interplay of Pesticides and Climate Change* (pp 173 – 230). Springer, Cham. https://doi.org/10.1007/978-3-031-81669-7_8

Babaniyi, G. G., Babaniyi, B. R., Olamide, G. I. & Akor, U. J. (2025b). Molecular mechanisms of plant-nanoparticle interactions. In Pabbati, R., Chepuri, K., Reddy, K. V., Maddela, N. R. (eds) *Plant-Based Nanoparticle Synthesis for Sustainable Agriculture (9781003477730)*. CRC Press. https://doi.org/10.1201/9781003477730-10

Babatunde, O. M., Akintayo, B. D., Emezirinwune, M. U., & Olanrewaju, O. A. (2024). Environmental impact assessment of a 1 kW proton-exchange membrane fuel cell: a mid-point and end-point analysis. *Hydrogen, 5*(2), 352–373.

Bahmani, Z., Nabizadeh, R., Yaghmaeian, K., & Yunesian, M. (2025). Evaluation of potentially toxic elements and pharmaceutical compounds in leachate and exhaust air from non-incineration medical waste treatment devices. *Scientific Reports, 15*(1), 6395.

Balaguer-Trias, J., Deepika, D., Schuhmacher, M., & Kumar, V. (2022). Impact of contaminants on microbiota: linking the gut–brain axis with neurotoxicity. *International Journal of Environmental Research and Public Health, 19*(3), 1368.

Boamah, B. B. (2020). *Identifying Target Organs and Candidate Contaminants Based on Adverse Outcomes Following Sub-Chronic Exposure in Rats to Contaminated Soil Extracts from a Pesticide Manufacturing Site* (Doctoral dissertation, University of Saskatchewan).

Briffa, J., Sinagra, E., & Blundell, R. (2020). Heavy metal pollution in the environment and their toxicological effects on humans. *Heliyon*, *6*(9), https://doi.org/10.1016/j.heliyon.2020.e04691

Čapek, J., & Roušar, T. (2021). Detection of oxidative stress induced by nanomaterials in cells—the roles of reactive oxygen species and glutathione. *Molecules*, *26*(16), 4710.

Cetinkaya, A. Y., Kuzu, S. L., & Bilgili, L. (2022). Development of an MFC-biosensor for determination of Pb+ 2: An assessment from computational fluid dynamics and life cycle assessment perspectives. *Environmental Monitoring and Assessment*, *194*(4), 245.

Chen, X., Wang, S., Mao, X., Xiang, X., Ye, S., Chen, J., ... & Wang, X. (2023). Adverse health effects of emerging contaminants on inflammatory bowel disease. *Frontiers in Public Health*, *11*, 1140786.

Darvishi, E., Ghasemi, F., Sadeghi, F., Abedi, K., Rahmati, S., & Sadeghzade, G. (2022). Risk assessment of the work-related musculoskeletal disorders based on individual characteristics using path analysis models. *BMC Musculoskeletal Disorders*, *23*(1), 616.

Devasena, T., Iffath, B., Renjith Kumar, R., Muninathan, N., Baskaran, K., Srinivasan, T., & John, S. T. (2022). Insights on the dynamics and toxicity of nanoparticles in environmental matrices. *Bioinorganic Chemistry and Applications*, *2022*(1), 4348149.

Duh-Leong, C., Maffini, M. V., Kassotis, C. D., Vandenberg, L. N., & Trasande, L. (2023). The regulation of endocrine-disrupting chemicals to minimize their impact on health. *Nature Reviews Endocrinology*, *19*(10), 600–614.

Ghafoor, I., Naz, T., Nawaz, S. N. S., Iqbal, M. M., Iqbal, S., Akhtar, T., & Shurjeel, H. K. (2023). A review on emerging strategies for heavy metal remediation from various sources. *Journal of Quality Assurance in Agricultural Sciences*, *3*(01), 38–46.

Ghuge, S. A., Nikalje, G. C., Kadam, U. S., Suprasanna, P., & Hong, J. C. (2023). Comprehensive mechanisms of heavy metal toxicity in plants, detoxification, and remediation. *Journal of Hazardous Materials*, *450*, 131039.

Gómez, M. M. (2020). Prediction of work-related musculoskeletal discomfort in the meat processing industry using statistical models. *International Journal of Industrial Ergonomics*, *75*, 102876.

Iqubal, A., Ahmed, M., Ahmad, S., Sahoo, C. R., Iqubal, M. K., & Haque, S. E. (2020). Environmental neurotoxic pollutants. *Environmental Science and Pollution Research*, *27*, 41175–41198.

Ishimoto, Y., Wulf, C., Schonhoff, A., & Kuckshinrichs, W. (2024). Life cycle costing approaches of fuel cell and hydrogen systems: A literature review. *International Journal of Hydrogen Energy*, *54*, 361–374.

Khamaj, A. M., Ali, A. M., & Alam, M. M. (2022). Investigating factors affecting musculoskeletal disorders: Predictive models for identifying caregivers at risk. *Work*, *72*(4), 1311–1320.

Khan, Y., Sattar, A., Ullah, S. A., Bibi, M., Sarfraz, H., & Mukhtiar, A. (2024). Chemicals that disrupt the endocrine system and their effects on behavior. *Kashmir Journal of Science*, *3*(2).

Kowalczyk, A., Wrzecińska, M., Czerniawska-Piątkowska, E., Araújo, J. P., & Cwynar, P. (2022). Molecular consequences of the exposure to toxic substances for the endocrine system of females. *Biomedicine & Pharmacotherapy*, *155*, 113730.

Kumar, M., Sarma, D. K., Shubham, S., Kumawat, M., Verma, V., Prakash, A., & Tiwari, R. (2020). Environmental endocrine-disrupting chemical exposure: role in non-communicable diseases. *Frontiers in Public Health*, *8*, 553850.

Kumar, V., Tyagi, S. K., Kumar, K., & Parmar, R. S. (2023). Heavy metal-induced pollution in the environment through waste disposal. *International Journal of Research Publication and Reviews*, *4*(7), 1205–1210.

Lotrič, A., Sekavčnik, M., Kuštrin, I., & Mori, M. (2021). Life-cycle assessment of hydrogen technologies with the focus on EU critical raw materials and end-of-life strategies. *International Journal of Hydrogen Energy*, *46*(16), 10143–10160.

Ma, C. S., Li, D. L., Wang, F., Wang, J. P., & He, M. T. (2024). Neurotoxicity from long-term exposure to 6-PPDQ: Recent advances. *Ecotoxicology and Environmental Safety*, *282*, 116689.

Manke, A., Wang, L., & Rojanasakul, Y. (2013). Mechanisms of nanoparticle-induced oxidative stress and toxicity. *BioMed Research International*, *2013*(1), 942916.

Mesnil, M., Defamie, N., Naus, C., & Sarrouilhe, D. (2020). Brain disorders and chemical pollutants: A gap junction link? *Biomolecules*, *11*(1), 51.

Mitra, S., Chakraborty, A. J., Tareq, A. M., Emran, T. B., Nainu, F., Khusro, A., ... & Simal-Gandara, J. (2022). Impact of heavy metals on the environment and human health: Novel therapeutic insights to counter the toxicity. *Journal of King Saud University-Science*, *34*(3), 101865.

Mohamed Qasim, A. F., Al-Doghachi, F. A., Abdullah, Q. K., Hammad, R. T., Rasedee, A., Ibrahim, W. N., ... & Al-Qubaisi, M. S. (2021). Oxidative stress cytotoxicity induced by platinum-doped magnesia nanoparticles in cancer cells.

Molot, J., Sears, M., Marshall, L. M., & Bray, R. I. (2022). Neurological susceptibility to environmental exposures: pathophysiological mechanisms in neurodegeneration and multiple chemical sensitivity. *Reviews on Environmental Health*, *37*(4), 509–530.

Mori, M., Gramc, J., Hojkar, D., Lotrič, A., Smeacetto, F., Fiorilli, S., & Fiore, S. (2025). New life cycle inventories for end-of-life solid oxide cells based on novel recycling processes for critical solid oxide cell materials. *International Journal of Hydrogen Energy*, *104*, 635–650.

Mori, M., Stropnik, R., Sekavčnik, M., & Lotrič, A. (2021). Criticality and life-cycle assessment of materials used in fuel-cell and hydrogen technologies. *Sustainability*, *13*(6), 3565.

Oladimeji, T. E., Oyedemi, M., Emetere, M. E., Agboola, O., Adeoye, J. B., & Odunlami, O. A. (2024). Review on the impact of heavy metals from industrial wastewater effluent and removal technologies. *Heliyon*, *10*(23), https://doi.org/10.1016/j.heliyon.2024.e40370

Pei, D. S., & Liu, Y. (Eds.). (2025). *Toxicological Assessment of Combined Chemicals in the Environment*. John Wiley & Sons.

Plunk, E. C., & Richards, S. M. (2020). Endocrine-disrupting air pollutants and their effects on the hypothalamus-pituitary-gonadal axis. *International Journal of Molecular Sciences*, *21*(23), 9191.

Puri, M., Gandhi, K., & Suresh Kumar, M. (2023). A global overview of endocrine disrupting chemicals in the environment: Occurrence, effects, and treatment methods. *International Journal of Environmental Science and Technology*, *20*(11), 12875–12902.

Raja, G. L., Subhashree, K. D., & Kantayya, K. E. (2022). In utero exposure to endocrine disruptors and developmental neurotoxicity: Implications for behavioural and neurological disorders in adult life. *Environmental Research*, *203*, 111829.

Reis, J., & Spencer, P. S. (2024). An introduction to environmental neurotoxicology: Lessons from a clinical perspective. *Journal of the Neurological Sciences*, *463*, 123108.

Scott, N. B., & Pocock, N. S. (2021). The health impacts of hazardous chemical exposures among child labourers in low-and middle-income countries. *International Journal of Environmental Research and Public Health*, *18*(10), 5496.

Shafique, M., & Luo, X. (2019). Nanotechnology in transportation vehicles: An overview of its applications, environmental, health and safety concerns. *Materials*, *12*(15), 2493.

Shi, T., Zhang, Q., Chen, X., Mao, G., Feng, W., Yang, L., ... & Chen, Y. (2024). Overview of deltamethrin residues and toxic effects in the global environment. *Environmental Geochemistry and Health*, *46*(8), 271.

Sun, Y., Wang, H., Long, X., Xi, H., Biao, P., & Yang, W. (2022). Advance in remediated of heavy metals by soil microbial fuel cells: Mechanism and application. *Frontiers in Microbiology*, *13*, 997732.

Talapatra, S., Parvez, M. S., Saha, P., & Kibria, M. G. (2024). Assessing the impact of critical risk factors on the development of musculoskeletal disorders: a structural equation modelling approach. *Theoretical Issues in Ergonomics Science*, *25*(3), 343–368.

Tawalbeh, M., Alarab, S., Al-Othman, A., & Javed, R. M. N. (2022). The operating parameters, structural composition, and fuel sustainability aspects of PEM fuel cells: a mini review. *Fuels*, *3*(3), 449–474.

Tompos, A. (2025). Nanomaterials applied to fuel cells and catalysts. *Nanomaterials*, *15*(5), 347.

Ubani, O., Uba, B. O., Modise, S. J., Okoye, E. L., Omeazu, S. C., Ndibe, C. R., ... & Dokubo, C. U. (2024). Responses of Selenastrum capricornutum, Eisenia fetida, Brassica nigra and Sorghum bicolor to sent phone battery toxicity. *Multidisciplinary Science Journal*, *6*(7), 2024107.

Uddin, M. N., Desai, F., & Asmatulu, E. (2020). Engineered nanomaterials in the environment: bioaccumulation, biomagnification and biotransformation. *Environmental Chemistry Letters*, *18*(4), 1073–1083.

Wu, Y., Wang, M., Luo, S., Gu, Y., Nie, D., Xu, Z., ... & Ge, X. (2021). Comparative toxic effects of manufactured nanoparticles and atmospheric particulate matter in human lung epithelial cells. *International Journal of Environmental Research and Public Health*, *18*(1), 22.

Xuan, L., Ju, Z., Skonieczna, M., Zhou, P. K., & Huang, R. (2023). Nanoparticles-induced potential toxicity on human health: applications, toxicity mechanisms, and evaluation models. *MedComm*, *4*(4), e327.

Yang, Z., Zhang, J., Wang, M., Wang, X., Liu, H., Zhang, F., & Fan, H. (2024). Prenatal endocrine-disrupting chemicals exposure and impact on offspring neurodevelopment: A systematic review and meta-analysis. *NeuroToxicology*, *103*, 335–357.

Yazdi, M., Moradi, R., Pirbalouti, R. G., Zarei, E., & Li, H. (2023). Enabling safe and sustainable hydrogen mobility: circular economy-driven management of hydrogen vehicle safety. *Processes*, *11*(9), 2730.

Yu, H., Zachman, M. J., Reeves, K. S., Park, J. H., Kariuki, N. N., Hu, L., ... & Cullen, D. A. (2022). Tracking nanoparticle degradation across fuel cell electrodes by automated analytical electron microscopy. *ACS Nano*, *16*(8), 12083–12094.

Zaynab, M., Al-Yahyai, R., Ameen, A., Sharif, Y., Ali, L., Fatima, M., ... & Li, S. (2022). Health and environmental effects of heavy metals. *Journal of King Saud University-Science*, *34*(1), 101653.

14 Fuel Cell in Sustainable Agriculture

Balwant Singh, Mukul Machhindra Barwant, Debasree Lodh, and Anand Kumar

14.1 INTRODUCTION

The world's need for sustainable agriculture has increased as a result of the urgent challenges posed by climate change, environmental degradation, and the increasing demand for food security. Agriculture, the backbone of human society, is by nature energy-hungry, dependent to a large extent on fossil fuels for powering irrigation, machinery, fertilizer manufacture, and food transport (FAO, 2021). But the over-reliance on these non-renewable sources of energy has not only resulted in greenhouse gas emissions but also caused pollution of the environment and depletion of resources. It is thus important to incorporate renewable and clean energy technologies into agriculture for the purpose of sustainability. Of these, fuel cells stand out as a viable alternative energy source owing to their high energy conversion efficiency, low environmental footprint, and ability to fit into decentralized energy systems.

Fuel cells are electrochemical converters that directly convert chemical energy into electrical energy by redox reactions, usually between hydrogen and oxygen, with water and heat as the sole by-products (Rao et al., 2022). In contrast to traditional combustion-based power systems, fuel cells are quiet, produce minimal pollutants, and have higher energy conversion efficiencies. These qualities make fuel cells particularly well-suited for off-grid and remote agricultural uses, including the operation of irrigation systems, greenhouses, sensors, and farm machinery in a clean way (Yamamoto et al., 2019).

In the last decade or so, advances in fuel cell technologies—mostly proton exchange membrane fuel cells (PEMFCs), solid oxide fuel cells (SOFCs), and microbial fuel cells (MFCs)—have made their implementation in agriculture systems possible. As an example, microbial fuel cells have the advantage of utilizing agricultural waste to create electricity in the process, and hence fulfilling principles of circular economy and green waste management practices (Logan et al., 2023). Also, solar-powered water electrolysis may be combined with hydrogen-powered fuel cells to generate off-grid rural energy solutions for farm use and cut down on dependence on traditional fuels and enhance energy access (Kalamaras & Efstathiou, 2013).

In addition, fuel cells will find applications in precision agriculture based on real-time feedback and automation for maximizing the utilization of resources. Low-power fuel cell systems for powering sensors, communication, and monitoring stations can promote data-based agriculture without raising the carbon burden (Mekhilef et al., 2020). As farming becomes more digital and sustainable, the contribution of fuel cells to enabling robust and environmentally friendly agricultural systems is gaining prominence.

This chapter discusses how fuel cells can promote sustainable agriculture by presenting their principles, technologies, environmental advantages, and field applications. It focuses on the synergy between clean energy and agriculture to show how fuel cells can help achieve a greener, more efficient, and climate-resilient agri-food system.

DOI: 10.1201/9781003645405-14

14.2 BASICS OF FUEL CELL TECHNOLOGY

Fuel cell technology is a promising area for providing clean and efficient energy solutions, especially in uses like sustainable agriculture. A fuel cell is an electrochemical device that converts the chemical energy of a fuel—usually hydrogen—directly into electricity, water, and heat, without combustion. This direct conversion technique leads to improved energy efficiency and much lower emissions of greenhouse gases than traditional fossil fuel-based power systems (Baroutaji et al., 2020). The fundamental operation of a fuel cell is through an anode, a cathode, and an electrolyte. Hydrogen is oxidized at the anode to yield protons and electrons. The electrons travel through an external circuit, delivering electricity, while the protons move through the electrolyte to the cathode, where they react with oxygen to produce water (Figure 14.1).

A number of fuel cell types are applicable to agricultural uses because they are scalable, fuel flexible, and have operational conditions. Among them, Proton Exchange Membrane Fuel Cells (PEMFCs) are the most widely utilized in portable and transportation-based agricultural systems. They work at comparatively low temperatures (between 60 and 80°C), possess high power density, and provide fast startup, and so are well suited for powering greenhouse systems, small tractors, or irrigation pumps (Kamarudin et al., 2021). Solid Oxide Fuel Cells (SOFCs), however, work under high temperatures (600–1000°C) and can be fueled by a range of fuels including hydrogen, methane, and bio-derived gases. This fuel flexibility is suitable for SOFCs for stationary agricultural power systems where heat and power co-generation is useful, e.g., in greenhouses or village-level agro-processing units (Zhang et al., 2020).

Another cutting-edge variant that is being considered in the context of agriculture is the Microbial Fuel Cell (MFC). These systems depend on the metabolic activity of microbes to degrade organic substances—usually from agricultural waste—and produce electricity along the way. MFCs not only represent decentralized energy resources but also represent a sustainable process for waste treatment, thus resolving two important demands in agriculture: waste management and energy

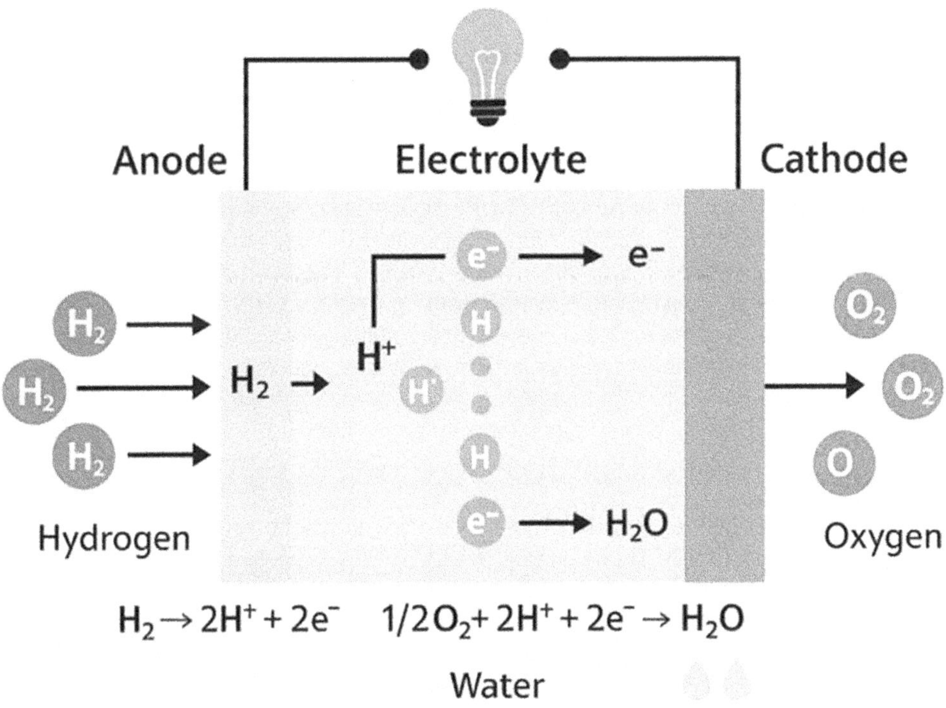

FIGURE 14.1 Fuel cell and their components.

supply (Logan et al., 2023). They are particularly well-suited for low-power applications such as environmental sensors, small lighting applications, or valve control systems for precision irrigation.

Fuel source selection is an important determinant of fuel cell system performance and sustainability. Hydrogen, the most widely used fuel, may be generated using a number of processes, including steam methane reforming, water electrolysis, and biomass gasification. For sustainable agriculture, focus is put on green methods of hydrogen production like electrolysis with renewable energy sources (wind, solar) and biomass hydrogen production, where crop residues, manure, or agro-industrial wastes are used (Staffell et al., 2019). These strategies allow farms to produce their own hydrogen from resources present in the immediate surroundings, and thus, the energy independence of farms increases, as well as lessens environmental stress.

In conclusion, the basic principles of fuel cell operation, together with developments in cell types and fuel sources, make this technology a key part of the shift toward energy-efficient, low-carbon agricultural systems. As research and deployment continue to grow, fuel cells are likely to play a central role in facilitating sustainable practice across a wide range of agricultural activities.

14.2.1 Principles of Fuel Cell Operation

Fuel cells are electrochemical converters that convert the chemical energy of a fuel directly into electricity by redox reactions without combustion. A common fuel cell includes an anode, a cathode, and an electrolyte membrane that enables ion transport without direct mixing of reactants (Barbir, 2020). In the anode, the fuel, usually hydrogen, is oxidized, releasing electrons and generating protons. These electrons move through an external circuit, producing an electric current, while the protons move through the electrolyte to the cathode. Oxygen molecules at the cathode combine with the protons and electrons to produce water as a by-product.

Fuel cell efficiency is generally 40–60% but is higher when in cogeneration schemes, with waste heat utilized as well (Kauranen et al., 2022). Fuel cells are also modular, scalable, and clean in nature, emitting only water and trace amounts of NOx or CO_2 on pure hydrogen (Kauranen et al., 2022). The low emission levels and silent nature of fuel cells make them amenable for deployment in agriculture systems, particularly where there are remote or off-grid locations.

14.2.2 Types of Fuel Cells Relevant to Agriculture

There are a number of fuel cells with possible agricultural applications because of their special operating conditions and fuel versatility. Among them, Proton Exchange Membrane Fuel Cells (PEMFCs), Solid Oxide Fuel Cells (SOFCs), and Microbial Fuel Cells (MFCs) are the most relevant.

PEMFCs run at comparatively low temperatures (60–80°C) and utilize a solid polymer electrolyte membrane, normally constituted of perfluorosulfonic acid (O'Hayre et al., 2016). The fuel cells are highly appreciated for their fast start-up time, high power density, and compactness, due to which they are perfectly suited for mobile farm equipment, UAVs, and portable power packs. The primary difficulty is the need for highly pure hydrogen and the membrane's contaminant sensitivity. In farming, PEMFCs power automatic machinery like tractors, irrigation systems, and greenhouse management equipment, to enhance energy efficiency and low emissions (Kamarudin et al., 2021).

SOFCs work at elevated temperatures (600–1000°C) with a solid ceramic electrolyte, such as yttria-stabilized zirconia (YSZ). They can operate with hydrocarbons like biogas, methane, or ethanol directly, which are usually by-products of agricultural residues (Zhang et al., 2020). The elevated operating temperature favors internal fuel reforming as well as high efficiency (up to 85% for combined heat and power applications). SOFCs are well-suited for stationary applications in rural farms where agricultural residues can be converted into syngas or biogas, supplying both electricity and heat for in-situ operations (Ishizaki et al., 2018).

MFCs employ the metabolic activity of microorganisms to transform organic substrates—like agricultural residues, animal waste, and wastewater—into electrical energy (Logan et al., 2023).

Bacteria in the anode oxidize the organic substrate and emit electrons, which travel to the cathode, where oxygen is reduced. MFCs have a twofold advantage: renewable electricity production and sustainable waste treatment. They have applications in agriculture such as powering low-power devices, treating wastewater from farms, and recycling nutrients (Pant et al., 2010). They may have low power output but have potential for decentralized waste-to-energy conversion in rural areas.

14.2.3 Fuel Sources and Hydrogen Production

The feasibility of fuel cell systems in agriculture is mostly based on fuel availability and sustainability. Hydrogen, the most widely used fuel for fuel cells, can be obtained from a variety of means-

Electrolysis of Water: Using renewable electricity (e.g., solar or wind), electrolysis provides a clean route to the production of green hydrogen. The process is most suitable for off-grid farms equipped with solar PV systems (Staffell et al., 2019).

Steam Methane Reforming (SMR): The most common approach, though it releases CO_2 unless carbon capture is used. Not suitable for sustainable agriculture because it relies on fossil fuels.

Biomass Gasification: Agriculture waste materials (e.g., straw, rice husks, animal manure) can be thermochemically transformed into syngas (a combination of H_2, CO, and CH_4), which can be processed to generate hydrogen (Caputo et al., 2005).

Dark Fermentation and Photobiological Processes: Hydrogen can be produced biologically through anaerobic digestion of organic waste, but with lower yields. Such ones are prospective for integrating hydrogen production into agricultural waste management systems (Ntaikou et al., 2010). In sustainable agriculture systems, integration of on-site renewable energy sources with hydrogen production facilitates energy self-sufficiency, minimizes carbon emissions, and adheres to circular bioeconomy principles.

14.3 APPLICATIONS OF FUEL CELLS IN AGRICULTURE

Fuel cells have transformative power in agriculture through the provision of clean, efficient, and decentralized energy solutions that solve both productivity and environmental sustainability. Agricultural processes, especially in remote or off-grid areas, tend to experience energy access, increasing fuel prices, and greenhouse gas emissions challenges. Fuel cells can solve these problems by providing constant electricity for various agricultural uses while greatly lowering the carbon footprint over conventional fossil-fuel-based systems (Kamarudin et al., 2021).

One of the leading uses of fuel cells in agriculture is in water management systems, where they drive irrigation pumps, drip systems, and water purification units. Proton Exchange Membrane Fuel Cells (PEMFCs) and Solid Oxide Fuel Cells (SOFCs) are particularly well-suited for such applications because of their flexibility of operation and energy efficiency. These fuel cells are capable of running on hydrogen or biofuels, so they are very flexible for integrating into farm-scale energy systems along with biomass- or solar-derived hydrogen production (Staffell et al., 2019). Microbial Fuel Cells (MFCs) also offer a sustainable solution through the use of organic agricultural wastes and wastewater for electricity generation and simultaneous waste treatment and energy harvesting (Logan et al., 2023).

Fuel cells have also been effectively utilized in driving farm equipment like tractors, harvesters, and transport vehicles. Fuel cell-powered vehicles produce only water vapor and run with lower noise levels as compared to conventional internal combustion engines, thus reducing environmental pollution and increasing the sustainability of agriculture. Portable and lightweight PEMFC units have been designed for small-scale horticultural and greenhouse equipment used for horticulture and greenhouse management, with the operational benefit in confined environments with limited ventilation.

In smart and precision agriculture, fuel cells deliver reliable power for sensors, weather stations, data loggers, and wireless modules of communication. Low but uninterrupted power is required by these devices, which can be easily delivered by small fuel cell systems, particularly in areas with unreliable access to grid electricity. This functionality contributes to the development of data-intensive farming practices that increase yields, maximize the use of resources, and enhance climate resilience (Mekhilef et al., 2020).

A second new use is the deployment of fuel cells within controlled agriculture environments, including greenhouses, where electricity and heat are both needed. SOFCs' high-temperature operation makes them a perfect choice for Combined Heat and Power (CHP) systems, providing thermal energy to maintain ideal growing conditions and electricity to power lighting, climate control, and CO_2 enrichment. The CO_2 produced as a by-product from SOFCs can be recycled back into the greenhouse to increase photosynthesis, leading to higher crop yields (Zhang et al., 2020).

Fuel cells also help to make sustainable waste-to-energy systems available on farms by leveraging agricultural residues, manure, and organic waste to generate electricity using MFCs. The systems provide a double advantage: they minimize the environmental impact of disposing of waste and produce on-site renewable energy. Additionally, by-products like treated water and nutrient-rich effluents are recyclable in farm operations, supporting circular economy principles (Yusuf et al., 2021).

In remote and rural farming villages, fuel cells can be integral to hybrid renewable energy systems consisting of solar photovoltaics (PV) and hydrogen production by electrolysis. Excess solar energy is captured as hydrogen and subsequently utilized to fuel cells during periods of non-sunshine, providing a stable energy supply for irrigation, lighting, and refrigeration. This improves energy security, minimizes dependence on diesel generators, and contributes to the overall objectives of rural electrification and sustainable agricultural growth (Staffell et al., 2019). In total, the use of fuel cell technology in agriculture is likely to improve energy efficiency, minimize environmental footprint, and facilitate the shift toward climate-smart agriculture. Ongoing research, policy incentives, and economies of scale will be essential to upscale these technologies and achieve their full potential in sustainable agricultural systems (Table 14.1).

TABLE 14.1
Applications of Fuel Cells in Agriculture

Application Area	Fuel Cell Type Used	Function/Benefit	Examples	References
Irrigation Systems	Hydrogen Fuel Cells (PEMFCs)	Provides off-grid, clean energy to operate pumps and control systems	Fuel cell-powered drip or sprinkler systems	IEA (2022); Kundu et al. (2018)
Greenhouses and Controlled Environments	Solid Oxide Fuel Cells (SOFCs)	Delivers both electricity and heat (CHP) for maintaining optimal growth conditions	Combined heating and power for temperature and CO_2 control	Brett et al. (2021); IRENA (2020)
Cold Storage and Post-Harvest Facilities	PEMFCs, SOFCs	Ensures uninterrupted refrigeration and humidity control using low-emission energy	Standalone fuel cell units for cold chains in rural areas	Zhang et al. (2021); Zhao et al. (2020)
Farm Equipment and Vehicles	PEMFCs, Ammonia-Based Fuel Cells	Replaces diesel with zero-emission energy for tractors, harvesters, and drones	Hydrogen tractors, fuel cell-powered autonomous machinery	Staffell et al. (2019); Lan et al. (2012)
Precision Agriculture Sensors	Microbial Fuel Cells (MFCs)	Powers low-energy sensors for monitoring soil moisture, nutrients, and pathogens	Soil-embedded MFCs for self-powered IoT systems	Santoro et al. (2017); Logan et al. (2023)
Agricultural Waste-to-Energy Conversion	MFCs, SOFCs	Converts biomass, manure, and organic residues into electricity	On-farm bio-waste digesters integrated with fuel cells	Logan et al. (2023); Santoro et al. (2017)

14.3.1 Powering Irrigation Systems

Fuel cells are proving to be a clean, efficient technology to power irrigation equipment in sustainable farming. Irrigation is among the most energy-using activities on farms, particularly in arid and semi-arid areas where water has to be pumped from deep aquifers or hauled in from far distances. Traditional irrigation pumps, frequently diesel-powered pumps, account for much of the greenhouse gas emissions and operating costs. Conversely, fuel cells provide a low-emission, efficient, and scalable solution to power both fixed and mobile irrigation systems, especially in remote or off-grid situations (Kamarudin et al., 2021).

Among all the different fuel cell types, Proton Exchange Membrane Fuel Cells (PEMFCs) and Solid Oxide Fuel Cells (SOFCs) are best appropriate for agricultural applications because of their efficiency of operation, modularity, and ability to be used with renewable energy systems. Low-temperature operation and rapid start-up capabilities of PEMFCs make them a good choice for efficiently powering electric pumps for sprinkler or drip irrigation. When combined with solar photovoltaic (PV) systems and hydrogen storage, PEMFCs provide energy autonomy and reliable irrigation schedules, even during low solar radiation or grid failure periods (Staffell et al., 2019). SOFCs, however, have the capacity to use multiple fuels such as biogas, methane, and agricultural wastes, providing a versatile fuel choice to rural farmers and increasing the on-farm circular utilization of waste products (Zhang et al., 2020).

Furthermore, microbial fuel cells (MFCs) have also shown promise for dual-purpose systems that purify agricultural wastewater while, at the same time, producing electricity to drive low-energy irrigation elements. MFCs utilize the metabolic activity of microorganisms to produce electricity from organic matter, representing a decentralized and sustainable source of energy for valve control or water lifting in precision irrigation systems (Logan et al., 2023). These systems are especially useful for smallholder farms looking for sustainable means to control both water and waste.

A number of pilot experiments and field demonstrations have confirmed the application of hydrogen fuel cells in operating submersible and surface irrigation pumps. For example, demonstration deployments in India and Africa have indicated that fuel cells powered by hydrogen can provide adequate energy for pumping water at depths of up to 60 meters, and thus they are viable alternatives to diesel engines for groundwater-based irrigation. These fuel cell systems also provide a huge maintenance cost and downtime reduction, as they have fewer moving parts and lower wear and tear than combustion engines.

14.3.2 Greenhouse and Controlled Environment Agriculture

Greenhouse and controlled environment agriculture (CEA) systems have become increasingly important in contemporary agriculture because they can maximize plant-growing conditions, increase resource utilization efficiency, and prolong growing seasons. Nevertheless, these systems consume lots of energy as they depend largely on electricity for climate control, lighting, irrigation, and automation. The application of fuel cell technology to greenhouse and CEA functions presents a viable solution to minimize carbon footprints and provide more energy reliability and sustainability. Fuel cells, through the direct transformation of hydrogen or other fuels into electricity and heat using electrochemical reactions, can be used as a decentralized and clean source of power consistent with the sustainability objectives of contemporary agriculture (Staffell et al., 2019).

Of the many types of fuel cells, Solid Oxide Fuel Cells (SOFCs) and Proton Exchange Membrane Fuel Cells (PEMFCs) are well suited for use in greenhouses. SOFCs work at high temperatures and can run on multiple fuels, such as biogas from agricultural waste. Their capability to produce electricity and heat simultaneously makes them particularly useful in greenhouses, where thermal power can be harnessed for the optimal heating of plants to grow (Zhang et al., 2020). PEMFCs, however, possess quick start times and effective low-temperature operations, which makes them well suited for use where flexible power delivery is needed, like LED plant lighting and building control systems (Baroutaji et al., 2020).

In areas of sparse grid networks or frequent outages, fuel cells can help improve the robustness of greenhouses. Integrating fuel cells with renewable sources like solar photovoltaics can provide on-site generation of hydrogen using electrolysis, which can then be stored to be utilized in times of low solar irradiation. This combined approach provides an uninterruptible power supply while limiting the use of fossil fuels (Kamarudin et al., 2021). Additionally, the carbon emissions with the use of biomass-derived hydrogen or biogas can be almost zero for fuel cell processes, thus a key component in a closed-loop agriculture system.

Aside from energy production, microbial fuel cells (MFCs) are researched for their integrated potential in greenhouse systems. The MFCs have the ability to treat hydroponic runoff or greenhouse organic waste while, at the same time, producing low amounts of electricity that can be utilized to power data loggers, sensors, or small actuators in precision-controlled systems (Logan et al., 2023). This bi-functional approach not only encourages waste reduction but also increases data-driven agriculture sustainability.

Pilot experiments have proven the feasibility of fuel cell-driven greenhouses in different climates. For instance, in Japan and the Netherlands, fuel cell systems have been employed to provide the entire energy requirement of sophisticated horticultural complexes, exhibiting enhanced energy efficiency and greenhouse gas emission reductions compared to traditional energy systems. These applications offer a useful example of how to scale up the use of fuel cells in greenhouse farming worldwide, particularly in urban and peri-urban areas where space and energy efficiency are critical.

Therefore, the incorporation of fuel cell technology into greenhouse and controlled environment agriculture is a visionary strategy to meet the twin challenges of energy needs and environmental footprint. By taking advantage of the co-generation capability, renewable fuel compatibility, and modularity of fuel cells, the agricultural industry can greatly improve its energy sustainability while maintaining high productivity and climate resilience.

14.3.3 Cold Storage and Post-Harvest Management

Post-harvest management and cold storage are critical elements of the agriculture value chain that play a great role in affecting the quality, safety, and marketability of food. Poor or unreliable supply of electricity in most areas, especially in developing nations, curbs the functionality of cold storage, resulting in huge post-harvest losses, particularly in perishable foods like fruits, vegetables, milk products, and meat. Incorporating fuel cell technology into cold storage systems presents an efficient and environment-friendly solution for meeting these needs. Fuel cells deliver constant, off-grid electricity with high efficiency and minimal emissions, which makes them especially fitting for rural and decentralized agricultural networks (Staffell et al., 2019).

Proton Exchange Membrane Fuel Cells (PEMFCs) and Solid Oxide Fuel Cells (SOFCs) are especially relevant in cold storage applications. PEMFCs have the advantage of fast startup and efficient operation at reduced temperatures and are therefore suitable for intermittent power supply and mobile refrigerated transport units. In contrast, SOFCs, being capable of operating at elevated temperatures and accepting a broad spectrum of fuels such as biogas and syngas from agriculture residues, are best suited for continuous operation and CHP applications. The SOFCs' waste heat can be reused for drying crops or sterilizing storage conditions, thereby increasing the overall energy efficiency of post-harvesting processes (Zhang et al., 2020).

Fuel cells also provide modularity and scalability, which are essential in scaling cold storage systems to suit different farm sizes and commodity quantities. Where grid access is poor or where there are frequent power outages, hybrid systems integrating fuel cells with renewable energy technologies like solar panels and battery storage can provide an uninterrupted cold chain, substantially lowering spoilage and enhancing food security (Kamarudin et al., 2021). Moreover, using green hydrogen generated through electrolysis from renewable energy or methane from biogas, such systems can be kept carbon-neutral, as per sustainable agricultural and climate change mitigation objectives (Baroutaji et al., 2020).

Microbial Fuel Cells (MFCs) also present nascent promise in low-power post-harvest management applications. For instance, MFCs can be used to supply power to small-scale sensors for monitoring inside cold stores, monitoring factors such as temperature, humidity, and gas content to maintain ideal storage conditions. The systems can be powered by organic waste streams from waste or trimmed produce, providing a closed-loop and waste-reducing solution for post-harvest energy needs (Logan et al., 2023).

Feasibility studies and field trials in countries such as South Asia and Sub-Saharan Africa have proven that the use of fuel cell-powered cold storage units can add a few days to weeks to the shelf life of crops, decrease post-harvest loss by up to 30%, and improve the profitability for small farmers by providing improved access to remote markets. These advantages highlight the technological transformative power of fuel cell integration in post-harvest supply chains and food supply chain resilience.

Hence, by meeting the essential energy requirements of cold storage and post-harvest management, fuel cell technology plays a role not just in agricultural waste reduction but also in improving food security, farmer income, and environmental responsibility throughout the agri-food industry.

14.3.4 Electrification of Farm Equipment and Vehicles

The electrification of agricultural vehicles and equipment is a revolutionary leap in contemporary farming, meant to improve the efficiency of operations, minimize the use of fossil fuels, and decrease greenhouse gas emissions. Conventional agricultural equipment, such as tractors, harvesters, and transport vehicles, rely heavily on diesel and gasoline, which in turn contribute greatly to air pollution, soil pollution, and global warming. Incorporating fuel cell technology as a source of power for these machines is a feasible and sustainable option, particularly in decarbonizing the agricultural industry and supporting energy self-sufficiency in rural areas (Staffell et al., 2019).

Hydrogen fuel cells, especially Proton Exchange Membrane Fuel Cells (PEMFCs), are highly applicable for mobile usage owing to their high energy density, rapid refueling capabilities, and stable operation under fluctuating load conditions. When used in agricultural vehicles, fuel cells transform hydrogen into electricity to power electric motors, providing emissions-free, noiseless, and efficient performance with water as the sole emission. This is specifically beneficial in environmentally sensitive agricultural situations, like greenhouses and livestock husbandry installations, where cleanliness of air and noise abatement are essential in maintaining plant and animal health (Baroutaji et al., 2020).

Fuel cell-powered tractors and autonomous farm robots are becoming popular in precision agriculture, where reliable, low-emission power sources are needed for data-hungry operations. Research institutions and companies have already showcased hydrogen-fueled tractor prototypes that are as capable as traditional machines in tilling, planting, and hauling but without toxic emissions (Kamarudin et al., 2021). In addition, fuel cells offer longer range operation than battery-electric competition, which is a critical factor for large farms where maximum runtime and reduced downtime are operational imperatives.

Aside from heavy equipment, light trucks like utility carts, sprayers, and drones can also gain from fuel cell technology. These small-scale applications, usually employed as intra-farm transportation and monitoring, can be more efficient utilizing portable and scalable PEMFC systems. The deployment of hydrogen refueling stations at farms using renewable energy and water electrolysis can provide a closed-loop energy system that can power the whole farm fleet, enhancing energy independence and climate resilience.

Solid Oxide Fuel Cells (SOFCs), though not as mobile because of their elevated working temperatures, can similarly be utilized in stationary battery chargers or combined heat and power (CHP) systems for electrically powered machinery. By producing electricity and available heat from biogas or ammonia derived from agricultural waste, SOFCs can decrease energy expenses and environmental problems stemming from the combustion of fuels (Zhang et al., 2020).

Microbial Fuel Cells (MFCs), while presently constrained in power output, hold promise for small-scale, self-sustaining sensor networks deployed in equipment telemetry and fleet management. They have the ability to boost predictive maintenance and logistics by providing real-time information on fuel use, mechanical wear and tear, and field efficiency, thus enhancing the overall sustainability and productivity of mechanized agriculture (Logan et al., 2023).

Generally, the use of fuel cell technology to electrify farm equipment and transport is a progressive approach to transforming agriculture in alignment with international sustainability objectives. Not only does it minimize environmental effects, but it also creates innovation, energy efficiency, as well as economic feasibility for both small-scale and industrial agriculture systems.

14.4 TYPES OF FUEL CELLS FOR AGRICULTURAL APPLICATIONS

Fuel cell technologies provide flexible and green energy solutions for agricultural use, offering sustainable alternatives to traditional fossil fuel-based systems. Of the different types, hydrogen fuel cells, microbial fuel cells (MFCs), and ammonia-based fuel cells have been found to be especially promising for meeting varied energy needs in contemporary agriculture, such as irrigation, transportation, greenhouse control, and post-harvest processing.

Hydrogen fuel cells, especially Proton Exchange Membrane Fuel Cells (PEMFCs), are extensively studied and utilized because of their high energy efficiency, low temperature operation, and zero-emission status, with water as the sole byproduct. These fuel cells are very well adapted to stationary and mobile agricultural uses, with fast start-up times and scalable output. The use of hydrogen fuel cells in tractors, irrigation pumps, and remote monitoring systems has been shown to lower operational costs and environmental effects considerably. In addition, the ability to generate hydrogen through renewable-powered electrolysis or biomass gasification increases their sustainability and usability in decentralized farming systems (Staffell et al., 2019; Baroutaji et al., 2020).

Microbial Fuel Cells (MFCs) are a new generation of bioelectrochemical systems that use microorganisms to generate electricity from organic matter, such as agricultural waste and manure. MFCs are especially appealing for small-scale, low-power applications in agriculture, such as energizing sensors, wireless monitoring equipment, and lighting systems in greenhouses or storage facilities. They have a twofold advantage of waste treatment as well as power generation, hence enabling the bioeconomy circle in rural and peri-urban agriculture settings. While their power density is as yet lower than that of hydrogen fuel cells, research and process optimization in the electrode materials as well as the microbial consortia are continuously enhancing their scalability and viability (Santoro et al., 2017; Logan et al., 2023).

Ammonia fuel cells are beginning to attract attention because ammonia contains high energy density, is easy to store, and is abundantly found in agricultural environments as a fertilizer. Ammonia can be both a hydrogen carrier and a direct fuel in some types of fuel cells, i.e., Solid Oxide Fuel Cells (SOFCs) and Direct Ammonia Fuel Cells (DAFCs). In farm use, ammonia-based fuel cells have the potential to provide closed-loop systems under which surplus or waste ammonia generated during fertilizer production and animal manure can be used to produce energy, minimizing environmental hazards due to ammonia volatilization and runoff. Ammonia-based SOFCs also have the ability to deliver combined heat and power (CHP) for on-farm usage, which could support operations such as drying crops and heating greenhouses with minimal carbon footprint (Lan et al., 2012).

These fuel cell technologies with distinct characteristics and fuel compatibilities play a considerable role in the decarbonization and modernization of agriculture. Their use sustains energy independence, enhances resource efficiency, and encourages sustainable approaches to support climate resilience and rural development objectives. Fuel cell technology has considerable potential for revolutionizing the energy dynamics in agriculture through the provision of clean, efficient, and decentralized energy solutions. Among the different fuel cells, hydrogen fuel cells, microbial fuel cells (MFCs), and ammonia-based fuel cells have particular relevance in various agricultural

TABLE 14.2
Types of Fuel Cells and Their Nature for Agricultural Applications

Fuel Cell Type	Working Fuel	Operating Temperature	Key Agricultural Applications	Advantages	References
Hydrogen Fuel Cells (PEMFC)	Pure Hydrogen	50–100 °C	Irrigation systems, farm machinery, electric tractors, power backup	Zero emissions, high efficiency, scalable	IEA (2022); Kundu et al. (2018)
Solid Oxide Fuel Cells (SOFC)	Hydrogen, methane, biogas	600–1000 °C	Combined heat and power (CHP) in greenhouses and cold storage	High efficiency, fuel flexibility, cogeneration capability	Brett et al. (2021)
Microbial Fuel Cells (MFC)	Organic agricultural waste	Ambient	Biosensors, low-power devices, waste-to-energy in precision agriculture	Waste valorization, low-cost, integration with smart farming	Santoro et al. (2017); Logan et al. (2023)
Ammonia-Based Fuel Cells	Anhydrous Ammonia	70–650 °C (depending on type)	Remote energy supply, fertilizer-derived fuel for equipment	Utilizes existing farm inputs, low CO_2 emissions	Lan et al. (2012); IRENA (2020)

contexts, ranging from providing power to farm machinery to treating waste and enabling controlled-environment agriculture (Table 14.2).

14.4.1 Hydrogen Fuel Cells

Hydrogen fuel cells, particularly Proton Exchange Membrane Fuel Cells (PEMFCs), are one of the most advanced and prevalent fuel cell technologies. They produce electricity from hydrogen and oxygen using electrochemical reactions, producing only water vapor as a waste product, hence a zero-emission technology best suited for agricultural use (Staffell et al., 2019). PEMFCs are relatively low-temperature operating devices (around 60–80°C), thus well-suited for mobile and distant applications such as driving tractors, irrigation pumps, and agricultural drones (Baroutaji et al., 2020).

Hydrogen may be generated locally by electrolysis with renewable power sources (solar, wind) or obtained from biomass gasification, which is also compatible with sustainable agriculture. This compatibility is in favor of off-grid energy supply in rural farms, which will increase energy autonomy and minimize the use of fossil fuels (Kamarudin et al., 2021). In addition, the high energy density and fast refueling feature of hydrogen fuel cells have unique advantages over conventional batteries, particularly for those applications that demand long hours of operation.

14.4.2 Microbial Fuel Cells (MFCs)

Microbial Fuel Cells (MFCs) harness the metabolic activity of microorganisms to oxidize organic substrates—like agricultural waste, manure, and wastewater—to produce electricity. The bioelectrochemical system has the dual benefit: waste treatment and renewable energy. MFCs are particularly effective in small-scale agricultural settings where low-power devices, like environmental sensors, remote monitoring systems, or greenhouse lighting, can be powered in a sustainable manner (Logan et al., 2023).

Recent developments in MFC technology involve advancements in electrode materials, biocatalysts, and reactor designs that have improved efficiency and scalability. The fact that they can work on locally sourced biomass makes them an attractive solution in circular farming systems and bio-integrated farms (Santoro et al., 2017). MFCs also have the potential to help counter methane emissions from livestock and composting plants by diverting organic waste into clean energy streams.

14.4.3 Ammonia-Based Fuel Cells

Ammonia is rapidly being seen as a high-potential biofuel alternative in agriculture because it has high energy density, existing infrastructure within the fertilizer sector, and simple storage and transportation. Ammonia fuel cells like Direct Ammonia Fuel Cells (DAFCs) and Solid Oxide Fuel Cells (SOFCs) can either oxidize ammonia directly or remove hydrogen from it to produce electricity (Lan et al., 2012).

In farm scenarios, ammonia fuel cells can fund off-grid electrical power generation to provide heating in greenhouses, pumping water, or even lighting cold storage systems. They are also a greener way to harness excess ammonia from fertilizer farms or animal production to minimize pollution risks like eutrophication and nitrogen leaching. SOFCs, which run at high temperatures, are best suited for combined heat and power (CHP) systems, thus enhancing the energy efficiency of agricultural processes such as crop drying or seed germination chambers.

Ammonia's potential as a carbon-free hydrogen carrier adds to its appeal, especially when generated through green synthesis routes such as electrochemical nitrogen fixation by renewable energy.

14.5 ENVIRONMENTAL BENEFITS OF FUEL CELLS IN AGRICULTURE

Fuel cells provide considerable environmental advantages when embedded in agricultural systems, providing a cleaner and more sustainable solution than traditional fossil fuel-based technology. With agriculture coming under increasing pressure to mitigate GHG emissions, save natural resources, and be more energy-efficient, fuel cells have proved themselves to be an effective solution congruent with environmental sustainability objectives.

The biggest environmental benefit of fuel cells for agriculture is that they can substantially lower GHG emissions. Hydrogen fuel cells, for example, produce no more than water vapor as waste, with zero carbon dioxide (CO_2), nitrogen oxides (NO_x), sulfur oxides (SO_x), and particulate matter common to diesel- and gasoline-run farm equipment (Staffell et al., 2019). The substitution of fuel cell systems for internal combustion engines in tractors, irrigation pumps, and transport vehicles can significantly lower the carbon footprint of farming activities. With hydrogen produced through renewable energy sources like solar or wind-powered electrolysis, the process is effectively carbon neutral, further improving the sustainability profile of fuel cell uses (Baroutaji et al., 2020).

Apart from the mitigation of emissions, fuel cells also help alleviate noise and air pollution. This is especially advantageous in closed environments like greenhouses or animal houses, whose crops' productivity and animals' health rely on indoor air quality and acoustic comfort. Fuel cells run quietly and do not release harmful pollutants upon operating, unlike diesel generators, which are not only loud but also a culprit of volatile organic compounds (VOCs) and fine particulates (Kamarudin et al., 2021).

Fuel cells are also key to the development of waste-to-energy technologies, especially microbial fuel cells (MFCs), that convert organic farm wastes into usable electricity. This ability not only minimizes the environmental impact of waste disposal but also provides an innovative route toward nutrient recovery and water reuse. For instance, livestock manure, crop wastes, and agro-industrial effluents can be employed as feedstocks in MFCs, hence preventing organic pollutants from entering landfills or water bodies and alleviating eutrophication threats (Logan et al., 2023; Santoro et al., 2017).

Additionally, fuel cells based on ammonia are environmentally friendly in the sense that they use excess or waste ammonia during fertilizer production or animal farming as a clean source of energy. This would aid in curbing ammonia volatilization and related atmospheric and water pollution and transform an undesired byproduct into an economic energy carrier (Lan et al., 2012). The coupling of fuel cell systems with the existing agricultural infrastructure promotes circular economy practices, in which waste streams are converted to energy inputs to create a closed-loop system with minimal environmental footprint.

The lifecycle environmental effects of fuel cells are also reduced as compared to fossil fuel systems, particularly as innovations in materials and production processes improve. Emerging green production methods for hydrogen and ammonia, along with the creation of recyclable or biodegradable fuel cell parts, further minimize the environmental impact of these technologies. Deployed on a large scale, fuel cells also support decentralized energy generation, lessening transmission loss and the necessity for widespread grid infrastructure in rural and distant agricultural regions.

Finally, the environmental advantage of fuel cells in agriculture is three-dimensional and includes reductions in emissions, pollution abatement, waste valorization, and resource efficiency. Such benefits place fuel cells as a central technology in shifting toward climate-resilient and environmentally sustainable agricultural production.

14.6 TECHNOLOGICAL ADVANCEMENTS AND INNOVATIONS

Advances in technology and innovations in fuel cell systems are quickly increasing their feasibility and performance for sustainable agricultural use. There have been tremendous advances in the efficiency, longevity, and cost-effectiveness of fuel cells, thus opening up their application across different aspects of farming activities. One of the leading advancements includes the improvement of proton exchange membrane fuel cells (PEMFCs), which currently have improved polymer electrolytes and platinum group metal-free catalysts. These have reduced material costs and improved tolerance to impurities in hydrogen fuel, thus becoming more viable for field use in applications like irrigation, electric tractors, and autonomous drones (Kundu et al., 2018).

Solid oxide fuel cells (SOFCs) with high operating temperatures and flexibility of fuel have also been optimized extensively. New ceramic materials based on advanced ionic conductivity have been fabricated, allowing for lower-temperature operation and minimizing thermal degradation. Such optimization enables SOFCs to be utilized efficiently in combined heat and power (CHP) systems for greenhouses and post-harvesting facilities, thereby enhancing energy efficiency and crop preservation at the same time (Brett et al., 2021). In addition, integration with alternative energy sources like solar panels has made it possible to have hybrid energy systems consisting of fuel cells and battery storage as well as smart grids so that there is no power disruption even in changing weather conditions.

Microbial fuel cells (MFCs) have also seen great progress in electrode design, biofilm control, and system design. Developments in nanostructured electrode materials, including carbon nanotubes and graphene composites, have added greatly to surface area and electron transfer rates, boosting the power output of MFCs. In farm applications, these advancements have enabled the use of MFCs for real-time biosensing and environmental monitoring, and for electricity generation from waste streams in livestock and aquaponic systems (Logan et al., 2023). Additionally, advances in scalable and modular MFC designs have created opportunities for distributed power generation in rural and remote regions.

Ammonia fuel cells are becoming a strategic technology for agriculture, especially given the infrastructure available for ammonia transport and storage within the fertilizer industry. Advances in catalyst technology have made more efficient cracking of ammonia and direct oxidation of ammonia possible, overcoming earlier limitations due to fuel purity and electrode poisoning (Lan et al., 2012). Such advances have placed ammonia fuel cells as a viable alternative for decentralized power generation in farms, particularly where hydrogen logistics are complicated or economically infeasible.

The incorporation of digital technologies and artificial intelligence (AI) in fuel cell systems is another innovation frontier. Predictive maintenance using AI, real-time monitoring of performance, and load balancing through automation have enhanced system reliability and minimized downtime. Smart controllers and Internet of Things (IoT) gadgets enable smooth energy management across farm processes, adjusting fuel cell production to match fluctuating energy demand and weather conditions (Zhao et al., 2020).

Cumulatively, these technologies are improving the performance, affordability, and flexibility of fuel cell technologies for agriculture. As the technologies continue to mature, they have the potential to be a key driver toward low-carbon, resilient, and energy-autonomous agriculture systems.

14.7 AGRICULTURAL FUEL CELL ADOPTION AND FUTURE PROSPECTS

The implementation of fuel cell technology in agriculture is a paradigm shift toward decentralized, sustainable, and clean energy systems. Although the use of fuel cells in agriculture is currently in the developmental phase, the world is increasingly looking at these technologies as the means to respond to increased energy needs without emitting greenhouse gases or depending on fossil fuels. The farming industry, especially in rural and off-grid areas, can gain significantly from the distinctive qualities of fuel cells, such as high energy efficiency, minimal emissions, silent operation, and flexibility to work with different fuels like hydrogen, ammonia, and organic waste.

The growing worldwide focus on climate-smart agriculture and carbon neutrality has provided a favorable climate for the use of fuel cells in agricultural activities. For example, the application of hydrogen fuel cells in tractors, irrigation systems, and other farm equipment can considerably reduce carbon footprints while creating greater energy independence. Japan, South Korea, and the United States have already started to pursue hydrogen-based agricultural technology, facilitated by government incentives, public–private partnerships, and strategic rural energy infrastructure investments (IEA, 2022). These trends are compounded by decreasing renewable electricity costs as well as evolving electrolyzer and hydrogen storage technology, which jointly increase the economic viability of producing hydrogen on the farm for local use (IRENA, 2020).

Microbial fuel cells (MFCs) that tap the metabolic processes of soil or wastewater microbes to power electricity present an exciting low-power, bio-interfaced farming application horizon. These systems have the ability to drive sensors for precision agriculture, allowing real-time tracking of soil nutrients, moisture, and pathogens with minimal environmental footprint (Santoro et al., 2017). With the growth of digital agriculture, the coupling of MFCs with Internet of Things (IoT) platforms will likely improve decision-making and minimize the energy footprint of data-driven farming.

Ammonia fuel cells, especially useful in fertilizer-high farming areas, have the added advantages of harnessing locally present ammonia as fuel and minimizing environmental emissions associated with its volatilization. Studies on high-efficiency catalysts for direct ammonia fuel cells and large-scale system designs will help facilitate their commercialization in stationary and mobile farm applications (Lan et al., 2012). Furthermore, efforts to green ammonia production with green hydrogen will also enhance the sustainability profile of ammonia-powered energy systems.

In spite of these encouraging developments, a number of hurdles hinder large-scale adoption. High capital requirements, low awareness levels among farmers, absence of infrastructure for hydrogen and ammonia supply, and techno-commercial challenges related to fuel cell durability under adverse field conditions are some of the major hindrances (Staffell et al., 2019). To facilitate solutions to these bottlenecks, concerted efforts must be made, such as focused subsidies, capacity-building initiatives, pilot demonstrations, and intersectoral coordination among agriculture, energy, and environmental groups.

In the future, the prospects of fuel cells in agriculture are very bright, with the impetus coming from improvements in materials science, bioelectrochemical engineering, and government support for clean energy. With the convergence of renewable energy sources, smart grid technologies, and sustainable waste management practices, fuel cells are set to become a pillar of environmentally sustainable and resilient agricultural systems. Their adoption will not only address energy security but also contribute to broader sustainability goals such as climate change mitigation, resource conservation, and rural development.

14.8 CONCLUSION

Fuel cell technology is a revolutionary solution to sustainable energy management in contemporary agriculture. As agricultural systems worldwide face rising energy needs, environmental degradation, and the need to limit greenhouse gas emissions, fuel cells constitute a viable option over traditional fossil fuel-based energy sources. Their capability to supply clean, efficient, and decentralized power fits neatly into the requirements of farm operation for a spectrum of farming activity from irrigation, greenhouse climate management, and cold storage to farm equipment and vehicle electrification. Additionally, fuel cells have several strengths, such as low emissions, efficient energy conversion, fuel flexibility, and low noise pollution, hence highly appropriate for rural and environmentally sensitive areas (Logan et al., 2023; Kundu et al., 2018).

Among the other fuel cell technologies, hydrogen fuel cells are finding favor because of their scalability and compatibility with renewable energy sources, including solar power and wind energy. Microbial fuel cells (MFCs) have a dual benefit of waste management as well as generation of energy and hence are the best for combined and sustainable agricultural operations. In contrast, ammonia-powered fuel cells reveal new opportunities for harvest expansion of already existing agricultural resources into the power industry, particularly in fertilizer-scarce areas. Advances in technology—from superior catalysts, nanomaterials, and hybrid systems to AI-based fuel cell management—keep coming down the cost ladder while enhancing system dependability, increasing the feasibility of the systems for agriculture adoption (Brett et al., 2021; Zhao et al., 2020).

In spite of these benefits, some challenges exist, such as the high initial capital, small-scale infrastructure for hydrogen and ammonia fuel supply, and the need for farmers to develop technical capacities. Policy assistance, research incentives, and public-private partnerships can be used to break these hurdles and speed up the deployment of fuel cells in agriculture. There are bright future prospects, especially with the increased focus on circular economy models and climate-smart agriculture. Fuel cells, when combined with renewable energy systems and smart technologies, can play an important role in energy self-sufficiency, environmental sustainability, and rural economic development.

REFERENCES

Barbir, F. (2020). *PEM Fuel Cells: Theory and Practice* (3rd ed.). Academic Press.

Baroutaji, A., Wilberforce, T., Ramadan, M., & Olabi, A. G. (2020). Comprehensive investigation on hydrogen and fuel cell technology in the aviation and aerospace sectors. *Renewable and Sustainable Energy Reviews*, 106, 31–40. https://doi.org/10.1016/j.rser.2019.02.024

Brett, D. J. L., Atkinson, A., Brandon, N. P., & Skinner, S. J. (2021). Future trends in solid oxide fuel cell technology. *Journal of Power Sources*, 500, 229983. https://doi.org/10.1016/j.jpowsour.2021.229983

Caputo, A. C., Palumbo, M., Pelagagge, P. M., & Scacchia, F. (2005). Economics of biomass energy utilization in combustion and gasification plants: effects of logistic variables. *Biomass and Bioenergy*, 28(1), 35–51. https://doi.org/10.1016/j.biombioe.2004.04.009

FAO. (2021). *Energy-smart food for people and climate*. Food and Agriculture Organization of the United Nations. Retrieved from http://www.fao.org

IEA. (2022). *Global Hydrogen Review 2022*. International Energy Agency. https://www.iea.org/reports/global-hydrogen-review-2022

IRENA. (2020). *Green Hydrogen: A Guide to Policy Making*. International Renewable Energy Agency. https://www.irena.org/publications/2020/Nov/Green-hydrogen

Ishizaki, A., Kishi, A., & Kojima, Y. (2018). Development of biogas-fueled SOFC systems for agricultural applications. *Journal of Power Sources*, 375, 191–198. https://doi.org/10.1016/j.jpowsour.2017.07.019

Kalamaras, C. M., & Efstathiou, A. M. (2013). Hydrogen production technologies: current state and future developments. *Energy & Environmental Science*, 6(4), 1095–1121. https://doi.org/10.1155/2013/690627

Kamarudin, M. Z. F., Daud, W. R. W., & Kim, B. H. (2021). PEM fuel cell system for agriculture: Current status and future prospects. *International Journal of Hydrogen Energy*, 46(23), 12248–12262. https://doi.org/10.1016/j.ijhydene.2020.03.162

Kauranen, P., Konttinen, J., & Reinikainen, M. (2022). Comparative analysis of fuel cell technologies for renewable energy systems. *Renewable Energy*, 181, 1284–1295. https://doi.org/10.1016/j.renene.2021.09.096

Kundu, A., Rauf, M., & Sahu, J. N. (2018). Recent developments in PEM fuel cell materials and their applicability for sustainable agriculture. *Renewable and Sustainable Energy Reviews*, 90, 107–125. https://doi.org/10.1016/j.rser.2018.03.005

Lan, R., Irvine, J. T. S., & Tao, S. (2012). Ammonia and related chemicals as potential indirect hydrogen storage materials. *International Journal of Hydrogen Energy*, 37(2), 1482–1494. https://doi.org/10.1016/j.ijhydene.2011.09.124

Logan, B. E., Rabaey, K., & Rozendal, R. A. (2023). Microbial fuel cells: current status and future prospects in sustainable agriculture. *Renewable and Sustainable Energy Reviews*, 175, 113146. https://doi.org/10.1016/j.rser.2023.113146

Mekhilef, S., Saidur, R., Safari, A., & Salam, Z. (2020). Fuel cell-based power generation systems for remote areas. *Renewable and Sustainable Energy Reviews*, 44, 877–884. https://doi.org/10.1016/j.rser.2015.01.005

Ntaikou, I., Antonopoulou, G., & Lyberatos, G. (2010). Biohydrogen production from biomass and wastes via dark fermentation: A review. *Waste and Biomass Valorization*, 1(1), 21–39. https://doi.org/10.1007/s12649-009-9001-2

O'Hayre, R., Cha, S. W., Colella, W., & Prinz, F. B. (2016). *Fuel Cell Fundamentals* (3rd ed.). Wiley.

Pant, D., Van Bogaert, G., Diels, L., & Vanbroekhoven, K. (2010). A review of the substrates used in microbial fuel cells (MFCs) for sustainable energy production. *Bioresource Technology*, 101(6), 1533–1543. https://doi.org/10.1016/j.biortech.2009.10.017

Rao, P. V., Kumar, S., & Satyanarayana, M. (2022). Advances in fuel cell technologies for sustainable development. *International Journal of Hydrogen Energy*, 47(6), 3554–3570. https://doi.org/10.1016/j.ijhydene.2021.11.113

Santoro, C., Arbizzani, C., Erable, B., & Ieropoulos, I. (2017). Microbial fuel cells: From fundamentals to applications. *Journal of Power Sources*, 356, 225–244. https://doi.org/10.1016/j.jpowsour.2017.03.109

Staffell, I., Scamman, D., Velazquez Abad, A., Balcombe, P., & Dodds, P. E. (2019). The role of hydrogen and fuel cells in the global energy system. *Energy & Environmental Science*, 12(2), 463–491. https://doi.org/10.1039/C8EE01157E

Yamamoto, T., Tanaka, S., & Watanabe, Y. (2019). Application of fuel cells in agricultural settings: energy and environmental performance evaluation. *Journal of Cleaner Production*, 235, 894–903. https://doi.org/10.1016/j.jclepro.2019.07.015

Yusuf, M., Rehan, M., & Ahmed, I. (2021). Bio-electrochemical energy production from agricultural wastes: A review of microbial fuel cell technologies. *Energy Reports*, 7, 7822–7837. https://doi.org/10.1016/j.egyr.2021.11.056

Zhang, X., Wang, Y., & Guo, X. (2021). Application of fuel cell-based systems in cold storage logistics. *Applied Energy*, 285, 116400. https://doi.org/10.1016/j.apenergy.2021.116400

Zhang, Y., Liu, M., Zhou, Y., & Shao, Z. (2020). High-performance SOFCs for rural energy: Challenges and prospects. *Journal of Materials Chemistry A*, 8(5), 2153–2179. https://doi.org/10.1039/C9TA11310H

Zhao, Y., Wang, X., Zhang, Y., & Liu, J. (2020). AI and IoT-enabled smart fuel cell systems for agricultural energy management. *Energy Reports*, 6(Suppl 2), 105–113. https://doi.org/10.1016/j.egyr.2020.11.004

15 Policy and Economic Considerations of Fuel Cells

Oluwafemi Christopher Igbasan, and
Olarewaju John Adunbi

15.1 INTRODUCTION

The United States Department of Energy (2023) defines fuel cell as an electrochemical device that converts the chemical energy of a fuel (usually hydrogen) and an oxidizing agent (often oxygen) directly into electricity, water, and heat, without combustion. Among the new energy technologies, fuel cells are a viable option for curbing greenhouse gas emissions. Hydrogen can be produced by water electrolysis, a process powered by electricity generated from renewable energy, nuclear, or fossil fuels (U. S. Department of Energy, 2018). Fuel flexibility is a primary advantage of fuel cells. Other advantages include low emissions, electric efficiency greater than 50%, and a diverse variety of applications (U.S. Department of Energy, 2018). In the United States, fuel cells are deployed across the stationary, transportation, and portable power sectors (U.S. Department of Energy, 2018). The widest application is small stationary fuel cells for commercial and government uses. Equipment for handling fuel cell materials had been commercially deployed in several states, such as Alabama, Arizona, California, and Colorado, as of 2017 (U.S. Department of Energy, 2018). On an international scale, the Intergovernmental Panel on Climate Change (IPCC) recognized the potential contribution of fuel cells to power generation, particularly decentralized stationary and mobile hydrogen fuel cells (IPCC, 2000). However, the development of fuel cells also encounters technical challenges and high costs (U.S. Department of Energy, 2015). To strengthen the advantages and reduce the disadvantages, the U.S. federal government has allocated funds and provided incentives to support fuel cell research and development. As of 2019, federal incentives included the Business Energy Investment Tax Credit (ITC) and the Residential Renewable Energy Tax Credit (Fuel Cell and Hydrogen Energy Association, 2019). State governments also have recently encouraged fuel cell development through initiatives and incentives (U.S. Department of Energy, 2018). As the world moves toward decarbonization and energy resilience, fuel cells have emerged as a promising clean energy technology. However, their adoption is heavily influenced by policy frameworks and economic viability. As of today, however, fuel cell technology has been overshadowed by the rise of hybrid and electric vehicles. One of the largest barriers to the continued deployment of hydrogen technologies in the automotive sector is that stringent laws and regulations are more complex for fuel cell electric vehicles (FCEVs) than for other technologies. Specifically, there are several legal barriers with respect to the fueling of FCEV. Firstly, inaccurate hydrogen dispensing options and non-standardized safety regulations are major issues in many European countries. For example, Belgium's fuel distribution system must adhere to very specific criteria that are different from those of its neighboring countries. Germany's fuel dispensing tolerance is very low, and current hydrogen technology is unable to achieve that threshold and, therefore, would require a relaxation of that rule to allow for higher hydrogen fueling tolerances (Dolci *et al.*, 2019). Nevertheless, current hydrogen fueling technologies across Europe do not fulfill the same criteria and would, therefore, require more intermediate regulations. The development of a regulatory framework that defines the

DOI: 10.1201/9781003645405-15

commercial production of FCEVs would allow this technology to advance to a level to more effectively compete with other technologies. However, without a clear regulatory framework that helps advance FCEVs and reduce cost factors such as filling stations, storage, and transportation cost, all efforts to introduce these vehicles to a larger commercial scale may be futile. To fully harvest the potential benefits of a clean energy economy, a successful business model and the best energy-economic policy practices are needed. Understanding these considerations is essential for stakeholders, including policymakers, investors, and industry leaders, to make informed decisions that support sustainable growth, innovation, and equitable energy transitions. This write-up explores the policy instruments and economic dynamics shaping the fuel cell sector to highlight pathways for effective deployment.

15.2 REGULATORY FRAMEWORKS

15.2.1 Hydrogen Policies for the United States

The United States is the world's largest producer of natural gas and oil and exports natural gas and oil to more than 35 countries. The United States has, therefore, a unique opportunity to reinforce and grow its energy leadership position in the world and create new jobs. As countries around the world look to hydrogen technology to reduce carbon emissions, the competitive and ample domestic supply of hydrogen would enable the United States to export even more fuel to markets around the world. With hydrogen technology, low-carbon electric power resources achieve a better power grid integration. Electrolyzers that produce hydrogen can significantly increase the flexibility for intermittent renewable energy resources when connected to the grid. This flexibility can provide long-term storage solutions that enhance and supplement the use of short-duration battery solutions. With these long-term storage solutions, hydrogen technology may complement other energy sources such as renewable and nuclear power. The current legislation and incentive programs for alternative fuel vehicles are quite complex in the United States. The legislations vary from state to state, which adds additional complexity. The Californian legislation appears to be among the most progressive within the United States.

15.2.2 Hydrogen Policies for Japan

In contrast to the United States, Japan has a different set of challenges as it relies heavily on imports of fossil fuels. Like most other industrialized countries, most of Japan's energy needs are dependent on imports. As currently approximately 94% of Japan's energy needs rely on fossil fuels to meet domestic energy demands, Japan is highly motivated to move more quickly to hydrogen in order to decrease its fossil fuel dependency. As a result of the 2016 Paris agreement, Japan formulated a plan to curb carbon emissions and reduce global warming. The Japanese government announced a plan to reduce greenhouse gas emissions by 26% by 2030 as compared to 2013. In order to achieve this ambitious goal, a mix of renewable energy, nuclear energy, and fossil fuels was proposed (Iida & Sakata, 2019). Japan also aims to reduce carbon emissions by 80% by 2050 (Sugiyama *et al.*, 2019). The Japanese government is aggressively promoting hydrogen fuel cell cars and is easing the legal framework to boost incentives and encouraging the use of the FCEV. Prime Minister Shinzo Abe showed his enthusiasm for hydrogen vehicles by suggesting that all the Japanese ministries and agencies should have fuel cell vehicles (Edelstein, 2015). Japan already has the highest incentives for fuel cell cars in the world, with some areas in Japan getting incentives of up to JPY 3 million (approximately USD 26,885 for a Toyota Mirai) that has an actual price tag in Japan of about USD 68,000 (Greenwood, 2019). As a leader in hydrogen technology, and since the enactment of the Paris agreement, the Japanese Ministry of Economy, Trade, and Industry (METI) published a strategic roadmap for hydrogen and fuel cells. These roadmaps are used to achieve a carbon-free society. This strategic roadmap is divided into two sections:

1. hydrogen use in mobility
2. hydrogen supply chain, and other applications for a global hydrogen society. The first two sections explain how hydrogen fuel and vehicle costs are being lowered by implementing specific action items detailed in the roadmap action plan.

15.2.3 Hydrogen Use in Mobility

The key objective of the action plan aims to reduce high FCEV prices and ultimately narrow the price gap between FCEVs and hybrid vehicles from JPY 3 million (USD 28,310) to JPY 700,000 (USD 6605) by 2025. This objective will be achieved by transparency and cooperation between all key stakeholders, such as government organizations, automobile industries, energy and power companies. The transparency and cooperation will facilitate more innovation in the field that could ultimately help lower the costs for the end-user (Japan Ministry of Economy, Trade and Industry, 2019).

15.2.4 Hydrogen Supply Chain

Japan's roadmap also includes a long-term initiative to lower hydrogen prices to a level similar to liquefied natural gas by 2030. This goal will be achieved by building a hydrogen supply network and initiating government-level agreements with countries rich in hydrogen resources. The Japan–Australia brown coal to hydrogen project, for example, will help to lower fuel costs by building a supply chain network and thus reduce costs by transporting and storing hydrogen in bulk. This project, also known as the HSEC (hydrogen energy supply chain) project, is one of the world's first to establish an integrated supply chain between Australia and Japan (Ishikawa, 2019).

15.2.5 Energy Efficiency Policies

Energy efficiency (EE) regulations are normally designed to decrease the energy intensity of an economy, i.e., to minimize "the amount of energy used per unit GDP". As fuel cells, when used with hydrogen, are relatively energy efficient (compared with traditional gasoline-powered internal combustion engines), legislation on EE may indirectly favor the development and market introduction of fuel cells. Since 2000, the Community has adopted several measures in the field of EE. Many legal instruments concern specific sectors such as buildings and the labelling of precisely defined products such as household appliances or electric ovens. Two relatively recent directives are already rather encompassing in their scope and set a framework for nearly all sectors:

- The directive on energy end-use efficiency and energy services of 2005 sets the target that every member state must on average improve by 1% its EE every year. These targets are indicative. The Member States are relatively free to choose the instruments by which they want to reach these goals (EC, 2006).
- In 2005, the EU also adopted a directive on the eco-design of energy-using products. This directive applies in principle to any energy-using product and aims to improve EE in the whole life cycle of the product. The text provides for EE standards and requirements for every stage of the production, beginning with the early design phase. The text itself does not set any binding targets (EC, 2005).

15.2.6 Hydrogen Policies for the European Union

Like Japan, the European Union is committed to decarbonizing energy systems throughout Europe in order to align with the targets defined in the Paris Agreement of 2016. The EU is planning to cut carbon emissions by 95% by 2050 (Jong *et al.*, 2015). To achieve these goals, the EU requires advancing and implementing hydrogen technologies on a wider scale, including both the commercial and private sectors (Fuel Cells and Hydrogen 2 Joint Undertaking, 2019). In the EU, the transport sector comprises one-third of the total carbon emissions. Decarbonizing the transport industry is, therefore, a vital step to meet the standards of the Paris agreement (Fuel Cells and Hydrogen 2 Joint Undertaking, 2019). In order to facilitate the use of hydrogen and the development of this technology, a total of 25 member states of the EU signed the Hydrogen Initiative even before the EU hydrogen roadmap was initiated.

The Hydrogen Roadmap Europe Report predicts that FCEVs could account for 1 in every 22 passenger vehicles and 1 in every 15 light commercial vehicles by the year 2030 (Fuel Cells and Hydrogen 2 Joint Undertaking, 2019). There are numerous benefits for the reduction of carbon emissions if the 2050 hydrogen vision is implemented in the EU. This roadmap can help meet 24% of the energy demand and reduce 15% of NOx emissions from road transport and will also reduce 560 Mt of CO2 (Fuel Cells and Hydrogen 2 Joint Undertaking, 2019). The EU hydrogen roadmap report published in early 2019 discusses that countries like Japan, China, and South Korea are aggressively pursuing hydrogen technologies. These countries made significant advances in hydrogen technology as they issued 55% of the fuel cell patents worldwide, whereas the EU only issued 16% of the patents. This discrepancy suggests that the EU does not have an equally strong decarbonization strategy. In response to this deficit, the EU hydrogen roadmap report addresses to take the following overarching steps to achieve the decarbonization goals:

1. Industry and regulators should work together to set clear, long-term objectives to achieve the decarbonization goals across different segments. Their objectives should not just include the end applications like zero-emission vehicles or decarbonization of houses but also include infrastructural developments necessary to support and sustain the end applications (Fuel Cells and Hydrogen 2 Joint Undertaking, 2019).
2. In order to remain competitive and attract emerging opportunities, the EU should invent hydrogen and fuel cell technologies. This would require alliances with fast accelerating hydrogen technology markets outside the EU like Japan, Korea, and China in order to reduce the market risk. They should also work with the regulators to build a strong home market within the EU (Fuel Cells and Hydrogen 2 Joint Undertaking, 2019).
3. In the transport industry, regulators and legislators should overcome the chicken-and-egg problem by developing a clear roadmap and policies that ensure a definite solution by unlocking proper investment for a hydrogen infrastructure. A roadmap with the goal of developing a basic infrastructure coverage across Europe will ensure that the automobile industries invest and scale up the development of FCEV, which would ultimately lead to overall cost reductions and more choices for the end-users (Fuel Cells and Hydrogen 2 Joint Undertaking, 2019).

As a response to the above roadmap, the interest in hydrogen technology increased in 2020. In July of 2020, the EU launched three key policy initiatives:

1. The EU Hydrogen Strategy established initial targets for the deployment of hydrogen, which will require up to EUR 470 bn by 2050.
2. The European Clean Hydrogen Alliance (ECHA), a government body, works with the leadership of energy giants like Shell, Siemens, Électricité De France (EDF), and Vattenfall, which comprise the ECHA.
3. A large clean technology innovation fund was set up with EUR 1 bn a year that focuses on investments in hydrogen energy (Table 15.1).

TABLE 15.1
Summary of Policy Review between the United States, Europe, and Japan

Policies	United States	Europe	Japan
National Strategy	The United States formed a systematic basket of laws, policies, and scientific research plans to promote hydrogen energy.	In 2019, Fuel Cells and Hydrogen Joint Undertaking released the Hydrogen Roadmap Europe, which proposed a roadmap for hydrogen energy development toward 2030 and 2050.	In 2014, Japan launched the fourth Strategic Energy Plan and published the Strategic Roadmap for Hydrogen and Fuel Cells, outlining an integrated approach to hydrogen production, storage, transportation, and applications.
Hydrogen Production And Distribution	In 2019, the Department of Energy issued a funding opportunity announcement for up to USD 31 million in funding to advance the H2Scale concept, including innovative concepts for hydrogen production and integrated production, storage, and fueling H2Scale pilot system	Relative high focus on clean production of hydrogen going forward. Hydrogen Roadmap Europe: a transition to a one-third ultra-low carbon hydrogen production by 2030.	Strategic Roadmap for Hydrogen and Fuel Cells: Building up a commercial-based domestic system for efficiently distributing hydrogen by the mid-2020s and fledged operation of manufacturing, transportation and storage of zero-carbon emission hydrogen by 2040.
Hydrogen Infrastructure	The DoE launched H2USA, a public–private partnership with FCEV OEMs, focusing on advancing hydrogen infrastructure.	Hydrogen Roadmap Europe: 3700 hydrogen refueling stations are expected by 2030.	From 2016 to 2018, the Ministry of Economy, Trade, and Industry provided a budget of approximately USD 88 million on R&D and approximately USD 539 million on construction subsidies of hydrogen fueling stations.
Support for Passenger Vehicles	The California Fuel Cell Partnership outlined targets for 1,000,000 FCEVs by 2030.	Hydrogen Roadmap Europe: 3.7 million fuel cell passenger vehicles on road by 2030.	The target of 800,000 FCEVs by 2030 Hydrogen Strategy 2017
Support for Commercial Vehicles	In 2018, the California Air Resources Board preliminary awarded USD 41 million for the shore-to-store project, developing 10 Fuel Cell (FC) class 8 drayage trucks	Hydrogen Roadmap Europe: 500,000 fuel cell Light Commercial Vehicles LCVs, 45,000 fuel cell trucks and buses on road by 2030	The target of 1200 FC buses and 10,000 forklifts by 2030 Hydrogen Strategy 2017.

15.3 GOVERNMENT SUPPORT AND INCENTIVES FOR FUEL CELL TECHNOLOGY

15.3.1 Tax Credits for the Alternative Fuel Infrastructure

The Federal Government provided a tax credit for hydrogen fueling stations installed through 31 December 2020. This tax credit included up to 30% of the infrastructure cost but could not exceed USD 30,000. However, permission and inspection fees were not included in these expenses. Nevertheless, for individuals who own multiple fueling stations and install qualified equipment at multiple sites, the tax credit could be applied to each location (U.S. Internal Revenue Service, 2021).

15.3.2 Tax Credits for Fuel Cell Motor Vehicles

A tax credit for up to USD 8000 is available for the purchase of qualified non-commercial fuel cell vehicles. Additional tax credits are available for commercially used vehicles, and the credit amounts are based on vehicle weight. Vehicle manufacturers must follow Notice 2008-33 (PDF) to certify to the Internal Revenue Service (IRS) that a vehicle meets certain requirements to claim the fuel cell vehicle credit (U.S. Internal Revenue Service, 2021). This incentive originally Sustainability 2021, 13, 5149 5 of 12 expired on 31 December 2017 and was retroactively extended through 31 December 2020 by Public Law 116-94.

15.3.3 Alternative Fuel Excise Tax Credit

Liquefied hydrogen enjoys a tax incentive of USD 0.50 per gallon that is available when used to fuel and operate an FCEV. The tax credit is based on the gasoline gallon equivalent (GGE) or diesel gallon equivalent (DGE). This incentive originally expired on 31 December 2017 and was retroactively extended through 31 December 2020 by Public Law 116-94 (U.S. Internal Revenue Service, 2021).

15.3.4 Improved Energy Technology Loans

The Department of Energy provides loans up to 100% of the project cost to eligible projects and programs. These projects must help reduce air pollution and carbon emissions by early adoption of technologies like fuel cell vehicles. These loans are not intended for research and development projects (U.S. Department of Energy, 2021).

15.3.5 Grant Agreement

A grant agreement was made between the Delaware Economic Development Office (DEDO) and Bloom Energy in 2012. It included a $12 million grant that could increase to $16.5 million. This grant required Bloom Energy to create 900 full-time jobs at its manufacturing plant and to spend $12 million on salaries over 12 months by September 2014. DEDO also provided a conditional incentive of $6250 for each additional employee (up to 600) created by suppliers that co-located to the site (Carper, 2011). DEDO also offered to provide 3% of Bloom's total capital expenditures up to the first $50 million (Carper, 2011). Combining all these incentives, DEDO offered Bloom Energy a potential $16.5 million grant. If the job target was unmet or if Bloom did not maintain the jobs, the state had the authority to recapture its investment (Carper, 2011).

15.3.6 Rent Benefit

A rent benefit, also called the lease execution grant, was another financial incentive. Bloom Energy planned to establish its manufacturing facility on 50 acres of land at the University of Delaware (UD). Chrysler had owned the land but had shuttered its vehicle manufacturing plant in 2008. UD acquired the property in 2009 and developed a new science and technology campus. Because Bloom Energy's innovative energy manufacturing image matched UD's science and technology planning (Carper, 2011), it received a $1 per year rent contract for 25 years (Mordock and Murray, 2016). The university also received a $7 million grant from the state to improve infrastructure throughout the entire site (Carper, 2011).

15.3.6.1 Legislative Amendment

Along with financial agreements, Delaware also amended its Renewable Energy Portfolio Standards Act (REPSA) to include fuel cells as an eligible resource. On July 7, 2011, Governor Markell signed these amendments into law. Because of the amendment, Delmarva, the only regulated utility in Delaware after electricity market deregulation (Chen, 2019), can fulfill its renewable credits requirements under REPSA by purchasing electricity generated from fuel cells. The amendments not only gave Delmarva Power a new way to meet its renewable-energy requirement, but they also secured a buyer for Bloom Energy, even though the primary energy source for the fuel cells is natural gas (Delaware PSC, 2019).

15.4 ECONOMIC CONSIDERATIONS FOR FUEL CELL TECHNOLOGY

15.4.1 Linkage between Clean Energy and Economic Development

Many studies have investigated links between new and renewable energy and national and local economic growth. These studies of clean energy and local economic growth vary in their analytical approach, period of data used, and study area. Most of them focus on renewable energy consumption and economics. For example, in studies of the so-called "Next 11" countries, economic growth was found to catalyze renewable energy consumption when it was evaluated using parametric and nonparametric causality tests, but bidirectional effects were identified using the Geweke Causality Test (Sinha *et al.,* 2018). Bidirectional causality between economic growth and renewable energy consumption was identified in Asia-Pacific Economic Cooperation countries as well. Bidirectional causality means that energy policies can influence economic growth and vice versa (Zafar *et al.,* 2019). Bidirectional causal associations between renewable energy and GDP, as well as between nonrenewable energy and GDP, have also been identified in 11 net oil-importing countries in the Middle East and North Africa. Furthermore, the long-term elastic effect of renewables on real GDP was found to be larger than that of nonrenewables (Kahia *et al.,* 2017). The direction of causality depends on the area studied, specifically on its economic structure and policies. In China, bidirectional causality was only identified in one province (Heilongjiang), where economic policies strongly supported renewable energy development and effectively guaranteed economic development. No link between renewable energy and economic growth was found in 16 other provinces (Bao and Xu, 2019). Similarly, no causality between renewable energy and GDP was identified in Turkey (Bulut and Muratoglu, 2018). The causality between economic growth and renewable energy has been primarily investigated by economists, mostly using time series or panel data causality tests that focus on renewable energy consumption and GDP (Bao and Xu, 2019). These types of studies mainly aim to identify the existence of an interlinkage and the causality between renewable energy consumption and economic development. Other approaches have been used to estimate the potential economic impact of the deployment of renewables (Ramos *et al.,* 2019). For example, Jenniches and Worrell (2019) evaluated the economic and environmental impacts of 280 PV plants with a total capacity of 3.7 MW. Using an input-output analysis, they found that solar PV would generate economic effects equivalent to €3.8 million from 2014 to 2034. These various approaches aimed at measuring the effects of renewable energy on socioeconomic development, such as GDP and employment, based on an assumption of one-way causality from renewable energy consumption to economic development. Also, many studies have investigated the effectiveness of policy measures such as feed-in tariffs or energy tax credits on the deployment of renewable energy (Frazier *et al.,* 2019).

However, as Cox *et al.* (2019) pointed out, the effectiveness of non-energy policies, including economic policies, has not attracted much attention from researchers or policymakers. Although a great deal of literature has investigated the interconnection between renewable energy consumption and economic growth, it has not scrutinized the effects of economic policies on clean energy deployment. While the effects of changes in the economic system, such as liberalization and energy-relevant taxes or incentives, were analyzed, the general tax regime was not dealt with from the perspective of this interlinkage (Cox *et al.*, 2019).

15.4.2 Linkage between Fuel Cells and Economic Development

When it comes to fuel cells, connections have rarely been investigated. Although Wang *et al.*'s 2019) study would seem to explore the links between "hydrogen-based renewable energy" and economic development based on its title, hydrogen-based renewable energy consumption actually refers to per capita renewable electricity generation, excluding hydrogen energy due to insufficient historical data. Similarly, Xu et al. (2019) investigated the links between economic development and "hydrogen-based renewable energy". Again, that term includes renewable energy and the share of hydrogen energy is negligible. When the regional or national scale impacts of fuel cell technologies have been analyzed, it has been for environmental impacts, not economic aspects (Kendall, 2018). Instead, most existing studies focus on the economic feasibility or cost analysis of specific fuel cell technologies (Owebor *et al.*, 2019). For example, they focus on the transportation sector (Hombach *et al.*, 2019) or on combined heat and power (Löbberding and Madlener, 2019). (Morrison et al. 2018) analyzed the competitiveness of fuel cell vehicles, together with electric vehicles. They focused on light-duty vehicles (LDV) and compared the total cost of ownership. They found that fuel cell vehicles would be cost-competitive in the long term (by 2040) with electric vehicles, especially for large cars with longer travel distances. (Chen and Melaina 2019) conducted a similar but more comprehensive study that compared the long-term cost and performance competitiveness of various powertrain technologies, including internal combustion, hybrid, plug-in hybrid, electric, and fuel cell vehicles. (Thompson *et al.*, 2018) estimated the manufacturing cost of a fuel cell system for LDVs in 2017, finding that the total system cost ranges from $45kWnet to $50kWnet when it is produced at a large volume, 100,000 to 500,000 units per year. However, a cost reduction to $30/kWnet would be needed to achieve cost parity with internal combustion systems. The effectiveness of policy measures for fuel cells and the impacts of policy measures on fuel cell deployment or the fuel cell industry have been studied less than the impacts of fuel cells on economic development. For example, (Upreti *et al.*, 2016) measured the impacts of the American Recovery and Reinvestment Act (ARRA) and the ITC on the North American fuel cell industry. The authors simulated the impacts of the policy measures using a model that considered learning-by-doing and economies of scale from increased sales of fuel cells. The ARRA will induce additional sales of about 5600 fuel cell units by 2025.

15.4.3 Infrastructure Challenges

One obstacle in advancing the use of FCEVs on a global scale is the lack of a global infrastructure through which to distribute fuel to the end-user. While this challenge may not be a direct result of complex policies and regulations for the technology itself, it manifests rather typical infrastructure issues. For example, according to the Hydrogen Delivery Technical Team Roadmap for the United States, the hydrogen pipeline network in operation expands to only 1600 miles nationwide and is almost exclusively used for delivering hydrogen to very large hydrogen clients such as chemical plants and petroleum refineries. Here, a change of regulations relating to the nationwide hydrogen pipeline network would indirectly impact a positive movement toward increasing the overall marketability of FCEVs.

BMW expert Axel Rücker stated that "As long as the network of refueling stations for hydrogen-powered cars is so thin, the low demand from customers will not allow for profitable mass production of fuel cell vehicles, and as long as there are hardly any hydrogen cars on the roads, the operators will only hesitantly expand their refueling station network" (Hale, 2020). On the other hand, the Japanese energy group Iwatani has started to establish a network of refueling stations. However, starting the process of establishing a network in Japan is time-consuming and expensive. In Japan, hydrogen is classified as an industrial gas, and as a result, unlike in many other countries, refueling stations need to comply with very strict safety regulations (Green Car Reports, 2017) and rigorous installation requirements.

15.4.4 Cost Challenges

One additional barrier to the commercialization of the FCEV continues to be the high cost of the technology itself. There is currently no lower-priced fuel cell vehicle available on the market that can compete with electric or combustion engine vehicles because the cost of fuel cell technology and hydrogen cylinders is still very high. However, comparing the Toyota Mirai (a fuel cell electric vehicle) priced at USD 49,500 to a Tesla Model 3 Performance (a battery electric vehicle, BEV), which is priced at USD 52,690, the Toyota Mirai appears to be quite competitive. This is particularly true since the driving ranges between the two vehicles are very similar. Nevertheless, expensive fuel cell technology hinders the development of lower-cost models. Nissan currently offers a battery electric vehicle for approximately USD 31,600, and that price is significantly lower compared to the lowest-price FCEV, the Toyota Mirai, which sells for USD 49,500. This is because the fuel cell technology used in the Toyota Mirai, also known as proton exchange membrane cell, utilizes platinum, a very expensive metal, in the catalyst layer of the actual cell. This platinum catalyst layer accounts for nearly half of the fuel cell cost (Proietti *et al.*, 2011). This makes the fuel cell stack by far the most expensive component in a fuel cell vehicle, accounting for approximately USD 11,000 (Tajitsu and Shiraki, 2018). Although the cost of the platinum loading of the fuel cells was reduced significantly over the last decade, it is still a very expensive technology and cannot yet be implemented in vehicles in a USD 20,000 to USD 30,000 price range (Litster and McLean, 2004).

15.5 CONCLUSION

The advancement and deployment of fuel cell technology are critical to achieving global climate goals, reducing dependence on fossil fuels, and fostering innovation in clean energy systems. While fuel cells offer significant environmental and operational advantages such as high efficiency, low emissions, and versatility across sectors, their widespread adoption hinges not only on technological improvements but also on supportive policy and economic structures. From a policy perspective, sustained government intervention remains essential. This includes regulatory clarity, direct financial incentives, public-private partnerships, and alignment with broader national decarbonization strategies. Countries that have implemented comprehensive hydrogen and fuel cell roadmaps, such as Japan, South Korea, and the EU, demonstrate how consistent and forward-looking policies can catalyze market development. In contrast, markets without coherent strategies risk falling behind in both innovation and competitiveness. Economically, while fuel cells are gradually becoming more cost-competitive due to advances in manufacturing and scaling, key challenges remain. High capital costs, limited hydrogen infrastructure, and price volatility of key materials (such as platinum and iridium) continue to constrain growth. Therefore, economic tools such as subsidies, carbon pricing, and green procurement policies are critical to leveling the playing field with incumbent fossil-fuel-based technologies.

Looking ahead, the future of fuel cells will depend on multi-stakeholder coordination among governments, industry players, investors, and civil society. Market confidence, technological innovation, and social acceptance will be driven by transparent governance, equitable investment, and inclusive policy design. Importantly, long-term policy certainty and economic signals must continue to prioritize clean energy deployment over short-term market fluctuations or political shifts. While fuel cells are not a silver bullet, they represent a vital component of a diversified and resilient clean energy portfolio. Realizing their promise requires deliberate, well-coordinated action to bridge policy goals and economic realities. By aligning regulatory support, investment flows, and infrastructure development, fuel cells can transition from niche applications to mainstream solutions in the global energy landscape.

15.6 RECOMMENDATIONS

To fully unlock the potential of fuel cells as a cornerstone of the clean energy transition, coordinated action across policy, economic, and industrial spheres is essential. The following recommendations provide a roadmap for governments, industry stakeholders, investors, and research institutions to accelerate the adoption and commercialization of fuel cell technologies.

Develop Clear National Roadmaps: Governments should create or update comprehensive fuel cell and hydrogen strategies aligned with long-term decarbonization goals. These roadmaps should set measurable targets, identify priority sectors, and outline implementation timelines.

Standardize Regulations: International cooperation is needed to harmonize safety, performance, and environmental standards for fuel cell systems and hydrogen infrastructure. Consistent regulation across borders will facilitate trade, investment, and cross-national projects.

Subsidize Early-Stage Deployment: Financial incentives should target both supply and demand sides to reduce upfront costs, encourage adoption, and stimulate production capacity. This includes tax credits, purchase subsidies, and low-interest loans.

Internalize Environmental Costs: Implement or expand carbon pricing schemes (e.g., carbon taxes or emissions trading) to reflect the true environmental costs of fossil fuel systems and improve the competitiveness of fuel cells.

Support Green Hydrogen Production: Since hydrogen is the primary fuel for most fuel cells, investments in electrolyzers, renewable energy integration, and hydrogen storage/distribution infrastructure should be prioritized.

Invest in Innovation: Governments and private sector partners should co-fund RD&D to lower fuel cell costs, improve durability, and reduce dependence on rare or expensive materials (e.g., platinum).

Pilot Projects and Demonstration Sites: Encourage the deployment of large-scale demonstration projects in diverse applications (e.g., transportation fleets, grid backup, industrial heat) to validate performance and build public and investor confidence.

REFERENCES

Bao, C., & Xu, M. (2019). Cause and effect of renewable energy consumption on urbanization and economic growth in China's provinces and regions. *Journal of Cleaner Production*, *231*, 483–493.

Bulut, U., & Muratoglu, G. (2018). Renewable energy in Turkey: Great potential, low but increasing utilization, and an empirical analysis on renewable energy-growth nexus. *Energy Policy*, *123*, 240–250.

Carper, J. (2011). *Bloom Energy and the future of clean power*. [Website/Publisher name if available]. https://www.example.com/bloom-energy-article

Chen, W.-M. (2019). The U.S. electricity market twenty years after restructuring: A review experience in the state of Delaware. *Utilities Policy*, *57*, 24–32. https://doi.org/10.1016/j.jup.2019.01.003

Chen, Y., & Melaina, M. (2019). Model-based techno-economic evaluation of fuel cell vehicles considering technology uncertainties. *Transportation Research Part D: Transport and Environment*, *74*, 234–244.

Cox, E., Royston, S., & Selby, J. (2019). From exports to exercise: How non-energy policies affect energy systems. *Energy Research & Social Science*, *55*, 179–188.

Delaware Public Service Commission, 2019. In the matter of the application of delmarva power and light company for approval of qualified fuel cell provider project tariffs. Docket # 19-0045 [WWW Document]. URL. https://delafile.delaware.gov/. accessed 2.15.19.

Dolci, F., Thomas, D., Hilliard, S., Guerra, C. F., Hancke, R., Ito, H., & Lucchese, P. (2019). Incentives and legal barriers for power-to-hydrogen pathways: An international snapshot. *International Journal of Hydrogen Energy*, *44*(23), 11394–11401.

EC, 2005. Directive of the European Parliament and of the Council establishing a framework for the setting of ecodesign requirements for energy-using products, 2005/32/EC, Official Journal, L 191/29–L191/58.

EC, 2006. Directive of the European Parliament and the Council of 5 April 2006 on energy end-use efficiency and energy services and repealing Council Directive 93/76/EEC, Official Journal, L 114/64.

Edelstein, S. (2015). *Japan eases laws, boosts incentives for hydrogen fuel-cell cars*. Green Car Reports. https://www.greencarreports.com/news/1096324_japan-eases-laws-boosts-incentivesfor-hydrogen-fuel-cell-cars

Frazier, A. W., Marcy, C., & Cole, W. (2019). Wind and solar PV deployment after tax credits expire: A view from the standard scenarios and the annual energy outlook. *The Electricity Journal*, *32*(8), 106637.

Fuel Cells and Hydrogen 2 Joint Undertaking. (2019). *Hydrogen roadmap Europe: A sustainable pathway for the European energy transition*. https://op.europa.eu/en/publication-detail/-/publication/0817d60d-332f-11e9-8d04-01aa75ed71a1/language-en

Green Car Reports. (2017, April 3). *All the challenges for hydrogen fuel-cell cars laid out*. https://www.greencarreports.com/news/1109684_all-the-challenges-for-hydrogen-fuel-cell-cars-laid-out

Greenwood, M. (2019, April 9). Japan sees big future in hydrogen cars. *Engineering.com*. https://www.engineering.com/japan-sees-big-future-in-hydrogen-cars/

Hale, A. (2020). A closer look at BMW's hydrogen fuel cell technology. HotCars. https://web.archive.org/web/20200808000000/https://www.hotcars.com/a-closer-look-at-bmws-hydrogen-fuel-cell-technology/

Hombach, L. E., Doré, L., Heidgen, K., Maas, H., Wallington, T. J., & Walther, G. (2019). Economic and environmental assessment of current (2015) and future (2030) use of E-fuels in light-duty vehicles in Germany. *Journal of Cleaner Production*, *207*, 153–162.

Iida, S., & Sakata, K. (2019). Hydrogen technologies and developments in Japan. *Clean Energy*, *3*(2), 105–113. https://doi.org/10.1093/ce/zkz003

IPCC. (2000). *Methodological and technological issues in technology transfer: Summary for policymakers: A special report of IPCC Working Group III*. Cambridge, University Press.

Ishikawa, K. (2019). *Hydrogen energy supply chain from Australia to Japan*. https://netl.doe.gov/sites/default/files/netl-file/19CCUSOGT-20190829-Kawasaki-Final.pdf

Japan Ministry of Economy, Trade and Industry. (2019). *The strategic road map for hydrogen and fuel cells*. https://www.meti.go.jp/english/press/2019/pdf/0312_002a.pdf

Jenniches, S., & Worrell, E. (2019). Regional economic and environmental impacts of renewable energy developments: Solar PV in the Aachen Region. *Energy for Sustainable Development*, *48*, 11–24.

Jong, W., Honselaar, G., Cebolla, O., & Institute for Energy and Transport (Joint Research Centre). (2015). *CEN—CENELEC sector forum energy management: Working group hydrogen: Final report*. https://op.europa.eu/en/publication-detail/-/publication/99f62cea-a877-11e5-b528-01aa75ed71a1

Kahia, M., Aïssa, M. S. B., & Lanouar, C. (2017). Renewable and non-renewable energy use-economic growth nexus: The case of MENA Net Oil Importing Countries. *Renewable and Sustainable Energy Reviews*, *71*, 127–140.

Kendall, M. (2018). Fuel cell development for New Energy Vehicles (NEVs) and clean air in China. *Progress in Natural Science: Materials International*, *28*(2), 113–120.

Litster, S., & McLean, G. (2004, March 2). PEM fuel cell electrodes. *Journal of Power Sources*, *130*(1–2), 61–76. https://doi.org/10.1016/j.jpowsour.2004.01.001

Löbberding, L., & Madlener, R. (2019). Techno-economic analysis of micro fuel cell cogeneration and storage in Germany. *Applied Energy*, *235*, 1603–1613.

Mordock, J., & Murray, M. (2016). The truth behind Bloom Energy's sweet deal in Delaware. *News J.*

Morrison, G., Stevens, J., & Joseck, F. (2018). Relative economic competitiveness of light-duty battery electric and fuel cell electric vehicles. *Transportation Research Part C: Emerging Technologies*, *87*, 183–196.

Owebor, K., Oko, C. O. C., Diemuodeke, E. O., & Ogorure, O. J. (2019). Thermo-environmental and economic analysis of an integrated municipal waste-to-energy solid oxide fuel cell, gas-, steam-, organic fluid-and absorption refrigeration cycle thermal power plants. *Applied Energy*, *239*, 1385–1401.

Proietti, E., Jaouen, F., Lefèvre, M., Larouche, N., Tian, J., Herranz, J., & Dodelet, J. (2011). Iron-based cathode catalyst with enhanced power density in polymer electrolyte membrane fuel cells. *Nature Communications*, *2*, Article 416. https://www.nature.com/articles/ncomms1427

Ramos, C., García, A. S., Moreno, B., & Díaz, G. (2019). Small-scale renewable power technologies are an alternative to reach a sustainable economic growth: Evidence from Spain. *Energy*, *167*, 13–25.

Sinha, A., Shahbaz, M., & Sengupta, T. (2018). Renewable energy policies and contradictions in causality: a case of Next 11 countries. *Journal of Cleaner Production*, *197*, 73–84.

Sugiyama, M., Fujimori, S., Wada, K., Endo, S., Fujii, Y., Komiyama, R., ... & Shiraki, H. (2019). Japan's long-term climate mitigation policy: multi-model assessment and sectoral challenges. *Energy*, *167*, 1120–1131.

Tajitsu, N., & Shiraki, M. (2018). *Toyota plans to expand production, shrink cost of hydrogen fuel cell vehicles*. Reuters. https://www.reuters.com/article/us-toyota-hydrogen/toyota-plans-to-expand-production-shrink-cost-of-hydrogen-fuel-cell-vehicles

Thompson, S. T., James, B. D., Huya-Kouadio, J. M., Houchins, C., DeSantis, D. A., Ahluwalia, R., & Papageorgopoulos, D. (2018). Direct hydrogen fuel cell electric vehicle cost analysis: System and high-volume manufacturing description, validation, and outlook. *Journal of Power Sources*, *399*, 304–313.

U.S Department of Energy. 2018. State of the States: Fuel Cells in America 2017.

U.S. Department of Energy. (2015). Quadrennial Technology Review 2015.

U.S. Department of Energy. (2021). Hydrogen Fueling Infrastructure Development. Available online: https://afdc.energy.gov/fuels/hydrogen_infrastructure.html

U.S. Department of Energy. (2023). *Fuel cells*. Office of Energy Efficiency and Renewable Energy. https://www.energy.gov/eere/fuelcells/fuel-cells

U.S. Internal Revenue Service. (2021). Fuel Cell Motor Vehicle Tax Credit. Available online: https://afdc.energy.gov/laws/350

Upreti, G., Greene, D. L., Duleep, K. G., & Sawhney, R. (2016). Impacts of the American recovery and reinvestment act and the investment tax credit on the north American non-automotive PEM fuel cell industry. *International Journal of Hydrogen Energy*, *41*(5), 3664–3675.

Wang, Z., Kamali, F. P., Osseweijer, P., & Posada, J. A. (2019). Socioeconomic effects of aviation biofuel production in Brazil: A scenarios-based Input-Output analysis. *Journal of Cleaner Production*, *230*, 1036–1050.

Xu, R., Chou, L. C., & Zhang, W. H. (2019). The effect of CO2 emissions and economic performance on hydrogen-based renewable production in 35 European Countries. *International Journal of Hydrogen Energy*, *44*(56), 29418–29425.

Zafar, M. W., Shahbaz, M., Hou, F., & Sinha, A. (2019). From nonrenewable to renewable energy and its impact on economic growth: the role of research & development expenditures in Asia-Pacific Economic Cooperation countries. *Journal of Cleaner Production*, *212*, 1166–1178.

16 Challenges and Future Directions of Fuel Cells

Falese Babatunde Abimbola, Olaewe Tunde Olaniyi, and Olafare Biola Blessing

16.1 INTRODUCTION

The world sees growing approval for fuel cells as a major stakeholder in efforts to make the world more sustainable, with a friendlier and sustainable energy system. This capability to produce electricity effectively from chemical energy, via the use of electrochemical reactions, makes fuel cells also more efficient with virtually no emissions, hence deserving to be considered as an alternative fuel source for power generation from fossil fuels (Sazali et al., 2020). The flexibility of fuel cells makes them appropriate for sectors comparable to transportation, power generation at fixed locations, and robust power for mobile gadgets (Lai and Michael, 2017).

Despite its significant benefits, the massive challenges fuel cells face slow them down when it comes to mass utilization. The high cost of production is one of the biggest challenges to the adoption of fuel cells, partly due to the higher costs of materials such as the platinum used in the cells' catalysts (Wang *et al.*, 2018). Sufficient system durability and longevity are also further challenges, as degradation of the performance of fuel cells is frequently noted as time elapses, particularly in varying operating conditions (Wang, 2017). Due to a lack of hydrogen refueling stations in various parts of the world, the commercialization of fuel cell vehicles is limited by major challenges (Hassan et al., 2023). Proving the safe handling of hydrogen in transport and use adds new complications to fuel cell incorporation.

Going forward, overcoming these stumbling blocks depends on continuous research and development projects. Innovations in material used for catalysts aim at reducing the consumption of expensive metals, thereby reducing costs considerably. Improved stability of membranes and electrodes is essential to cater to fuel cell systems that can last this long (Jamal et al., 2023). It is essential to make available effective and cost-effective methods of production, storage, and distribution of hydrogen to create an infrastructure on which fuel cell uptake can be made. Furthermore, the direct participation and financial support of the public and private sectors are also crucial for the rapid development of technologies and the comfortable joining of fuel cells to the sphere of energy.

Eventually, even though fuel cells' promotional potential for sustainable energy exists, one must address the persistent technical, economic, and infrastructural barriers. How this can be realized through fuel cells is continuous innovation and convergence of various stakeholders.

16.2 CHALLENGES OF FUEL CELLS

Even though fuel cells are capable of providing clean and efficient energy, there are numerous obstacles that prevent their mass use. Research targeting the removal of these technical impediments toward the commercialization of fuel cells has been supported by the Department of Energy (DOE). The poor energy costs, durability of the polymer electrolyte membrane (PEM) fuel cells, challenges with hydrogen production and supply, fuel cells' performance efficiency under variable conditions,

DOI: 10.1201/9781003645405-16

environmental and regulatory factors to the deployment of fuel cells, and water and heat management issues were among the main issues with the use of fuel cells.

16.2.1 The High Price of Production and Materials

Materials used in catalysis, including platinum, are central to the fabrication of Proton Exchange Membrane Fuel Cells (PEMFC) and other fuel cells. Because these materials are expensive, and their mining practice may harm the environment, they are considered environmentally unsustainable (Okonkwo et al., 2021). Though pilots have broken traditional cost trends in industry, fuel cells tend to cost more than both didactic energy sources and batteries in typical comparisons, as seen by Xun et al. (2022). In order to replace traditional solutions, fuel cells must be developed at a similar cost to the internal combustion engine (ICE) system. At the current price, fuel cells are costly and require reduced expense. The transportation direct hydrogen 80-kW fuel cell system could be expected to cost $73 per kilowatt when produced at a volume of 500,000 per year, according to Martin et al. (2010). The DOE's cost target for fuel cell systems is $30/kW by 2015. The fuel cell stack turns out to be more than half the cost of the entire system when cost estimates are examined. Specific activity is reached by utilizing finely dispersed Pt particles in the widely used catalysts. At higher production levels, catalyst ink makes up about 50% (47 %) of the fuel cell stack's cost owing to the high price of Pt (Liu et al., 2022). Under these production figures (1,000 systems/year), the membrane is the most substantial cost element in the fuel cell stack, according to Martin et al. (2010). In addition, currently used PEMs must be humidified and face a restriction on the maximum operating temperature of the cell (Ge et al., 2022). If membranes can work efficiently at low humidity and high operating temperatures, it would be possible to reduce the complexity of the system by either removing or decreasing humidification steps, and minimizing the size of the thermal management system.

16.2.2 Durability and Longevity Issues

In the long run, fuel cells may corrode and suffer contamination by matter such as carbon monoxide (CO) or high temperatures. As Tellez-Cruz et al. state, efficiency and effectiveness may suffer; more maintenance and replacement are needed. Therefore, fuel cells, in particular for automotive propulsion, need to have endurance in operating and cyclic environments. The system is to function under temperature range below freezing to above boiling, which is accompanied by humidity range from ambient to saturation and a half-cell potential from 0 to more than 1.5 V. (Rajalakshmi et al., 2017). The rapid changes, along with the repeated patterns, in the operating conditions significantly increase the need to cope with such conditions. Variations in voltage, temperature, humidity, and gas composition cause rapid changes in cell and stack performance at elevated rates (Schüttoff et al., 2024). Various physical and chemical changes occur in repetitive cycles with frequent adverse effects manifested (Martin et al., 2010). Additionally, the anode side of the cell, under start/stop conditions when exposed to hydrogen and air, produces potentials that are higher than 1.5 V. DOE has designated 40,000 hours for stationary and 5,000 hours for transportation fuel cells under realistic conditions of load cycling and start/stop when in operation (Madheswaran et al., 2023). For transport fuel cells transient working includes:

1. 17,000 start/stop cycles
2. 1,650 freeze cycles
3. 1,200,000 load cycles.

(Martin et al., 2010).

These cycles lead to the following effects: nutrition and the broad-ranging environmental consequences of anthropogenic land use.

1. Up-transient – hydrogen starvation
2. Down-transient – differential pressure imbalance

3. Load cycling intensifies the corrosion risks and contributes more mechanical stress to the membrane in dynamic cell operation.
4. Low power operation – high voltage (catalyst and support corrosion potential
5. Off – transit of oxygen to the anode, and rusting of the support.

(Martin et al., 2010)

Not only do normal operations cause cycles, but system failures caused by non-stack components can generate additional unplanned operating cycles. Martin et al. (2010) report that 85–90% of system failures are caused by unplanned shutdowns. The ability of fuel cells to operate when system failures are unexpected is essential, according to Wang (2017). The most troublesome source of degradation consists of starting and shutting down the system. When the system is subjected to repeated start-stop cycles, local potentials may rise to 1.5 V, which encourages oxidation in carbon supports. Martin et al. (2010) state that catalysts and supports must remain stable despite both periods of fuel scarcity and the changing potentials that accompany starting and stopping operations.

16.2.3 Hydrogen Production and Supply Challenges

Hydrogen is used as the principal fuel in a number of fuel cell systems, the Proton Exchange Membrane Fuel Cells (PEMFCs) being widely employed for transport and stationary applications (Aminudin et al., 2023). Although they have enjoyed a reputation for clean operation, the actual long-term sustainability of the systems relies on methods for sourcing and transporting hydrogen. Currently, about 95% of hydrogen is produced from fossil fuels, likely by heavy-emitting processing such as steam methane reforming and coal gasification (Yan et al., 2020). The application of "grey hydrogen," which is primarily obtained from fossil fuels, directly contradicts the objective of low-emission fuel cell technology, thus leaving it questionable whether it can be sustainable (Azadnia et al., 2023). The production, which is realized using renewable energy through electrolysis, is more of a green proposition but, Analyses by the International Energy Agency (IEA) claim green hydrogen is financially uncompetitive due to its higher costs, which are 2–5 times higher than those of grey hydrogen, in the absence of massive policy incentives or carbon pricing systems (Hermesmann and Müller, 2022).

Greater production of green hydrogen might not overcome the huge obstacles created by the lack of supporting infrastructure. The obstacles like lack of hydrogen pipelines, lack of refueling stations, and compression and storage costs are conducive to increasing the cost of hydrogen delivery (Wang, 2015). Also, the safe and efficient transportation of hydrogen, in high-pressure gas or cryogenically, in a liquid form, is a consideration in itself. In the arrangement of renewable energy resources and clean water, an inadequate balance exposes energy security and supply reliability to risk, and some regions are more exposed to risk (Carley et al., 2017. Such problems of fabrication, cost, and operation in transporting the fuel cells limit the fuel cells' ability to support sustainable energy. The immediate necessity is for green hydrogen production investments, supply chain development, and international regulation calibration to completely harness the benefits of fuel cell technologies.

16.2.4 Performance Efficiency of Fuel Cells under Variable Load Conditions

Fuel cells are known for their outstanding efficiency and environment-friendly operation, particularly in respect to their operation under optimal conditions, which are steady state (Qasem and Abdulrahman, 2024). However, in normal activity associated with vehicles, mobile power sources, and power regulation systems, fuel cells face variable demands (Qasem and Abdulrahman, 2024). The changing loads represent significant challenges to effective operation. Rapid load swings are not one of the initial engineering features of PEMFCs and SOFCs (Zhou et al., 2022). The electrochemical and thermal inertia in fuel cell systems cause a lagging reaction to occur upon sudden power demand variation (Peng et al., 2008). Eventually, these delays could interfere with the system, creating a state of short-term power mismatches that would incorporate weakening the efficiency and against the system balance (Machowski et al., 2020).

Besides, performance decreases in fuel cells when efficiency declines under partial loads instead of it operating at full capacity. Running in dynamic load conditions tends to see the fuel cell fluctuating between maximum, partial, and idle performance, compromising its efficiency (Zhang et al., 2021). This issue is most critical for hybrid vehicles, which are dominated by the requirement for frequent changes in the power demand depending on the driving pattern. Besides, the repeated fire-up and shut-down episodes experienced during intermittent use hasten the erosion of fuel cell components, including catalysts and membranes, as indicated by Yu et al. (2012). The aggregated impact of this operational stress declines the system's operational life and boosts maintenance needs. For SOFCs where operating temperature is high (typically between 600 and 1000°C), the long, complex startup and shutdown procedures are especially important and pose serious challenges to system durability due to long durations and many steps (Corigliano et al., 2022).

Thermal stability management is essential to fulfill these operational requirements. Under variable load, fuel cells require optimum heat management approaches to reduce the threat of thermal cycling and the stability of performance. Poor temperature regulation in systems such as SOFCs can lead to structural damage on ceramics, seal failure, and a continued inefficiency drop (Peng et al., 2021). This drawback in the load following ability of existing fuel cell technologies hinders the widespread use of such technology in dynamic operational application scenarios (Yang et al., 2017). In an attempt to overcome these problems, new fuel cell designs will need to employ state-of-the-art control schemes, effective hybrid combinations with batteries or supercapacitors, and substantial thermal management strategies to maximize.

16.2.5 Challenges to the Deployment of Fuel Cells brought on by Environmental and Regulatory Factors

The expanding attention given to fuel cell technologies has not alleviated the substantial regulation and environmental problems that limit their deployment at scale. The problem of uncoordinated international rules for hydrogen production, storage, and use is one of the main impediments to rapid deployment. The standards for building hydrogen fueling stations, achieving safety certifications, and defining handling methods vary substantially both among countries and internal regions (Kumar et al., 2024). Such inconsistencies make international cooperation more difficult for manufacturers and developers planning to bring fuel cell systems to the market (Kumar et al., 2024). A lack of global standards not only delays advancements in the field but also drives up compliance expenses, thereby discouraging general market uptake.

Regulatory inconsistency is compounded by the fact that fuel cell projects receive smaller financial allocations than renewables like wind, solar, or Li-ion battery storage (Chalk and Miller, 2006). Hydrogen and fuel cell systems receive few subsidies and infrastructure funds from governments, especially in countries in the Global South. The absence of these support mechanisms causes fuel cell technologies to have a higher cost than more mature alternatives, thereby reducing private investment (Nyangon and Darekar, 2024). Besides, poor financial support for R&D and the scarcity of public–private alliances in developing nations slow down the pace of innovation and infrastructure investment. This gap in technology means that fuel cell integration is less consistent and more fragmented globally (Rumiantsev et al., 2024).

When taken together, these types of barriers generate a major obstacle for the large-scale adoption of fuel cells. If regulatory frameworks lack clarity, both investors and other stakeholders must deal with high levels of risk, thus hindering their ability to plan over the long term and encourage innovation. Additionally, insufficient attention to policy making constrains the establishment of the essential hydrogen infrastructure, including refueling stations and storage units upon which fuel cells are dependent (Apostolou and Xydis, 2019). Appropriate regulation, added investment in research, and effective public–private alliances are needed for nations to surpass these challenges. Fuel cell technologies will reach their maximum potential in driving the global energy transition solely when supported by robust and harmonized policies.

16.2.6 Water and Heat Management Issues

The efficient operation of fuel cells, particularly PEMFCs, is complicated by the need to balance water and heat, since these cells need water for ion transfer and run at relatively low temperatures (Zhang et al., 2024). High proton conductivity depends on the hydration of the membrane within a PEMFC (Tellez-Cruz et al., 2021). Balancing water effectively is hard, since too much water can inundate the MEA and inhibit reactant gas flow, leading to lower cell performance, while too little water can dry the membrane, greatly decreasing proton conductivity and hastening membrane breakdown (Ous Hence, reliability and durability of operation require exact control of humidity and water balance.

Control of heat is as vital in fuel cells as they generate notable heat over the course of operation (Xu et al., 2022). A lack of effective cooling systems enables temperature differentials within the cell, which then causes inconsistent chemical reactions and faster material wear. Moreover, SOFCs operating at high temperatures confront extra complications, especially the slow initiation and strain on internal components induced by repeated thermal fluctuations. A lack of proper thermal control in both kinds of fuel cells can reduce overall efficiency, harm internal elements, and decrease the system's overall lifespan (Tarancón, 2009). As a result, advanced techniques for cooling and intelligent control are indispensable for keeping operating temperatures and water content steady, which is necessary for continuous high performance and system robustness.

16.3 FUTURE DIRECTIONS OF FUEL CELLS

The pace of progress for fuel cell technology is directly tied to increases in oil prices worldwide, declines in accessible oil supplies, poor success in new oil discoveries, and more funding for developing hydrogen energy systems (Basu, 2007). Environmental problems linked to vehicular, power plant, and oil refinery emissions will not drive progress toward development unless users and manufacturers recognize the cost reduction benefits of pollution reduction measures. Constructing a hydrogen energy infrastructure requires activities such as fuel production, supply chains, safe dispensing, and safety regulation. Hydrogen, alcohols, esters, naphtha, and natural gases are among the gases supplied to fuel cells and fuel processors without any additional conversion. While some energy sources are produced using renewable techniques such as wind and solar energy, water electrolysis, gasification of biomass, and fermentation, others are produced from fossil fuels, especially hydrogen and alcohol (Ous and Arcoumanis, 2013). According to Basu (2007), greenhouse gas emissions in the first case are very small and may be treated as insignificant.

A highly attractive goal is to suppose that, except for biomass gasification, neither cars nor stationary power plants produce air pollutants or greenhouse gases. With this solution, all the air pollutant emissions will occur at one specific place; so, cities will not have to manage pollution from car emissions that derive from internal combustion power. The hydrogen fuel cell vehicle (H_2 FCV) and H_2FCV-hybrid electric vehicle cause the least environmental impact when compared to other advanced options. Basu (2007) shows that choosing hydrogen from natural gas as a fuel would minimize environmental pollution damage costs to one eighth and one fifteenth compared to current gasoline internal combustion engines, if there is no or with CO_2 capture. In spite of the fact that the economic situation is not supportive of PEMFC vehicles or SOFC stationary plants due to insufficient hydrogen availability, growing crude costs, limited reserves, and technological advances in fuel cell performance and price allow for optimism. Increased hydrogen energy infrastructure, along with this, will independently raise the profitability of FCVs and fuel cell distributed power generation.

Optimally sizing the fuel cell, according to Stambouli (2011), is important for placing it in a distributed system to handle utility peak loads. For markets ranging from 0.5 kW to 2 MW,

Farooque and Maru (2001) have compiled a selection of intermediate sizes for fuel cells. Scientists from many nations are working to find better materials for fuel cells, which have been approaching commercial readiness for years. This combined work is targeted at solving the main technical issues that current FCs encounter. Many approaches in Japan and the USA make use of an uncomplicated ceramic method to fabricate thin-film electrolytes.

As a result, this advancement decreases cell resistance, raises power output by 100%, and greatly decreases the expense of SOFCs. According to Le Goff et al. (2009), a nickel bisdiphosphine catalyst has been shown for use in fuel cells. The effort continues to link FCs together, create innovative stack arrangements, and increase the lifetime of the system to surpass 60,000 hours. Because it was shown that fuel cells run on methane at reduced expense, the prospect has greatly expanded for the simple and affordable design of power stations (Farooque and Maru, 2001). Further significant changes in standard FC design have been developed by companies dedicated to FC manufacturing. Since 2000, a 48.6% rise in single-cell power densities has been attained, which is the highest value for commercial-sized fuel cells reported globally.

Such improvements in power density are brought about by changes in both cell design and component materials. That higher power densities lead to lower system weights, volumes, and costs is well established in fuel cell systems. At the projected cost of $1000/kW, fuel cells could in future be considered for small residential applications (Stambouli, 2011).

16.4 CONCLUSION

As soon as the oil, coal, and gas reserves are depleted, a transition to a hydrogen economy will establish fuel cells as the preferred technology. Because all three generations of fuel cells, AFC, PAFC, and MCFC, have been adequately developed, there are presently no unresolved research and development questions. Both kinds of fuel cells find themselves in the segment of the S-curve with the greatest rate of improvement. Yet, improvements to SOFC, PEMFC, and DMFC continue to gain momentum on the S-curve, and many development milestones must still be reached before they are ready for consumer use. Therefore, the focus of ongoing fuel cell research and development has been limited to SOFC, PEMFC, and DMFC. The strong fiscal support by leading countries, such as the USA, Europe, Canada, and Japan, for fuel cell research does not prevent smaller countries from specializing in PEMFC and DMFC and introducing these systems to niche markets (Sopian and Wan-Daud, 2006).

Fuel cells, along with other substitutes, are capable of significantly increasing the economic efficiency and environmental effectiveness of existing energy systems. In consequence, the use of fuel cells must be part of a broader strategy devoted to efficient energy use and a higher share of renewable energy sources (Pehnt and Ramesohl, 2003). Realizing the objective to switch from fossil fuel systems and infrastructures to a renewable hydrogen alternative will require considerable time (Hassan et al., 2024). Advancing and evaluating pioneering approaches in major areas could guide the development of the proposed energy system. The projects are thus designed to work together with existing policies to drive the development and consumption of sensible and renewable power. To move away from fossil fuels, policy settings and industrial support by governments should favor a portfolio of environmentally sound energy alternatives. Fuel cells that operate with natural gas as a transitional source of energy are critical for moving toward both a renewable energy system and zero-emission electricity (Pehnt and Ramesohl, 2003).

REFERENCES

Aminudin, M. A., Kamarudin, S. K., Lim, B. H., Majilan, E. H., Masdar, M. S., & Shaari, N. (2023). An overview: Current progress on hydrogen fuel cell vehicles. *International Journal of Hydrogen Energy*, *48*(11), 4371–4388.

Apostolou, D., & Xydis, G. (2019). A literature review on hydrogen refuelling stations and infrastructure. Current status and future prospects. *Renewable and Sustainable Energy Reviews*, *113*, 109292.

Azadnia, A. H., McDaid, C., Andwari, A. M., & Hosseini, S. E. (2023). Green hydrogen supply chain risk analysis: A European hard-to-abate sectors perspective. *Renewable and Sustainable Energy Reviews*, *182*, 113371.

Basu, S. (2007). Future directions of fuel cell science and technology. In *Recent trends in fuel cell science and technology* (pp. 356–365). Springer New York.

Carley, S., Baldwin, E., MacLean, L. M., & Brass, J. N. (2017). Global expansion of renewable energy generation: An analysis of policy instruments. *Environmental and resource economics*, *68*(2), 397–440.

Chalk, S. G., & Miller, J. F. (2006). Key challenges and recent progress in batteries, fuel cells, and hydrogen storage for clean energy systems. *Journal of Power Sources*, *159*(1), 73–80.

Corigliano, O., Pagnotta, L., & Fragiacomo, P. (2022). On the technology of solid oxide fuel cell (SOFC) energy systems for stationary power generation: A review. *Sustainability*, *14*(22), 15276.

Farooque, M., Maru, H. C. (2001). Fuel cells – the clean and efficient power generators. *IEEE Proceedings*, *89*, 1819–1829.

Ge, X., Zhang, F., Wu, L., Yang, Z., & Xu, T. (2022). Current challenges and perspectives of polymer electrolyte membranes. *Macromolecules*, *55*(10), 3773–3787.

Hassan, N.S., Jalil, A.A., Rajendran, S., Khusnun, N.F., Bahari, M.B., Johari, A., Kamaruddin, M.J. and Ismail, M. (2024). Recent review and evaluation of green hydrogen production via water electrolysis for a sustainable and clean energy society. *International Journal of Hydrogen Energy*, *52*, 420–441.

Hassan, Q., Azzawi, I. D., Sameen, A. Z., & Salman, H. M. (2023). Hydrogen fuel cell vehicles: Opportunities and challenges. *Sustainability*, *15*(15), 11501.

Hermesmann, M., & Müller, T. E. (2022). Green, turquoise, blue, or grey? Environmentally friendly hydrogen production in transforming energy systems. *Progress in Energy and Combustion Science*, *90*, 100996.

Jamal, T., Shafiullah, G. M., Dawood, F., Kaur, A., Arif, M. T., Pugazhendhi, R. & Ahmed, S. F. (2023). Fuelling the future: An in-depth review of recent trends, challenges and opportunities of hydrogen fuel cell for a sustainable hydrogen economy. *Energy reports*, *10*, 2103–2127.

Kumar, S., Nanan-Surujbally, A., Sharma, D. P., & Pathak, D. (2024). Hydrogen safety/standards (national and international document standards on hydrogen energy and fuel cell). In *Towards hydrogen infrastructure* (pp. 315–346). Elsevier.

Lai, J.-S., and Michael W. E. (2017). Fuel cell power systems and applications. *Proceedings of the IEEE*, *105*(11), 2166–2190.

Le Goff, A., Artero, V., Jousselme, B., Phong-Dinh, T., Guillet, N., Métayé, R., Fihri, A., Palacin, S., Fontecave, M. (2009). From hydrogenases to noble metal–free catalytic nanomaterials for H_2 production and uptake. *Science*, *326*(5958), 1384–1387.

Liu, H., Ney, L., Zamel, N., & Li, X. (2022). Effect of catalyst ink and formation process on the multiscale structure of catalyst layers in PEM fuel cells. *Applied Sciences*, *12*(8), 3776.

Machowski, J., Lubosny, Z., Bialek, J.W. and Bumby, J.R. (2020). *Power system dynamics: Stability and control*. John Wiley & Sons.

Madheswaran, D. K., Thangamuthu, M., Gnanasekaran, S., Gopi, S., Ayyasamy, T., & Pardeshi, S. S. (2023). Powering the future: Progress and hurdles in developing proton exchange membrane fuel cell components to achieve department of energy goals a systematic review. *Sustainability*, *15*(22), 15923.

Martin, K. E., Kopasz, J. P., McMurphy, K. W. (2010). Status of fuel cells and challenges facing fuel cell technology today. *Hydrogen Fuel Cells and Infrastructure Technology*, *1*, 1–13.

Nyangon, J., & Darekar, A. (2024). Advancements in hydrogen energy systems: A review of levelized costs, financial incentives and technological innovations. *Innovation and Green Development*, *3*(3), 100149.

Okonkwo, P. C., Ige, O. O., Uzoma, P. C., Emori, W., Benamor, A., & Abdullah, A. M. (2021). Platinum degradation mechanisms in proton exchange membrane fuel cell (PEMFC) system: A review. *International Journal of Hydrogen Energy*, *46*(29), 15850–15865.

Ous, T., & Arcoumanis, C. (2013). Degradation aspects of water formation and transport in proton exchange membrane fuel cell: A review. *Journal of Power Sources*, *240*, 558–582.

Pehnt, M., Ramesohl, S. (2003). Fuel cells for distributed power: Benefits, barriers and perspectives. *World wide Fund for Nature*, 1–8.

Peng, J., Huang, J., Wu, X. L., Xu, Y. W., Chen, H., & Li, X. (2021). Solid oxide fuel cell (SOFC) performance evaluation, fault diagnosis and health control: A review. *Journal of Power Sources*, *505*, 230058.

Peng, J., Shin, J. Y., & Song, T. W. (2008). Transient response of high temperature PEM fuel cell. *Journal of Power Sources*, *179*(1), 220–231.

Qasem, N. A., & Abdulrahman, G. A. (2024). A recent comprehensive review of fuel cells: History, types, and applications. *International Journal of Energy Research*, *2024*(1), 7271748.

Rajalakshmi, N., Imran Jafri, R., & Dhathathreyan, K. S. (2017). Research advancements in low- temperature fuel cells. In *Electrocatalysts for low temperature fuel cells: fundamentals and recent trends* (pp. 35–74).

Rumiantsev, A., Morozov, V., & Panikar, G. (2024). Innovation at the crossroads: How public- private partnerships drive technological development in Europe. *Actual Problems of International Relations*, *1*(159), 94–108.

Sazali, N., Wan Salleh, W. N., Jamaludin, A. S., & Mhd Razali, M. N. (2020). New perspectives on fuel cell technology: A brief review. *Membranes*, *10*(5), 9.

Schüttoff, M., Wachtel, C., Schlumberger, R., Wilhelm, F., Scholta, J., & Hölzle, M. (2024). Development of accelerated durability test protocols for polymer electrolyte membrane fuel cell stacks under realistic operating conditions. *Fuel Cells*, *24*(5), e202300263.

Sopian K., Wan-Daud, W. R. (2006). Challenges and future developments in proton exchange membrane fuel cells. *Renewable Energy*, *31*, 719–727.

Stambouli A. B. (2011). Fuel cells: The expectations for an environmental-friendly and sustainable source of energy. *Renewable and Sustainable Energy Reviews*, *15*, 4507–4520.

Tarancón, A. (2009). Strategies for lowering solid oxide fuel cells operating temperature. *Energies*, *2*(4), 1130–1150.

Tellez-Cruz, M. M., Escorihuela, J., Solorza-Feria, O., & Compañ, V. (2021). Proton exchange membrane fuel cells (PEMFCs): Advances and challenges. *Polymers*, *13*(18), 3064.

Wang, J. (2015). Barriers of scaling-up fuel cells: Cost, durability and reliability. *Energy*, *80*, 509–521.

Wang, J. (2017). System integration, durability and reliability of fuel cells: Challenges and solutions. *Applied Energy*, *189*, 460–479.

Wang, J., Wang, H., & Fan, Y. (2018). Techno-economic challenges of fuel cell commercialization. *Engineering*, *4*(3), 352–360.

Xu, J., Zhang, C., Wan, Z., Chen, X., Chan, S. H., & Tu, Z. (2022). Progress and perspectives of integrated thermal management systems in PEM fuel cell vehicles: A review. *Renewable and Sustainable Energy Reviews*, *155*, 111908.

Xun, Q., Murgovski, N., & Liu, Y. (2022). Joint component sizing and energy management for fuel cell hybrid electric trucks. *IEEE Transactions on Vehicular Technology*, *71*(5), 4863–4878.

Yan, Y., Manovic, V., Anthony, E. J., & Clough, P. T. (2020). Techno-economic analysis of low- carbon hydrogen production by sorption enhanced steam methane reforming (SE-SMR) processes. *Energy Conversion and Management*, *226*, 113530.

Yang, J., Qin, S., Zhang, W., Ding, T., Zhou, B., Li, X., & Jian, L. (2017). Improving the load- following capability of a solid oxide fuel cell system through the use of time delay control. *International Journal of Hydrogen Energy*, *42*(2), 1221–1236.

Yu, Y., Li, H., Wang, H., Yuan, X. Z., Wang, G., & Pan, M. U. (2012). A review on performance degradation of proton exchange membrane fuel cells during startup and shutdown processes: Causes, consequences, and mitigation strategies. *Journal of Power Sources*, *205*, 10–23.

Zhang, C., Zeng, T., Wu, Q., Deng, C., Chan, S. H., & Liu, Z. (2021). Improved efficiency maximization strategy for vehicular dual-stack fuel cell system considering load state of sub-stacks through predictive soft-loading. *Renewable Energy*, *179*, 929–944.

Zhang, Z., Mao, J., & Liu, Z. (2024). Advancements and insights in thermal and water management of proton exchange membrane fuel cells: Challenges and prospects. *International Communications in Heat and Mass Transfer*, *153*, 107376.

Zhou, S., Fan, L., Zhang, G., Gao, J., Lu, Y., Zhao, P., Wen, C., Shi, L. and Hu, Z. (2022). A review on proton exchange membrane multi-stack fuel cell systems: Architecture, performance, and power management. *Applied Energy*, *310*, 118555.

Index

Pages in **bold** refer to tables.

For Product Safety Concerns and Information please contact our EU
representative GPSR@taylorandfrancis.com
Taylor & Francis Verlag GmbH, Kaufingerstraße 24, 80331 München, Germany

www.ingramcontent.com/pod-product-compliance
Ingram Content Group UK Ltd.
Pitfield, Milton Keynes, MK11 3LW, UK
UKHW052219190926
475702UK00006B/1109

* 9 7 8 1 0 4 1 0 8 2 7 4 3 *